Dryland Agriculture

NIPA® GENX ELECTRONIC RESOURCES & SOLUTIONS P. LTD.
New Delhi-110 034

About the Authors

Dr M.V.R. Prasad worked as the Director of the Directorate of Oilseed Research (Currently Indian Institute of Oilseeds Research) and held positions relating to dryland agriculture in India and in several international assignments in Brazil, Panama and Africa. Dr Prasad possess vast experience in handling diverse crop production problems of dryland agriculture in the arid and semi-arid regions of India and Brazil.

He has trained number of scientists and post-graduate students on researches relating to breeding and production of dryland crops in India and abroad. He possesses over 137 publications to his credit on oilseed crops, annual and perennial legumes and dryland agriculture.

Dr. G. Subba Reddy worked as the Head Division of Crop Sciences and the Project Coordinator of Dryland Agricultural Research at the ICAR- Central Research Institute for Dryland Agriculture (CRIDA), Hyderabad. He also worked as the Principal Investigator for assessment and refinement of Indigenous Technologies. His has specialized in the Intercropping Systems for dryland agriculture. Dr Subba Reddy's contributions in technology transfer in the areas of rainfed cropping systems, rainwater management, drought management strategies, integrated nutrient management, participatory watershed Management, sustainable soil management practices and rainfed farming systems are amply recognized. Documentation of the strategies for contingency crop planning in around 100 districts of the country and as well as the modalities of establishment of integrated Rural Bio Resource centers carried out by Dr Reddy are well documented and appreciated.

He has 84 publications in the national and international journals and 20 Books as co-author to his credit.

Dryland Agriculture

M.V.R. Prasad
Former Director
ICAR- Indian Institute of Oilseed Research (ICAR- IIOR)
Hyderabad, Telangana State, India

G. Subba Reddy
Former Project Coordinator and Head
Division of Crop Sciences
Central Research Institute for Dryland Agriculture (ICAR-CRIDA)
Hyderabad, Telangana State, India

NIPA® GENX ELECTRONIC RESOURCES & SOLUTIONS P. LTD.
New Delhi-110 034

NIPA® GENX ELECTRONIC RESOURCES & SOLUTIONS P. LTD.

101,103, Vikas Surya Plaza, CU Block
L.S.C.Market, Pitam Pura, New Delhi-110 034
Ph : +91 11 27341616, 27341717, 27341718
E-mail:newindiapublishingagency@gmail.com
www: www.nipabooks.com

For customer assistance, please contact
Phone: + 91-11-27 34 17 17
Fax: + 91-11- 27 34 16 16
E-Mail: feedbacks@nipabooks.com

ISBN: 978-81-19072-15-6

Composed and Designed by NIPA®.

Trust for Advancement of Agricultural Sciences

Avenue II, Indian Agricultural Research Institute, New Delhi - 110 012
Phone : 011-25843243; +91-8130111237
E-mail : taasiari@gmail.com Website : www.taas.in

Dr. R.S. Paroda
Founder Chairman

Foreword

Dryland agriculture is plagued by constraints such as shortage of water for crop growth due to inadequate rainfall, low soil organic matter, poor biodiversity and increased biotic stresses which limit the dryland agricultural production. Under dryland farming, crops are grown in areas having annual rainfall generally <1,150mm and hence, these areas need an improved system of farming where maximum water can be conserved. Degraded soils with low water holding capacity along with multiple nutrient deficiencies and depleting ground water table contributes to low crop yields and also lead to further land degradation. About 55 per cent of the cultivated area in India is under dryland which contributes about 44 per cent of the total food production and plays a critical role in India's food security. However, despite scientific advancements in dryland farming, farmers are still unable to meet the needs of their families. Therefore, concerted efforts are needed for site-specific research to put dryland farming on a stronger pedestal.

To ensure sustainability of dryland agriculture, conserving our natural resources, integrated water shed development, improving water use efficiency, diversification of agriculture including livestock farming, alternative land uses and integrated soil-nutrient-water-crop management need to be given priority' attention. There is urgent need to fill the knowledge gaps, particularly to benefit the young researchers, students, and agricultural extension workers for better understanding of complexities of the dryland farming.

The book entitled 'Dryland Agriculture' is quite comprehensive and covers all aspects of dryland agriculture, namely, drought management, efficient plant types, soil and water management technologies, crops and cropping/ inter-cropping/ farming systems, crop production challenges, climate resilient practices, alternate land use systems, biotic stresses, and biodiversity' of dryland farming, as well as economic condition as smallholder farmers.

I congratulate Drs MVR Prasad and G Subba Reddy for writing this book which is indeed timely. I am sure this publication will be useful to dryland researchers, agricultural scientists, post-graduate students, agricultural extension workers, farmers and all those interested in dryland agriculture.

(R.S. Paroda)

Chairman TA AS
Former Secretary', DARE
& Director General, ICAR, Govt, of India

Acknowledgement

The prevailing situation of poverty and indebtedness of dryland farmer was a constant poke on the minds of the authors that despite several scientific advancements, dryland farmer is unaware of the skills to improve his job.

The difficulties encountered by the dryland farmers though are not new, the play of climate change together with degradation of soil and ambience in general have added to the complexity of the perennial intricacies of dryland farming.

There is a feeling among research and extension workers that the complexities of dryland agriculture need to be documented.

In this context the suggestion from Shri Sumit Shobha Pal Jain of New India Publishing Agency, New Delhi to write the book acted as a stimulus to the authors to go ahead with the job.

We are grateful to Dr R.S. Paroda, Former Secretary to the Govt. of India (DARE) and Director General ICAR and Chairman of the Trust for Advancement of Agricultural Sciences (TAAS), New Delhi to have consented to our request and written a befitting Foreword to the book.

We thank Dr V. Ranga Rao former Director of the ICAR-Directorate of Oilseeds Research (Currently known as Indian Institute of Oilseeds Research) for his useful suggestions.

Thanks are also due to Dr Y.S. Ramakrishna, former Director of the ICAR-Central Institute of Dryland Agriculture and Dr ARG Ranganatha for suggesting certain references.

The authors thank Shri V. Siva Prasad for formatting the entire contents of the book.

The encouragement and support received from our families boosted our efforts in writing the book.

One of us (MVRP) is grateful and indebted to Shri-Hari-Sai for the abundant Grace showered without which it would have been practically impossible to accomplish the task.

Preface

The diversity in Indian dryland agriculture is of special interest in the sense that the rainfall dependent farming system varies remarkably based on soil types, latitude, quantum and distribution of rainfall and variety of animals and crops and agronomic factors including crop growing conditions. The variability of the system is also dependent upon the factors of temperature and humidity that bestow the special characters to each of the dryland ecosystems. Biodiversity of dry-farming systems is of added significance in relation to the sustainability and stability of the complex ecosystem.

The gamut of complexities of dry farming is spread out in 68% of India's cultivated area and contributes to 44% of Indian food requirement. Hence the Indian dryland agriculture forms a mega ecosystem, which is the very important factor in Indian agricultural economy and food security. Additionally, the dryland agriculture supports 60% of the livestock population, a crucial factor in small holders' agriculture.

Disasters of crop failures are not unusual under dry-farming conditions not only due to prolonged dry spells during the crop growth period but also due to the biotic stresses which take a heavy toll of the meagre production. The duration of moisture availability to crop growth in dryland agriculture is more or less a constant factor restricted to less than 5 calendar months and this is the most critical factor dictating the crop production in this ecosystem.

The farming communities dependent on dryland agriculture too are rather varied in terms of the crops grown and cropping intensity. Nevertheless, the dryland farmer in general is poor with his income varying from IRs. 3000 to 5000 per month, with which he fails to meet the maintenance expenditure of his family dependent on agriculture. Most of the dryland farmers fall under the category of small farmers with an average land holding size of less than two hectares. All the intricacies that the dryland farmer faces push him into a never ending monetary debt trap.

Despite the dismal picture of Indian dryland agriculture, the efficiency of the system can be enhanced by harnessing science and technology in terms of in situ moisture conservation, rainwater harvesting connected to micro-irrigation, genetically superior breeds of crop varieties combining resistance to biotic stresses, soil amelioration, organic nutrient management, conservation agriculture and finally farming systems approach. The escalating cost factor in dryland agriculture could be controlled to a great extent if the farmer learns to generate inputs from his own farm. Added to the above, there exists an urgent need to policy interventions in terms of market support to agricultural commodities.

The book on Dryland Agriculture while introducing the reader to the special features of dry-farming in India, covers the topics on drought management, soil and water management, crop production technologies, crops and cropping patterns, intercropping systems, efficient plant types, biodiversity, climate resilient practices to meet the challenges of climate change and dryland based farming systems.

In view of the pivotal role of Intercropping in dryland production system, a separate chpter on the subject in addition to the general discussions of the intercropping practice in the other chapters. The theme of the book is towards finding workable solutions to the problems plaguing the system of dryland agriculture.

The book caters to the needs of agricultural scientists, students, teachers, extension workers and literate farmers.

M.V.R. Prasad
G. Subba Reddy

Prologue

Earth's land surface is covered by drylands forming about 41 per cent which is inhabited by 2 billion people forming about one third of world population.

Drylands are areas with low soil moisture, high evapotranspiration which results in water deficit prevailing throughout the year. The drylands are not equally distributed between the rich and the poor nations. About 72 per cent of the global drylands are in the developing nations and rest 28 per cent falls in industrialized nations (www. millenniumassessment.org). In a country like India, 44 per cent of the total food production is being supported by drylands, thereby playing a critical role in nation's food security.

Rain-fed farming constitutes 80% of the world's cropland and produces more than 60% of the world's cereal grains, generating livelihoods in rural areas while producing food for cities. In temperate regions with relatively reliable rainfall and good soils, rain-fed agriculture generates high yields.

Most of the countries in the world depend primarily on dryland agriculture for their food grain production. Despite large strides made in improving productivity; environmental conditions in many developing countries, a great number of poor families in Africa and Asia still face poverty, hunger, food insecurity and malnutrition where rain-fed agriculture is the main agricultural activity.

With the increasing population, food production has to be increased. A real need to achieve the second green revolution has been envisioned, which can be a reality, largely by improving the dryland agriculture. Geographically dryland agricultural area in India includes the north western desert regions of Rajasthan, the plateau region of central India, the alluvial plains of Ganga Yamuna river basin, the central highlands of Gujarat, Maharashtra and Madhya Pradesh, the rain shadow regions of Deccan in Maharashtra, the Deccan Plateau of Telangana, Andhra Pradesh and the Tamil Nadu highlands (Singh et al., 2004).

Rain-fed farming and dryland agriculture account for more than 95% of farmed land in sub-Saharan Africa; 90% in Latin America; 75% in the Near East and North Africa; 65% in East Asia; and 60% in South Asia. This system includes both permanent crops (such as rubber, tea, and coffee) as well as annual crops (such as wheat, maize, and rice). For example, tubers, a staple crop for sub-Saharan Africa, have been all but uninfluenced by the technological developments of the green revolution. In rain-fed farming, the amount of rainfall converted into plant-available soil water is determined by the amount and intensity of rainfall, topography, infiltrability and water retentivity of soil, depth of root zone, and soil depth.

Rainfed agriculture is generally overlooked by development investors, researchers and policy makers due to limited confidence in its ability to increase agricultural production and development. However, research undertaken by a team of leading scientists from global organizations demonstrates its potential in achieving food security, improving livelihoods and most importantly addressing issues of equity and poverty reduction in dryland areas, the hot spots of poverty. On the basis of case studies from varied agricultural and ecological regions in Asia and Africa, the need for adopting new paradigms between rainfed and irrigated agriculture, catchment/micro-watershed management approaches, upgrades in science-based development and more investments in rainfed areas. Yield gaps for major rainfed crops are analysed globally and possible ways and means including technological, social, and institutional options to bridge the yield gaps are discussed in detail. Covering areas such as rainwater harvesting and its efficient use, the rehabilitation of degraded land and assessment methods for social, environmental and economic impacts are most relevant currently, considering the increasing concerns over the high cost of expanding large-scale irrigation and the environmental impacts of large dams, upgrading rain-fed agriculture is gaining increased attention (Rockström et al., 2010).

Rain-fed farming and dryland farming are often used interchangeably, but this is an error. They both exclude irrigation, but beyond that, they can differ significantly. Dryland farming is a special case of rain-fed agriculture practised in arid and semiarid regions in which annual precipitation is about 20–35% of potential evapotranspiration. Conditions of moderate-to-severe moisture stress occur during a substantial part of the year, greatly limiting yield potential, and in which farming emphasizes water conservation in all practices throughout the year. Rain-fed systems, although they include dryland systems, can also include systems which emphasize disposal of excess water, maximum crop yields, and high inputs of fertilizer.

The major contributing characters impeding normal productive agriculture in dryland zones are as follows. –

(i) Low rainfall: In the dryland regions, low rainfall is the primary factor within a range of from 375 mm to 1125 mm which is unevenly distributed, highly erratic and uncertain. The crop production on drylands is mainly dependent on the frequency and intensity of rainfall making it a less productive.

(ii) Soil related problems: The major causes for land degradation include the chemical degradation of soil, loss of soil structure and texture, depletion of soil organic matter and loss of natural vegetation leading to soil erosion. Sequestration of carbon also turns out to be a major problem, which further degrades the soil making it less productive. (ftp://ftp.fao.org).

(iii) Occurrence of drought: The extensive climatic hazards are seen in drylands as the soils are weak and can be subjected to environmental stress to a higher level, leading to further land degradation. Drought is a common scenario in drylands as water availability is less; further leading to low productivity.

(iv) Extensive agriculture: Prevalence of mono-cropping extensively makes farm lands lack of nutrients and result in reduction of yield.

(v) Crops grown: The traditional crops grown in a particular region will be similar and are not much remunerative compared to the major crops like rice and wheat. When similar crops are grown, all mature at the same time and a large quantity of produce will reach the market leading to glut in the markets. This situation is severely exploited by the traders and the middlemen in the markets. The issue of marketing turns out to be a big problem in dryland agriculture.

(vi)Poor economy of farmers: Economic status and of living of farmers is low in drylands, due to the less choice of the crops that are grown in these areas.

There are three components of a successful dryland farming system: (1) retaining the precipitation on the land, (2) reducing evaporation from the soil surface to increase the portion of evapotranspiration used for transpiration, and (3) utilizing crops that have drought tolerance and that match the precipitation patterns. Although these components have been known for centuries, new technologies continue to be developed that increase crop production in water-short areas.

Crop productivity depends on the cumulative effects of water deficits on plant performance. Therefore, it is critical to understand the processes of water supply and demand in dryland crop communities and the role of adaptation in mitigating the detrimental effects of drought on crop yields. It is now known that drought occurs over an almost infinite variation of space and time under drought prone environments.

Water productivity (Gowda et al., 2009) 'the amount of crop produced per drop', tends to be low in dryland farming systems, while losses from evaporation are high. Land is often degraded; crops frequently die because of drought or floods; but certain methods are in place for managing water more effectively. In parts of sub-Saharan Africa and South Asia, productivity is particularly low, which results in food insecurity and poverty for rural communities.

In its broadest aspects, dryland farming is concerned with all phases of land use under semiarid conditions.

Considering the present rate of development of irrigation facilities in India, it is estimated that around 50% of the cropped area shall remain under rain-fed farming system.

Nevertheless, the major finding of a Comprehensive Assessment (CA) of rainfed agriculture for food security is that the vast untapped potential of rainfed agriculture needs to be tapped as the current farmers' crop yields are lower by two- to fivefold than the potential, based on the clear documented evidence (Wani *et al.*,2009)

Use of modern technological tools such as Nano-technology for efficient water and nutrient management and plant protection and biotechnology to enhance the recovery of genetic potential of the plant species suited to dryland agriculture and conservation of productive biodiversity in arid and semi-arid zones are expected to open new avenues in dryland agriculture

Contents

1

Drought and Drought Management in Dryland Agriculture

The world's drylands include hyper-arid, arid, semi-arid and dry sub-humid areas where rainfall is highly scanty and variable. Droughts are common in areas / zones where water is the principal limiting factor for agriculture (https://www.fao.org/3/t0122e/t0122e03.htm) . Dryland soils are characterized by low levels of moisture, organic matter, and biological activity, consequently such soils are very poor in fertility devoid of plant nutrients. When, these soils suffer further degradation in fertility, resulting in erosion, desertification, and salinization when crops are grown on such soils inappropriately without any scientific basis. However, the major factor limiting agricultural production on drylands is the lack of adequate moisture for crop growth or drought.

Drought is characterized by the moisture deficits, due to atmospheric factors such as deficient rainfall and /or soil factors such as low soil moisture. Drought is a phase of dry weather when there is primarily shortage of rainfall. The shortage of precipitation of various degrees could be detrimental to the communities, resulting in damage to agricultural production and a shortage of drinking water. These effects can trigger economic and social disasters, such as famine, forced migration away from drought-stricken areas, and conflict over remaining resources. (https://education.nationalgeographic.org/resource/understanding-droughts). Drought Many times drought may be considered well-defined dry season, which is a regular feature of climate. Drought in contrast is a recurrent, yet sporadic feature of climate, known to occur under diverse climatic regimes, exhibiting variability in terms of its spatial expanse, severity and duration. The spread and severity of this kind of adverse aberation is contingent on several factors, including the status of surface and ground water resources, agro-climatic features, cropping patterns, socio-economic vulnerabilities of the population etc. It is very difficult to determine the initiation and termination of a drought phase because of the slow, 'creepy' onset, silent spread and gradual. In India, the drought is considered to be coterminous with the monsoons (MOA, 2021)

The drought phase can be variable from a short spell of 15 days to some months and may be of one or more years. The longest period ever reported of drought was in the central Chile from 1770 to 1782 with a three non-consecutive years of the highest degree of drought (Astaburuaga and Ricardo 2004)

The adverse impact created on the ecosystem and agriculture of the region, results in immense harm to the entire region. The preponderance of continuous dry seasons with least quantum of rainfall in the semi-arid zones significantly enhances the chances of incidence of drought accompanied by calamities such as bush fires.

According to the Irrigation Commission of India, the probability of incidence of drought is higher in the regions where annual rainfall is less than 75% of the average precipitation. The acute paucity of adequate moisture levels needed for satisfactory crop growth would have a devastating effect on crop yields leading to the agricultural drought. In other words, the acute ecological imbalance among the factors of crop growth of which the moisture deficit is the most important adverse factor triggering the damage to crop productivity.

Regular droughts spells do occur in all most all parts of several countries, which have become more extreme and more unpredictable due to climate change. As back as the decade of 1900's the negative play of global warming in accentuating drought to the detriment of crop growth was noticed. Many plant species, such as Cacti and certain perennial trees with xerophytic attributes shown in Fig.1 have drought tolerance adaptations like reduced leaf area and waxy cuticles in order to conserve moisture in the plant system and cut down evapotranspiration. The plant types pertaining to the xerophytic category including cacti and the bushes and trees with higher leaf density and reduced leaf size escape the adverse effects of drought by conserving water in their tissues and reducing the transpiration levels. Some plant species survive long dry spells as dormant seeds. Throughout history, humans have usually viewed droughts as "disasters" due to the impact on food availability and the rest of society.

Fig. 1: Dry soil without moisture in different layers of soil and presence of bushes and trees with xerophytic attributes.

Academically droughts may be defined in three main ways (https://en.wikipedia.org/wiki/Drought) :

1. Meteorological drought is observed with extended time period with below average rainfall (Swain *et al.*, 2017). Meteorological drought initially sets in earlier than the onset of agricultural drought (NOAA, 2007)

2. Agricultural drought alters the ecology negatively which in turn has the adverse effect on crop performance. The factors such as negatively altered rainfall pattern, deterioration of soil condition, including its erosion and poor soil nutrient content and faulty crop management practices too accentuate the drought. The key factor, nevertheless is the incidence of prolonged spells of inadequate rainfall. (Wang-Qianfeng *et al.*, 2015)

3. Hydrological drought occurs with the marked decline in water levels of structures with stored water such as lakes and reservoirs. This kind of drought happens gradually with time as it involves slow decline of water levels of the above structures without replenishment. As in the case of agricultural drought, hydrological drought can happen due to several other factors in addition to low rainfall. (BBC 2004)

The duration of the moisture stress being the key factor, the phenomenon of drought is defined as follows.

1. Permanent drought: This is characterised by aridity with the least moisture availability, due to which no agriculture can't be practised without regular water supply. The ecosystem of the arid zones with permanent drought consists of scanty vegetation with reduced green parts in their size.
2. Seasonal drought: As the name indicates this kind of drought is a feature of the zones with well demarcated wet and dry seasons in terms of the moisture availability.
3. Contingent drought: This kind of drought may not follow any regularity in it occurrence; but could be the result of absence of rainfall for a short spell of time in other wise non-drought areas with regular humid weather. This kind of 'stoppage' of moisture availability for a short time-zone being irregular, may not result in adverse crop performance.
4. Invisible drought: The nature of the crop season in this case is marked by obvious rainfall perhaps without breaks or spells of drastic shortage of rains of long dry spells. Nevertheless, the quantum of rains received might be inadequate in terms of meeting the evapo-transpiration needs of the crop. The invisible drought is a feature of shortage of soil moisture with the concomitant result of reduced crop yields.

Based on the time of occurrence, droughts may be classified as follows

1. Initial phase of drought: Moisture shortage caused by the absence of rainfall initially or lag in the onset of rainfall or stoppage of rains after initial soaking showers, causing a dry-spell is the initial drought
2. Mid-season drought: This is characterised by shortage of soil moisture for more than 15 days in the mid-crop season when the crop is in its peak or grand vegetative growth or initial flowering phase and this is the drought spell occurring in between two rainy spells. The length and intensity of this kind of drought spell determines success or failure of the crop. If the rainless period, causing soil drought continues beyond two weeks, it would result in deficient soil moisture, adversely affecting the crop performance.
3. End season drought: This is characterised by the stoppage of rains in the fag-end of the crop season, resulting in deficient soil moisture during the grain maturity phase of the crop. In other words, it can be called 'pre-empting drought' due to early withdrawal of the seasonal rainfall adversely affecting crop maturity and grain development.

The adverse effects of drought alter negatively the agricultural performance and also gradually worsen the living conditions of rural communities. Pantuliano and Pavanello (2009) reported that during the periods of 1984–85, 2006 and 2011 desertification had set in East Africa due to continuous droughts, resulting in adverse ecology leading to hunger in rural populations due to shortage of food grains.

What are the factors leading to Drought?

(i) Precipitation deficiency

Convective processes resulting in rain (Emmanoui and Anagnostou 2004) are due to intense linear motions leading to the capsizing of the troposphere of the area at that location resulting in heavy precipitation (Pearce, 2002), while strati-form processes cause weaker upward motions and less intense rains over a longer time duration. (Houze,1993).

Precipitation falls into three types, depending upon whether it is of (i) normal watery nature, or of (ii) water, which might solidify as it touches the surface, or of (iii) frozen solid type. Incidence of droughts is in the areas of normal rainfall patterns, that are of lower magnitude. If the factors described above do not manifest for the ecosystem to meet the needs of plant species and crops, it would result in drought.

Table 1 gives an account of deficient precipitation for the drought years that caused damage to crops in India. India receives most of its rainfall (73%) from the South-West or "summer" Monsoon i.e., the rainfall received in the monsoon season comprising the months of June to September.

Table 1: Distribution of Deficient (%) Rainfall in Meteorological Sub-divisions of India during Major Drought Events (Number of meteorological sub-divisions = 36)

Drought year Mid-July Mid-August Drought Year in India	Mid-July	Mid-August	Mid-September
1966	19	14	16
1972	13	21	21
1979	17	15	15
1987	25	25	21
2002	25	25	21
2009	15	19	16
2014	16	14	13
2015	23	23	14

Source: MOA (2021)

(ii) Low moisture holding capacity of soils a consequence of depletion of organic matter

Organic matter in the soil is a decomposed mix of materials- fragments of previous year's stalks and roots, earthworm casts, and living microbes and invertebrates, to name just a few. These materials are broken down by physical and biological processes. For example, freezing and thawing causes plant residue to lose its structure. Tiny dissolved molecules flow deep into the soil with rainwater. Hungry invertebrates, fungi, and bacteria consume complex living and dead organic material and excrete nutrients they don't need in a smaller, simpler form. These small organic molecules can stick to clay surfaces. Clay surfaces covered with organic material grow like snowballs, and soil aggregates are formed. Soil aggregates are critical for moisture retention in the soil for two reasons. First, a well-aggregated soil has large pores between aggregates to let water enter the soil profile. Second, small pores within aggregates hold water tightly enough to keep it around, but loosely enough for plant roots to take it up. It's critical that soil both let water flow through and hold water for later. If your soil doesn't let water infiltrate, you'll have ponding, runoff and soil loss, and lower plant water supply. If your soil doesn't hold water, plants suffer from drought.

Hence, organic matter is critical for forming aggregates, and aggregates are critical for holding water. Because of that link, there is organic matter is positively correlated to moisture-retentive capacity of soil. Thus, the quantum of water stored is based upon on soil texture, also in addition to the role of organic matter. If soil doesn't hold water, plants suffer from drought. Hence, organic matter is critical for forming aggregates, and aggregates are critical for moisture retention in the soil. The ideal quantum of organic matter of soil should be from 3 to 6%. Regrettably, considerably in a large proportion of Indian soils the current organic matter content is reported to be 1% or even less.

(https://www.covercropstrategies.com/articles/1206-soil-organic-matter-is-key-to-soil-water-)

(ii) Dry season

The rainy and dry seasons of the tropical zones arise due to the movement of the Monsoon Trough, which is portrayed as the line on a weather map exhibiting the sites of minimum sea level pressure, which is a zone of convergence between the wind patterns of the southern and northern hemispheres (Wang, 2006). Drought gets enhanced by the rainless dry season, causing gradual reduction in soil moisture (Vijendra *et al.*, 2005). This adverse phenomenon forces the animals to migrate in search of fodder and water. A simple soil-water balance using long term values of monthly rainfall and potential evapotranspiration gives some indication of the available soil water (Thornthwaite, 1948). The falling rate soil-water budget could be used to calculate daily soil water (Thonthwaite and Mather 1955)

(iii) El Niño

According to the world Meteorological organization, the El Niño/La Niña Southern Oscillation (ENSO) exert marked influence on the world climate patterns. El Nino occurs with rapidly fluctuating ocean temperatures in the central and eastern equatorial Pacific, together with variations in the atmosphere.

Unlike the normal climatic patterns, incidence of El Nino brings about a change with warmer winters and dry atmosphere as it happened in the Northwest and Northern Midwest of United States of America. This in turn results in lower degrees of snowfall in the winters. This kind of reversal of climatic pattern was observed in African countries viz., Mozambique, Botswana, Zimbabwe and Zambia too.

Southeast Asia and Northern Australia have been the in the category of victims of El Niño. Consequently, there have been increased bush fires together with worsening haze, and deterioration of atmospheric quality. Even in the developed country like Australia, it was observed that drier conditions prevailed in Queensland, Victoria, New South Wales, and eastern Tasmania during the months of June to August.

Wide spread drought prevailed due to heating up of oceanic waters in the western Pacific. Thus, Singapore experienced the driest climate in February in 2014 for the first time since the year1869 (Channelnewsasia.com,2010).

According to Kumar *et al*., (2013), El Niño-Southern Oscillation (ENSO) events resulted in El Niño-related droughts leading to severe crop loses in several parts of India (Anonymous, 2014)

Soil Erosion due to human activities

The human intervention with erroneous and unscientific agriculture together with bad water management causing flooding, and deforestation, result destruction of delicate balance of the fragile soil ecosystems leading to decrease in the moisture holding capacity of soil due to continuous erosion(Forest.org, 2006). In arid ecosystems, the erosion resulted by high speed wind (Hofman and Franzen, 1997) causing due to the upward movement and flight of small particles away from the initial sites. This process is known as deflation.

Such suspended particles create erosion by abrasion. The zones with scanty vegetation cover together with very little rainfall have always been prone to Wind erosion (USGS,2004).

The most practically sustainable practice to check wind erosion is by installation of wind breaks employing suitable tree species such as *Leucaena leucocephala* K636 (Giant leucaena), *Sesbania sesban*, *Calliandra calothyrsus*, and *Gliricidia sepium* in tropical countries (Hofman and Franzen,1997). The other tree species such as Western Red Cedar, Pacific Crabapple, Red Alder, Shore Pine, Black Hawthorn and Bitter Cherry are used in western countries including United States of America to control wind erosion (https://mrtreeservices.com).

Soil loss due to wind erosion is estimated to be of the tune of 6100 times greater in drought years than in wet years. (Wiggs and Giles, 2011)

(iv) Climate change

The IPCC Sixth Assessment Report published in (2021) projected multiplicative enhancement of extreme events in relation to the pre-industrial era for heat waves, droughts and heavy rainfall, for various global warming scenarios (Climate Change 2021) According to Adam and Richard (2013) the agricultural will be adversely affected by drought due to the negative impact of global climate change, throughout the world, and especially in developing nations (USDS,2005). While drought is of known occurrence in certain zones, excessive down-pours and soil erosion are reported in others zones. According to Brahic (2007) technologies of solar-energy could be of utility in this context.

Climate change leads to multiple factors together with droughts. Increase in the atmospheric temperatures over land results in enhancement of atmospheric evaporative demand, leading to preponderance of droughts (Benjamin *et al.*, 2018). Though it is currently not possible to attribute drought to human induced climate deterioration, Mukherjee *et al.*, (2018) reported the 'man-made' adverse situations of the above order in the Mediterranean and California.

The report of Intergovernmental Panel on Climate Change issued in 1990 (IPCC,1990) revealed that the area of drylands subjected to drought, increased by 1% per year. In the year 2015 around 500 million people lived in areas that was impacted by desertification in the years 1980s – 2000s. People who live in the areas affected by land degradation and desertification are "increasingly negatively affected by climate change".

Adverse effects of incidence of drought in dryland agriculture

Drought can have serious health, social, economic, and political impacts with far-reaching consequences, apart from crop failure on drylands. Monsoon plays an important controlling role in Indian Agriculture, providing the needed quantum of water for crop and animal production. In parts of India, failure of seasonal rains has got accentuated due to the altered climate particularly in the Deccan plateau zones of south-eastern Maharashtra, northern Karnataka, and Andhra Pradesh, and Telangana, together with Odisha, Gujarat, and Rajasthan, which are subjected to periodical droughts leading to crop losses (Swain *et al.*, 2017).

Wildfires

The low moisture and precipitation that often characterize droughts can lead to wild fires destroying all vegetation around such as forests, range lands and farm lands adding misery to already prevailing food shortages. In addition, even plants generally adapted to dry conditions will drop needles and leaves during a drought, contributing to a layer of dead vegetation on the ground. This dry duff then becomes a dangerous fuel for damaging wildfires.

Wildlife

Wild plants and animals suffer from droughts, even if they have some adaptations to dry conditions. In grasslands, sustained lack of rain decreases forage production, affecting herbivores, grain-eating birds, and indirectly, predators and scavengers. Droughts will lead to increased mortality and reduced reproduction, which is especially problematic for populations of at-risk species whose numbers are already very low. Wildlife needing wetlands for breeding (for example, ducks and geese) experience drought as a decline in available nesting sites. Biodiversity is absolutely essential for dryland agricultural productivity. Regrettably, continuous levels of drought take a heavy toll of the valuable biodiversity.

Electricity Generation

Many areas in the world rely on hydroelectric projects for electricity. Drought will reduce the amount of water stored in reservoirs behind dams, reducing the amount of power produced. This problem can be very challenging for the numerous small communities relying on small-scale hydro, where a small electric turbine is installed on a local creek.

Desertification: Continuous drought of acute order over a considerable period of time leads to deterioration of soil fertility and aggregation resulting in destruction of prevailing positive ecosystem. The above entire gamut of adverse changes leads to the process of desertification. Desertification also causes progressive loss of valuable vegetation and fauna consisting of several adapted animal species, damaging the prevailing biodiversity. It goes without saying that the pangs of phenomenon of desertification hit agriculture drastically leading to several total crop failures for want of minimum levels of moisture needed for plant growth and survival of fauna.

Migration or Relocation

Faced with the other impacts of drought, many people will flee a drought-stricken area in search of a new home with a better supply of water, enough food, and without the disease and conflict that were present in the place they are leaving.

Disease

Drought often creates a lack of clean water for drinking, public sanitation and hygiene, which can lead to life-threatening diseases. The problem of water access is critical: every year, millions are sickened or die due to lack of clean water access and sanitation, and droughts only make the problem worse.

Hunger and Famine

Drought conditions are characterised by paucity of minimum levels of water for the sustenance of crops, provided by either natural precipitation or irrigation using reserve water supplies. The same problem affects grass and grain used to feed livestock and poultry. When drought undermines or destroys food sources, people go hungry. When the drought is severe and continues over a long period, famine may occur. The 1984 famine in Ethiopia was due to a deadly combination of acute and prolonged drought and a dangerously ineffective government, resulting in the tragic death of hundreds of thousands died.

While drought is inevitable consequence of several factors including deficient moisture availability, the management of drought through available technologies and certain innovative procedures should be put in place, to minimise the adverse impact of drought.

Drought Environment in relation to plant growth

Plants' use of soil water could be divided into three physiologically distinct phases (Sinclair and Ludlow, 1986). In the first stage the soil moisture content is fairly high facilitating germination and proper crop establishment. The availability of water to roots generally causes no inhibition of transpiration rates and photosynthesis. Stomata are open in the first stage and the transpiration is mostly determined by the weather parameters.

In the subsequent second stage characterised by the intermittent drought environment, starting with paucity of water to plant roots that cannot support the transpirational demands.

In this case the stomatal conductance is decreased to maintain a balance between the transpirational rate causing depletion of moisture and rate of soil moisture availability. Largely the second stage begins with soil bound 0.2 to 0.3 fraction of transpirable water. This phase is characterised by the imminent moisture stress in respect of dryland crops. Crop establishment subsequent to germination is one crucial phase that may face sever water stress by the second stage, when the intermittent drought can be damaging to the performance of dryland crop. In such cases the crop genotypes with faster establishment rate by which the root system grows rapidly shall have an advantage.

The above follows the condition, that may be called the third stage, when little additional decreases in possible stomatal conductance, resulting in the transpirable water getting practically exhausted (Sinclair, 1988). In this case the crop faces the survival mode. The amount of stored moisture in the initial phase of the drought and duration of the period without rainfall dictate as how quickly the third stage is reached and how long the crop is subjected to this condition. The prolonged rainless period will take the crop under this kind of drought to the next fourth stage, under which the crop's survival is the big question (Sinclair, 1 988).

Drought stress during the crop growth stages of peak leaf area indices might suffer pronounced decrease in crop yield. At high leaf area indices, the crop gas exchange rates are the highest and hence the moisture lost at the highest rate, which put the crop at the critical third stage of drought indicated above. At this point, the stage of high leaf area indices coincides with the greatest potential of CO2 accumulation rates, so that the inhibited gas exchange at this period results in significant loss of productivity (Sinclair, 1988). The anthesis or the mid reproductive phase, as is commonly believed to be the sensitive to drought. According to Sinclair (1988), the crop at anthesis stage will be adversely affected only if it is subjected to very severe drought stress. In order to assess whether drought induced yield decreases are due to overall restricted biomass accumulation or are attributable to some unique sensitivities of reproductive phase such as the anthesis, harvest index could be an indicator (Sinclair, 1988). If harvest index is relatively unaffected by intermittent drought the plant cannot be sensitive to drought, beyond the effect on overall biomass accumulation (Sinclair, 1988).

In order to avoid the adverse effects of predictable intermittent drought to the reproductive phase, it should be possible to conserve the water for the predicted drought phase. If the timing of the drought is predictable, the management practice of deferring the planting time or reducing the plant population densities would effectively conserve water to minimise the severity of the intermittent drought. The genotypes possessing plant characters such as lower stomatal conductance, or less vegetative attribute with reduced leaf size or decreased leaf loss could be chosen for the above specific environment (Ludlow and Muchow, 1988). Also, the increased rooting depth is an effective attribute to maintain plant turgor during drought (Jones and Zur,1984).

The next phase of drought is that of post rainy season, in which crops are grown on conserved soil moisture. The quantum of water available to support crop production under the above situation is defined at the initial phase of post rainy season by the amount of transpirable water stored in the soil. Using the approach of Tanner and Sinclair (1983), it is possible to estimate the maximum crop biomass that can be produced from stored soil water. By assuming a post rainy season of uniform vapour pressure deficit, the following relationship can be used to arrive at the probable maximum crop biomass (Tanner and Sinclair, 1983)

$$B \leq (W-E)\, k/(e^{*}-e)$$

Where

B= Crop biomass produced (g per meter squire)

W= total available water (g per meter squire)

E= total soil water evaporation (g per meter squire)

K= explicit coefficient defined by physiology of crop species or variety (Pa)

$(e^{*}-e)$ = average daily atmospheric vapour pressure deficit. (Pa)

Water deficit and Crop water use: The dry matter production per unit of water transpired by crop plants (WUE) is more or less constant at a given environment (Sinclair *et al.*, 1984). Also the vapour pressure deficit of the air is also constant for a particular crop plant (Squire *et al.* 1986). These kind of relationships show us a feasible means to model potential crop productivity, explaining as to why several high yielding varieties get more vulnerable to severe drought.

Crop growth, in terms of accurately estimated dry matter production taking into consideration (i) amount of water transpired and (ii) water use efficiency (Landsberg,1988) would provide clues regarding the drought resistance of the crop plant genotype.

According to May and Milthrope (1962) drought resistance as applied to crop plants is normally used as an all-embracing term to describe those crop varieties or species, which are able to yield satisfactorily in area liable to periodic drought. Thus it covers an extensive complex of properties, which can be appreciated by considering the ecological situations, which lead to, and shortage of water within the plant.

Currently, however, it is possible to examine these ecological situations more precisely. Additionally, it is possible to consider effects of, and adaptation to drought at a number of operational levels viz., cells, individual plants and the crop as a whole and the system within which crops are grown (Bunting and Kassam, 1988)

A field crop consumes not more than about 1 to 2% of all the water it takes up during its life cycle. The rest of the water is transpired through the leaves. The small amount of water that is retained in the plant's system is of great significance (Sutcliffe, 1968). The transpiration rate depends not only on the 'evaporative demand' of the atmosphere, but also on proportion of each day during which the stomata are open, and the size of the evaporating surface area (leaf area) which intercepts radiant energy or receiving reflected heat. If this rate is greater than the rate the rate at which water can be taken up, the plants lose water, with decrease in leaf water potential.

As water shortage develops, the water potential becomes smaller or more negative due to dehydration and at some point changes in the turgor potential of different leaf cells lead to partial or complete closure of stomata. If the water supply shortage together with associated plant water deficits continue to increase, then the proportion of each day the stomatal opening decreases, leaf temperature rises, and osmoregulation of solute-osmatic potential occurs. Initially the *decrease in solute potential maintains positive cell turgor potential as water potential continues to decrease*. However, later it serves to avoid irreversible dehydration and to withstand desiccation (Bunting and Kassam1988).

By and large the cultivated leguminous crops do not have a large working range of water potential. The typical figures at their zero turgor potential (i.e., wilting) range from -1.0 to-2.5 MPa. In cereals the values range from -2.5 to -7.0 MPa. Therefore, these crops can withstand greater dehydration levels and can extract more water from soil (Bunting and Kassam1988).

According to Bunting and Kassam (1988), the rate of water use is influenced by the following three sets of conditions.

1. The evaporative demand from the air,
2. The size of the canopy cover and
3. The water supply.

The total amount of water used by the crop depends on the length of the life of the crop and the time course of the rate at which it uses water.

Different crop varieties grow and develop their canopies at different rates and hence exhibit different economic yield levels per unit time, surviving in environments with different evaporation conditions. Thus their water use efficiencies too differ for total dry matter production as well as for their rain yield levels (Bunting and Kassam, 1988). Hence, the different growth rates between crops are due to the differences in the leaf area or assimilatory surfaces per unit area of land. Additionally, the differences due to given Leaf Area Index and level of light interception determine the rate of canopy photosynthesis per unit area of assimilatory surface. These growth rate differences affect the time course of water use efficiency.

Therefore, it is important that the physiology of the crop fit approximately into the time available for crop growth, so that the crop is able to adjust the life cycle to match the unpredictable year to year variations in the length of the growing period. This facilitates proper dry matter accumulation. The portioning of the dry matter is governed the plant's inherent genetic nature.

These observations have bearing in developing crop varieties suited for moisture stress conditions.

Crop Adaptation to Drought

There are three mechanisms given below by which a crop can offset the adverse effects of drought (Bunting and Kassam1988).

1. The crop can *escape drought if its life cycle is short enough* to enable it to attain maturity before the onset of dry period or drought.
2. The crop can *withstand or endure dry period by extracting more stored soil moisture,* with its larger working range in water potential of leaves and other plant parts and by storing water in its tissues so that the wilting is delayed. Thus the plant can continue to assimilate CO2 facilitating its growth.
3. The crop may survive and recover from dry period by losing water, so that much of the canopy wilts and dies, following which the crop recovers by producing new leaves from the dormant buds.

The performance of certain grain legumes such as green gram (mung bean), moth bean (*Vigna aconitifolia*) and groundnut under dryland conditions reveals the operation of the above phenomena. In the areas that receive a meagre annual rain fall of the order of 300 to 400 mm with evapotranspiration (ET) exceeding 2mm, very short season legume crops viz., mung bean and moth bean tend to avoid drought at crucial developmental phases of flowering and fruiting by maturing in less than 70 to 75 days.

Also in the case of groundnut, spanish-valencia genotypes mature fairly early in 85-90 days. The primary root of groundnut plant grows rapidly reaching a soil depth of around 120 cm. Groundnut leaves are reported to contain a layer of water storing cells, which presumably help the plant to overcome the adverse effects of water loss by delaying the day-time closing of the stomata and wilting of leaves (Bunting and Kassam, 1988).

Farmer's adaptation to drought in traditional dryland areas with meagre rainfall

In traditional areas that depend on meagre rainfall for crop production, over centuries, the farmers have come up with procedures to overcome or minimise the effect of drought on crop production (Bunting and Kassam, 1988).

Crops are planted at wider spacing with the receipt of monsoon rains so as to make the best use of the limited and certain availability of moisture. When the rains appear to be assured, the staple food crops like sorghum, pearl-millet and finger millet are sown.

In case monsoon rains are delayed beyond a reasonable limit, the above mentioned cereal crops fail. The farmers, however, would chose grain legumes like cowpeas, green gram etc. and a photo-insensitive crop sunflower, which provide tangible insurance against crop failure.

In traditional groundnut growing areas of Gujarat in India, the adaptation mechanism to ward of adverse effects of uncertain rainfall and drought, farmers of the region grow two types viz., sequential branching early maturing *Spanish* and alternate branching late maturing *Virginia* types together in intercropping system. In recent years, farmers have chosen to grow deep rooted sunflower as intercrop with groundnut.

In the arid zones of western Rajasthan and Gujarat, the practice of fallow to accumulate water in soil profile together with growing of early maturing millet and /or grain legumes planted at wider spacing in known to combat to certain extent the adverse effects of drought.

In Northern Nigeria which is more arid, the production systems are based on more of day neutral plant types, flowering during the determined period, irrespective of day length (Bunting and Kassam, 1988).

In the drier regions of Asia and Africa, farmers capture run-off water from higher ground by wide ranging methods (Bunting and Kassam, 1988)

Strategies to minimise the adverse effects of drought and enhance crop productivity levels in dryland agriculture

Crop growth on drylands depends apart from its regular management practices, on how efficiently rainfall is used to achieve economic yield levels. Rainfall potentially available for crop evapotranspiration (ET) is that, which is received during the period from harvest of the most recent crop to harvest of the crop under study. Water loss from the system other than the crop ET lowers rainfall-use efficiency.

Factors such as low infiltration and high run-off, low soil water storage capacity, evaporation (E) of soil water before crop establishment, and ET by non-crop plants all contribute to water loss. Added to the above, crops with inefficient or limited root system may not use water from soil profile efficiently, thus contributing to low rainfall-use efficiency (Unger *et al.*, 1988). Thus, the crop varieties with deep and robust root system are to be preferred for dryland agriculture.

Rainfall, soil and crop characteristics influence water infiltration and runoff. Runoff occurs when rainfall rates and amounts exceed the surface storage capacity and infiltration rate of soil, which happens with intense rain storms or when rainfall occurs very frequently. Low infiltration rates occur due to steep slopes, soil aggregate dispersion, soil surface sealing and impervious horizons in the soil profile. This problem also occurs with smooth soil surface devoid of crop residues before crop establishment. In some cases, despite good infiltration rates, the storage capacity of the soil might be low leading to low ratios of crop ET to rainfall levels. This phenomenon of low soil storage capacity has been observed in the case of shallow soils, which exhibit bed rock or other unfavourable soil conditions at low soil depth. Also the low storage capacity has been observed in sandy soils and soils with high clay content. In the case of permeable soils with good soil depth, infiltered water gets lost due to deep percolation. It is necessary to enhance the water storage capacity of soil. In order to achieve this soil should possess good depth without any impediments. Initial deep ploughing will improve the storage capacity of heavy soils.

Though evaporation of soil water is a natural process, its occurrence before crop establishment cuts down the amount of water available for subsequent ET by the crop. Evaporation as a part of total rainfall may be especially high in arid and semi-arid regions. Evaporation also occurs when inversion tillage exposes moist subsurface soil to atmosphere.

Losses due to transpiration by non-crop plants may occur before crop establishment or during crop growth due to weeds, volunteer crop plants, trees and shrubs.

Ratio of crop transpiration (T) to evapotranspiration (ET) may be relatively low when crop canopy coverage is incomplete to due to low and erratic plant populations. This may be rectified using adequate rates of good and viable seeds and satisfactory seeding methods.

Good germination, seedling emergence and survival are critical for success of dryland crop production. Some of the negative soil features such as unfavourable pH, poorly prepared seed beds, high salt or alkali levels and aluminium toxicity should be rectified.

Transpiration use efficiency (TUE) level

In crop production systems, transpiration use efficiency (TUE) is often based on the proportion of the crop that is of economic importance rather than total dry matter production. To achieve yields that are above the average level requires major changes in the crop such as increase in the proportion of partitioning of dry matter into the yield of the economic product or superior yield quality. Such changes are generally achieved through genetic improvements or major modification of growing conditions such as mist irrigation for temperature modification, which are of practically far reaching nature in dryland agriculture. Crop yields below the average occur when drought stress occurs during the vulnerable crop growth phases or due to lack of nutrients, biotic stresses that adversely affect the crop growth and development.

The important factor is that the amount of crop transpiration (T) rather than evaporation (E) which influences the crop yield. Thus transpiration or near potential transpiration for the prevailing conditions is desirable (Unger *et al.*, 1988). The TUE could also decline by delaying crop harvest beyond physiological maturity, which results in continued water use without increasing the crop yields.

Crops experience stress during the growing season either due to drought or high temperature in the arid and semi-arid ecosystems. Stress at any time may reduce plant growth and harvestable yield. However, stress at critical crop growth stages, even for short periods, can drastically reduce the yield.

Many grain crops viz., wheat (*Triticum aestivum* L), rice (*Oryza sativa* L) and sunflower (*Helianthus annuus*, L.) reach physiological maturity beyond which no further yield increase occurs despite the continuance of transpiration (T) until the completion of the crops life cycle.

Management techniques to improve Rainfall-use Efficiency

Soil and water considerations: Under conditions of dryland agriculture, soils are refilled with water from top i.e., through rains. On clay soils which have the property of swelling when get wet, fully wetting the soil profile is difficult, particularly when layers adjoining the surface is loaded with clay content. Additionally, the problems of clay pans and semi-hard pans in the soils too interfere with infiltration of rain water. Practices such as profile modification and deep ploughing, vertical mulching or para ploughing, which loosens the soil and increase water infiltration. Thus deeper and more uniform soil wetting that facilitates better root penetration and proliferation could be achieved. Deep tillage also enhances the water holding capacity of even sandy soils and cut down the erosion. The benefits due to better infiltration, crop yield and water use efficiency from soil profile modification and deep tillage are greater under conditions of limited precipitation which is common in dryland agriculture (Unger, 1979, Eck and Unger 1985). Research carried out on coarse-textured soils in India exhibited that deep tillage improved plant rooting by reducing soil mechanical resistance (Chaudary *et al.*, 1985).

When rainfall intensity exceeds the soil infiltration rate, storm run-off and soil erosion might occur. Soil and water conservation practices, such as land levelling and grading, furrow diking, contour tillage and terracing could be used to increase surface storage, reduce the slope gradient and manage and conduct water at no-erosive velocities. Ru-off conservation in water deficit areas could greatly enhance crop yields on rain dependent lands (Unger *et al.*, 1988). Soil and water conservation practices are explained in greater detail in a separate chapter.

The water holding capacity of coarse textured soils can be improved by adding organic materials to the soil or by deep tillage to bring finer soil materials to surface (Miller and Aarstad, 1972).

Evaporation accounts for major loss of water in arid and semi-arid zones (de la Paix Mupenzi *et al*, 2012; Unger, 1988; Bertrand, *et al.*,1966) Evaporation decreases and transpiration (T) increases as plant canopies develop. Also, Evaporation could be decreased by maintaining a crop residue or mulch cover on soil surface (Unger *et al.*, 1988). Low density straws such as wheat straw are much more effective on weight basis than sorghum or cotton residues in reducing evaporation (E) (Unger and Parker 1976).

Another lacuna in rainwater use is for dryland agriculture is losses by the way of 'water exploitation' by weeds and voluntary crop plants before the crops are planted and subsequently the competition for soil moisture and nutrients by the above during the crop cycle. Some common annual weeds growing with crops transpire about four times more water than a crop plants and use up to three times as much water to produce a pound of dry matter as do the crops. Under water stress condition annual weeds can cut crop yields more than 50% through moisture competition alone. The competition between weeds and crops is depending on weed density, the plant's physical characteristics rather than the aboveground biomass (Abouziena *et al.*, 2015). Effective weed control is essential pre-requisite for stabilising dryland crop production. According to Wiese (1983) perennial weeds can reduce crop yields as much as 75%. Weed populations can be controlled by effective tillage operation or by using herbicides. The most effective weed control, however, is through exploitation differences in biological characteristics of crops and competing weeds (Wiese, 1983).

In areas of less than 250 mm of seasonal rainfall, water harvesting using *salted sealed beds* to shed water and skip-row planting to concentrate water in the micro-zones of plants is reported to have considerable potential to enhance crop yields on soils with relatively high clay content (Unger *et al.*, 1988).

Crop Considerations: Cropping strategies to enhance precipitation use efficiency depend on climate, resources available and farmers' needs. Higher precipitation use efficiencies could occur using effectively annual crops, inter-cropping, double cropping. Non-cropped or fallow periods should be kept to a minimum. The improved management system developed at ICRISAT for crop production on Vertisols in semi-arid tropics, envisages dry sowing of sorghum in graded broad-bed-furrow (BBF) system immediately prior to the onset of monsoon. After harvest of the monsoon grown crop, a post-rainy season crop of chick pea is grown utilising the stored soil moisture.

Alternatively, pigeon pea is inter-cropped with the rainy season crop, which is the preferred means of extending cropping during the post-rainy season on Vertisols, because it eliminates the need for land preparation between the crops (El-Swaify *et al*, 1985).

A practical method of utilizing soil water and nutrients stored deep below the normal rooting zone of most of the annual crops is to rotate occasionally to a deeper rooted crop such as sunflower which has roots extended to a depth of over 3 meter. Alternatively, one could choose alfalfa (*Medicago sativa* L), whose roots go as deep as 6 to 7 meters

Ratio of Crop Transpiration (T) to Evapotranspiration (ET) Level

It is always desirable to maximise T to ET ratio so as to minimise the E (Evaporation levels) from soil surface (Unger *et al.*, 1988) to ensure appropriate crop water use efficiency. The E component can be minimised ensuring the use of well-adapted soil and crop management practices.

Soil Considerations: In order to achieve minimisation of E, the soil should be covered with crop at all times, which could be possible with perennial crops; but not with annual crops. Since E can't be avoided till the crop is sown, it is very important that soil conditions are favourable for timely and rapid seed germination and seedling establishment. This could be achieved with well-prepared seed bed backed up by efficient moisture conservation practices, which are fairly well known. Thus favourable seed-soil contact should be ensured for good germination and seedling establishment. The importance of checking biotic stresses together with avoiding unfavourable soil chemical conditions can't be over emphasised.

Crop Considerations: Once the crop canopy coverage is achieved the ET takes over T. Nevertheless, under dryland conditions total crop coverage is not achieved many times. As long as an appreciable amount of the soil surface is exposed to radiation, causing soil-air interchange with the atmosphere, E from the soil surface can be large. Therefore, to maximise ET, early plant growth and establishment are important requisite. In this context good and high seed quality together with good seed bed preparation backed up by moisture conservation are of paramount importance.

Another equally important factor is adoption of reduced tillage together with crop residue treatment to the soil surface, which changes the micro environment positively since soil temperature is reduced by such treatments (Ashton and Fisher 1986). Though closer or narrow-row spacing could be thought of (Steiner, 1986) in this context, narrow spacing could lower the crop yields (Bond *et al.*, 1964) when the soil water content at planting is low.

When the crop depends on water stored in the soil at sowing time, the depletion rate of water with time is important. This factor is much more relevant in respect of dryland crops grown on conserved soil moisture in the post rainy season, when no rains are received during the crop cycle. If the depletion of soil water is excessively early in the growing season, the crop is likely to undergo extreme stress during the subsequent reproductive and seed filling stages. Passioura (1976, 1983) proposed that plants with a high axial resistance to flow in roots and reduced partitioning of assimilates to roots would allow the crop to extend the period of water use over the growing season. Other researchers (Wright and Smith 1983) propose that root systems that extract water more efficiently from the soil profile, through deeper rooting depth can expand the soil water supply.

Use of efficient crop varieties with drought resistance and high moisture efficiency is one of the most spectacular means of combating drought on drylands. This has been discussed in greater detail in the chapter on 'Efficient Plant types'.

Where the cases of successful cropping depend both on stored soil water and highly variable seasonal rainfall, the two management options viz., rapid development of plant canopy and slow depletion of the soil-water supply are in operation. Unger *et al* (1986) showed that high residue levels from previous crops improve the response of dryland crops to growing season precipitation. Such mulching practices reduce E as a portion of growing season ET without excessively early season depletion stored soil moisture.

Well managed crop residues are highly effective to control erosion by water and wind. The residue mulches also conserve water, thereby assist in enhancing the crop yields. Crop residues also have value as livestock feed.

Though there are recommendations on the use of certain chemical agents, the recommended anti-transparent chemicals have not been found to be effective as these have shown toxic effects on field applications (Unger *et al* 1988). However, kaolinite treated soybeans showed a large increase a large increase in reflectance (up to 300%) in photosynthetic range and little effect on reflectance at other wave lengths (Doraiswamy and Rosenburg, 1974).

The highest T:ET ratios are generally in crops grown under a high level of management. If crop growth is limited by factors other than water or if water gets used by non-crop species, then the efficiency of the system is reduced. Timely harvesting is important to maintain high Transpiration Use Efficiencies (TUEs). Once the physiological maturity is reached, it is important to stop T of the crop to conserve soil moisture for the subsequent crop.

Under semi-arid dryland production, a range of management options is needed. Since the farmer operates under marginal conditions, the crop type, variety and management need to be carefully matched to the conditions of planting time as well as to probable conditions during the crop growing season (Unger *et al*., 1988)

Crop rotations that match precipitation patterns to crop water requirements and critical growth stages during the growing season are generally best suited to dryland systems. These often include crops with shorter growing seasons and crops that are able to predictably produce harvestable grain or forage with less water. In most semi-arid areas, including a fallow period in the rotation is necessary prior to crop with a higher water requirement. Crop Rotations are used to combat biotic stresses too and to achieve better use of the stored soil moisture. Crop rotations with different nutrient requirements could improve the nutrient availability to plant stand thus increase production with the same amount of water. Crop rotation can increase farmers' income while ensuring soil health and resisting the effects of climate change. Crop rotation reduces the negative effects of extreme weather on the planting system and reduces the fragility of the agricultural system. Some such examples of crop rotation are (1) sesame followed by gram or chickpea, (2) Sorghum followed by chickpea and (3) Pearle millet followed by groundnut.

Conservation tillage is any tillage system that leaves at least 30% of crop residue covering the soil surface. However, this is the minimum amount of residue generally considered to reduce soil erosion to acceptable levels. To reduce significantly reduce surface evaporation, 50% of crop residue cover is generally considered necessary. Crop residue performs a multitude of functions for water conservation: First it protects the soil surface from sealing caused by raindrop impact; it can provide small pockets or dams of crop residue to provide temporary water reservoirs during precipitation events; it captures snow – particularly effective with standing crop residue; and it reduces evaporation by shading the soil surface from the sun and reducing the wind velocity at the soil surface.

Capture of Runoff in Water Ponds

In many semi-arid regions, excess water is lost as storm runoff. It should be possible to collect the runoff water and store in the farm ponds. Such quantum of water may be used to irrigate part of the crop during moisture deficit periods. For efficient storage, the ponds must be constructed in low lying zone of the farm in soils with low permeability.

A water harvesting scheme will only be sustainable if it fits into the socio-economic context of the area as described in the previous chapter and also fulfils a number of basic technical criteria.

Figure 2 contains a flowchart with the basic technical selection criteria for the different water harvesting techniques. (https://www.fao.org/3/u3160e/u3160e07.htm) The following details are taken from the FAO.

SLOPE: The ground slope is a key limiting factor to water harvesting. Water harvesting is not recommended for areas where slopes are greater than 5% due to uneven distribution of run-off and large quantities of earthwork required which is not economical.

***SOILS*:** Should have the main attributes of soils which are suitable for irrigation: they should be deep, not be saline or sodic and ideally possess inherent fertility. A serious limitation for the application of water harvesting are soils with a sandy texture. If the infiltration rate is higher than the rainfall intensity, no runoff will occur.

***COSTS*:** The quantities of earth/stonework involved in construction directly affects the cost of a scheme or, if it is implemented on a self-help basis, indicates how labour intensive its construction will be.

Table 1 can be used as a quick reference to check the quantity of necessary earthworks required for different systems. A more quantified breakdown is given in the following sections where each system is described in detail.

Negarim microcatchment

Negarim micro-catchments are diamond-shaped basins surrounded by small earth bunds with an infiltration pit in the lowest corner of each. Runoff is collected from within the basin and stored in the infiltration pit. Microcatchments are mainly used for growing trees or bushes. This technique is appropriate for small-scale tree planting in any area which has a moisture deficit. Besides harvesting water for the trees, it simultaneously conserves soil. Negarim micro-catchments are neat and precise, and relatively easy to construct.

Table. 1: Earthwork/Stonework for Various Water Harvesting Systems

	Earthwork (m^3/ha treated)						Stonework (m^3/ha treated)	
System Name and Number	Negarim micro-catch-ments (trees)	Con-tour bunds (trees)	Semi cir-cular bunds (grass)	Conr ridges (crops)	Trapezoi-dal bunds (crops)	Water spread-ing bunds (crops)	Con-tour stone bunds (crops)	Per-meable rock dams (crops)
Slope %	(1)	(2)	(3)	(4)	(5)	(8)	(6)	(7)
0.5	500	240	105	480	370	305	40	70
1.0	500	360	105	480	670	455	40	140
1.5	500	360	105	480	970	N/R*	40	208
2.0	500	360	210	480	N/R*	N/R*	55	280
5.0	835	360	210	480	N/R*	N/R*	55	N/R*

*Not recommended

Notes

- Typical dimensions are assumed for each system: for greater detail see relevant chapters.
- For micro-catchment systems (1, 2, 3 and 4), the whole area covered (cultivated and within-field catchment) is taken as "treated".
- Labour rates for earthworks: Larger structures (e.g. trapezoidal bunds) may take 50% more labour per unit volume of earthworks than smaller structures (e.g. Negarim microcatchments) because of increased earthmoving required. Typical rates per person/ day range from 1.0 to 3.0 m3.
- Labour rates for stone-works: Typical labour rates achieved are 0.5 m3 per person/day for construction. Transport of stone increases this figure considerably.

Although the first reports of such microcatchments are from southern Tunisia (Pacey and Cullis, 1986) the technique has been developed in the Negev desert of Israel. The word "Negarim" is derived from the Hebrew word for runoff - "Neger". Negarim microcatchments are the most well-known form of all water harvesting systems.

Israel has the most widespread and best developed Negarim microcatchments, mostly located on research farms in the Negev Desert, where rainfall is as low as 100-150 mm per annum. However the technique, and variations of it, is widely used in other semi-arid and arid areas, especially in North and Sub-Saharan Africa Because it is a well-proven technique, it is often one of the first to be tested by new projects.

Technical details

i. Suitability

Negarim microcatchments are mainly used for tree growing in arid and semi-arid areas.

Rainfall: can be as low as 150 mm per annum.

Soils: should be at least 1.5 m but preferably 2 m deep in order to ensure adequate root development and storage of the water harvested.

Slopes: from flat up to 5.0%.

Topography: need not be even - if uneven a block of microcatchments should be subdivided.

ii. Overall configuration

Each microcatchment consists of a catchment area and an infiltration pit (cultivated area). The shape of each unit is normally square, but the appearance from above is of a network of diamond shapes with infiltration pits in the lowest corners.

iii. Limitations

While Negarim microcatchments are well suited for hand construction, they cannot easily be mechanized. Once the trees are planted, it is not possible to operate and cultivate with machines between the tree lines.

iv. Microcatchment size

The area of each unit is determined, depending on the plant (tree) water requirement (see Chapter 4) or, more usually, an estimate of this.

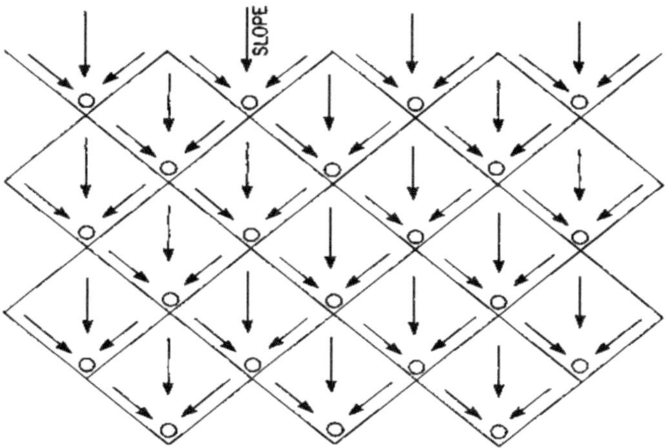

Fig. 2: Negarim micro-catchments - field layout

Size of micro catchments (per unit) normally range between 10 m2 and 100 m2 depending on the speciec of tree to be planted but larger sizes are also feasible, particularly when more than one tree will be grown within one unit.

The Table 2 below gives the bund heights in cm on higher ground slopes.

Table 2.

Size Unit Microcatchment	Ground slope			
(m^2)	2%	3%	4%	5%
3x3	even bund height of 25 cm			
4x4				30
5X5			30	35
6X6			35	45
8X8		35	45	55
10X12	30	45	55	
12X12	35	50	not recommended	
15 X 15	45			

The bund height is dependent on the selected size of the micro-catchment and the ground slope. It is suggested that bunds be constructed with a minimum height of 25 cm in order to forestall the damage by over-topping.

With the 2% ground slope or more, bund height near the infiltration pit may be increased . Table 2 presents the suggested figures for different sizes and ground slopes.

The 25 cm width of top of the bund and 1:1 range of the slopes, help in cutting down soil erosion in stormy rains. The bunds should have a cover of vegetation preferably of a grass to prevent any damage due to erosion.

This subject is discussed in greater detail in a separate chapter on soil and water and water management on drylands.

Role of Biotechnology in combating drought in dryland agriculture.

The traditional plant breeding methods rely on the available genetic variability for traits to be incorporated in the breeding programmes to achieve the needed genetic recombination involving the gene(s) of the identified promising germplasm lines. These approaches have served the needs of breeding for plant disease resistance, grain quality and of course higher seed / grain yield. Despite the marked success achieved through the conventional plant breeding methods, the time taken for the final recovery of needed genetic recombinants is rather slow and prolonged. Also the traditional methods depend on the naturally available genetic variability within the related plant genera and species for the needed traits.

On the other hand, the advances made in genetic research through gene manipulations and molecular-marker technologies facilitate exploitation of the unrelated or remote biological systems for the genes for abiotic stresses such as genetic resistance against moisture stress or drought leading to the recovery of trans-genetic recombinants. Thus, the trans-genetic recombinants offer structural genes for key enzymes that promote stress resistance and LEA proteins which play crucial role in drought / stress resistance together with certain regulatory genes. Thus, a number of drought/ stress-induced genes have been identified and some of these have been cloned.

Through the use of *Agrobacterium* spp., transgenic recombinant derivatives possessing different novel genes for drought tolerance have been developed in rice, wheat, maize, sugarcane, tobacco, groundnut, tomato, and potato. Field trials are being conducted to evaluate the GM cotton for drought tolerance and maize in the USA. Molecular markers have been used to identify drought-related quantitative trait loci (QTL), which can be used in crop varieties of rice, wheat, maize, pearl millet, and barley (Gosal *et al*., 2009)

The group of genes for drought resistance are those that produce proteins, which provide protection to the cells from the adverse effects of desiccation (Shinozaki and Yamaguchi-Shinozaki 2007). Umezawa *et al.*, (2006) and Shinozaki and Yamaguchi-Shinozaki (2007) reported the regulatory proteins for the transduction of the stress signal and modulation of gene expression.

With modern biotechnological tools, it has become possible to identify and isolate the specific genes / DNA segments controlling the trait and transfer them to another genetic system with impeccably perfect method. Biotechnology plays a vital role through development of crops resistant to drought, salinity, alkanity and iron toxicity, leading to increased yield levels from less available arable lands.

Arabidopsis thaliana of cruciferous group has been used extensively in understanding the functioning of the novel genes for stress tolerance, which are in use for transferring to crop plants grown on drylands.

Drought tolerance could be positively enhanced by the ABA levels, through manipulation of the functioning of stomata towards their closure to prevent excessive transpiration, which could be reversed with the dropping of ABA levels with the replenishment of water resulting in re-opening of stomata.

Wang *et al.*, (2005) identified the gene *ERA1* in *Arabidopsis* for certain enzymes, involved in ABA signalling. This is being used to advance further the genetic improvement for drought resistance. Transgenic derivatives for drought tolerance have exhibited promise under the conditions of water stress.

It is noteworthy that the performance between transgenic and traditional controls under the conditions of sufficient water demonstrated that superiority of transgenic derivatives (Wang *et al*, 2005). Multi-location trials have resulted in yield increases due to enhanced protection of the order of 15-25% to drought extended by the improved trnsgenics compared to non-transgenic controls (http://www.performanceplants.com).

The Canadian plant biotechnology company viz., *Performance Plants Inc,* is involved in developing the technologies for commercialization of superior transgenic maize, soybean, cotton, ornamentals and turf grass under the name *Yield Protection Technology™ (YPT™)* (https://www.bionity.com/en/companies/1039673/performance-plants-inc.html)

Despite significant and perceptible progress achieved in expounding the genetic mechanisms for drought tolerance, there is an urgent need to refine the genetic engineering technologies further in order to convert transgenic and genome modification technologies to suit field conditions.

The field conditions prevailing under agriculture are harsh and varied. In addition to drought, crops encounter several other kinds of stress such as biotic, salinity, alkalinity etc., to name a few. Hence it becomes necessary to test the available transgenics and other newly developed crop varieties for their response to multiple stresses under field conditions of dryland agriculture. (Mittler 2006).

2

Efficient Plant Types for Dryland Agriculture

Moisture deficit causing shortfalls in crop production is the characteristic feature of dryland agriculture. Other adverse factors viz., decline in soil fertility together with play of biotic stresses add to the complexities of dryland crop production.

Added to the proven agronomic management to combat the problem of drought in dryland agriculture, there is the need to tailoring of suitable crop varieties which can yield reasonably well under conditions of moisture deficit (Varshney *et al.*, 2021). In this context, the appropriate developmental rhythm along with the combination of the constellation of characters of the crop plant matching with the rainfall and the soil moisture availability patterns of the zone are the important considerations together with stability of yield (Hanson and Nelsen, 1980). This task being a complex one, the problem should be approached through multidisciplinary strategy that should buffer the genetic and breeding efforts.

The reproductive efficiency of a crop type depends upon the conversion of sun's energy into economic end-product, which in most cases happens to be fruit or grain particularly in crops used for human consumption and recovery of commercial product from grain or fruit. Portioning of photosynthates, the differential distribution of assimilates among organs, tissues and cells of plant should be more efficient (Good and Bell,1980).

Synthesis, translocation, partitioning and accumulation of photosynthetic products with in the plant are under genetic control and influenced by the environment in which, the plant is grown (Passioura,1981). The leaves and other green parts of the plant, which produce the photosynthates are called source and the plant parts that receive the products of photosynthesis through translocation for conversion into the end-product of economic value are known as sink. Source- Sink relationship in crop plants has immense bearing on crop productivity and production (Smith *et al.*, 2018).

The plant breeding advances that have been made to achieve sustainable productivity levels under conditions of dryland environment are due to not only specific drought adaptation; but also due to generalised genetic enhancement that expresses equally well under wide range of agro-ecological conditions (Hanson and Nelsen, 1980).

The genotype X environment interactions under drought conditions are large and non-systemic. Hence there is a need to widen the environmental situations under which the breeding materials are to be tested. In other words, evaluation of the plant breeding material over a wider range of drought-prone situations would be essential to arrive at consistently tangible results. Thus the selection of environments used by the breeder have to be progressively refined to differentiate between terminal drought and intermittent or mid-season drought of varying intensity during crop growth.

There is a need to describe clearly the probability of occurrence of a particular kind of drought regime in the environment under consideration. Thus the breeder should be in a position to tailor the character combination of the plant type to match the environment.

The characters involved may have to be considered in combination rather than in isolation.

The genotype X environment interaction that the breeder has to face in the task of breeding suitable crop varieties for dryland conditions is rather complex, considering the gamut of plant physiological attributes and the intricacies of the drought patterns. In other words, the crux of the problem remains in combining higher yield potential with stability of performance under variable conditions of dryland agriculture.

Thus designing suitable productive plant types for dryland situations demands a thorough integration of drought related plant physiological and breeding researches.

Understanding of the plant physiology of genotype X environment interaction helps in refining the breeding programme. For example, the complex objective of breeding for 'drought resistance' can be more precisely partioned into specific physiological processes (Lawn, 1988) viz., sink capacity or pod-development on one hand and root activity in terms of osmotic adjustment on the other. Selection criteria involving plant phenological attributes to match the duration of crop growth with the duration of adequate moisture availability is another powerful, but complex attribute to find place in crop breeding. Variation for phenological attributes among the segregates with in the breeding population contributes for selection for the interaction of phenology X drought pattern.

Gowda *et al*., (2009) described response of most crops to soil water deficit as a sequence of the following three successive stages of soil dehydration.

Stage I occurs at high soil moisture when water is still freely available from the soil and both stomatal conductance and water vapour loss are maximal. The transpiration rate during this stage is therefore determined by environmental conditions around the leaves.

Stage II starts when the rate of water uptake from the soil cannot match the potential transpiration rate. Stomatal conductance declines to keep transpiration rate similar to the uptake of soil water for maintaining the water balance of the plant.

Stage III begins when the ability of stomata to adjust to the declining rate of water up-take from the soil has been exhausted, and stomatal conductance is minimal. due to restriction of the assimilate supply rather than due to reduction of the grain storage capacity (Bieler *et al*., 1993)

Under very low water potentials, stomatal closure, and a consequent reduction in photosynthetic activity, have been reported in pearl millet (Henson *et al* 1984)

However, the supply of assimilates through the mobilization of stored soluble sugars can compensate for the impaired photo-synthetic activity (Fussell *et al*. 1991).

The transfer of assimilates from the leaves, with stems serving as a buffer during the grain development, appears to be the main adaptation trait during terminal drought stress in pearl millet (Winkel and Do, 1992). From a study involving normal and extended day length, Mahalakshmi and Bidinger (1985) suggested that photoperiod control of floral-initiation can provide an escape mechanism to avoid the coincidence of mid-season water stress with sensitive stages of millet growth.

Drought incidence can be of broadly three kinds in relation to crop growth. (a) Initial or early-season drought (b) Midseason drought and (c) End season drought (Prasad, 1973; Alagirisami,2016).

(a) *Initial phase or early drought* is the moisture stress that the crop encounters from the seedling to pre-flowering phase. Normal recommendation is that the irrigation be withheld for about 21 days soon after emergence. No doubt, the moisture stress that the crop encounters is not considered to adversely affect the growth and development of the plant or the final yield. It has been observed that the initial with holding of moisture supply for the first three weeks does help the crop in acquiring deeper root system and also promote condensation of the internodes towards the pod-bearing zone of groundnut plant (Prasad and Sudhakarbabu, 1997) resulting in profuse flowering, soon after availability of moisture, following during the initial drought of 20 to 23 days (Nageswara Rao, *et al.*, 1988, Puangbut *et al.*, 2010) leading to better yield.

(b) *Midseason drought:* This is the kind of moisture stress that crop commonly encounters under conditions of dryland agriculture, since it occurs at the stages of peak flowering, initial sink development into pods and grains. If this kind of drought is prolonged for more than 15 to 20 days, it shall have an adverse effect on pod or grain development and results in decline in yield. While the root system could tap moisture from deeper layers of the soil, the crop canopy may not be much affected. The crop genotypes possessing a kind of extended mid vegetative phase, are known to tolerate this kind of drought rather than early flowering and early maturing types, which get caught up in mid-season drought. For example, among the sub-specific forms of *Arachis hypogaea,* the *virginia* types can withstand the midseason drought while precocious *spanish* types in general are more susceptible to this kind of drought. The major difference between the above two types is that *virginia* types recover rapidly from the prolonged midseason drought, while *spanish* types fail to do so (Prasad and Reddy, 1992)

(c) *End season drought:* This kind of drought occurs at pre-maturity stage. Normally the end season drought occurs at the final phase of the crop growth, coinciding with the seed filling and pod hardening stage. The early flowering and precocious crop varieties like *spanish* and *virgina* bunch types of groundnut can tide over the ill effects of the end season drought. The *virgina* runner types of groundnut, late maturing varieties of other crops are more vulnerable to the end season drought due to their staggered or spread-out pod formation and late maturity. The end season drought does affect the oil and protein contents and fatty acid composition in groundnut to certain extent (Dwivedi *et al.*, 1996), in addition to the total yield.

Since the kind of drought is site specific, it is always advisable to choose the right kind of crop variety / genotype depending upon the location with characteristic drought pattern.

Drought-resistant plants share a mechanism known as Crassulacean Acid Metabolism (CAM), which allows them to survive despite low levels of water. According to Xiaohan Yang *et al* (2017) of the US Department of Energy's Oak Ridge National Laboratory CAM is a proven mechanism for increasing water-use efficiency in plants. The research paper published in the journal *Nature Communications*, brings out that CAM is essentially a form of photosynthesis in which the pores in a plant's leaves only open to let in carbon dioxide at night.

To dilate the process further, Crassulacean acid metabolism (CAM) is a metabolic adaptation of photosynthetic CO_2 fixation that enhances plant water-use efficiency (WUE) and associated drought avoidance/tolerance by reducing transpirational water loss through stomatal closure during the day, when temperatures are high, and stomatal opening during the night, when temperatures are lower.

In the face of the rapidly increasing human population and global warming predicted over the next century, the WUE of CAM plants highlights the potential of the CAM pathway for sustainable food and biomass production on semi-arid, abandoned, or marginal agricultural lands (Borland *et al* 2009, Cushman *et al.*, 2015), CAM photosynthesis can be divided into two major phases: (1) nocturnal uptake of atmospheric CO_2 through open stomata and primary fixation of CO_2 by phosphoenolpyruvate carboxylase (PEPC) to oxaloacetate (OAA) and its subsequent conversion to malic acid by malate dehydrogenase; and (2) daytime decarboxylation of malate and CO_2 re-fixation via C_3 photosynthesis, mediated by ribulose-1,5-bisphosphate carboxylase/oxygenase (RuBisCO). Malic acid is stored in the vacuole of photosynthetically active cells reaching a peak at dawn and can be used as a reference point to divide the two phases. CAM is found in over 400 genera across 36 families of vascular plants and is thought to have evolved multiple times independently from diverse ancestral C_3 photosynthesis lineages. The core biochemical characteristics of the CAM cycle are similar in all the plant lineages in which CAM has evolved, with some variation in the enzymes that catalyse malate decarboxylation during the day, and in the storage carbohydrates that provide substrates for malic acid synthesis at night (Holtum *et al.*, 2005). During the day, when the sun is out, the pores remain closed in order to prevent water escaping through them. This means they are better able to tolerate dry conditions.

The relevant plant attributes that need consideration in breeding suitable plant types for dryland agriculture are as below.

Water Use Efficiency (WUE): Among the various characters that contribute to drought escape, avoidance or tolerance to drought and through minimising the transpirational losses, total water use efficiency and harvest index are important criteria in breeding.

Grain yield is the function of water used (WU), water use efficiency (WUE) and Harvest Index(HI) (Passioura 1977). Also, it is reported (Fischer and

Turner1978, Tanner and Sinclair1983) that biomass accumulation is linearly related to cumulative transpiration. In other words, theoretically in order to recover maximum productivity, soil evaporation should be minimised and crops should extract as much water as possible. This theory is beset with risk in the sense that the crop is likely to exhaust available soil water before maturity, if it takes place in the field.

Alternatively, a more practical strategy would be where water use is less than the estimated availability, the plant should utilise it in a manner that would lead to higher yield stability. In other words, the efficiency of the genotype lies in its ability to adjust its cycle so as to accomplish grain development before the soil moisture gets exhausted. That leads to water use efficiency which is the ratio of biomass to the total amount of water transpired. Hence, it is suggested that water use efficiency should be an important and efficient selection criterion in breeding.

Rooting is an important character in dryland agriculture where moisture is the limiting factor. The simulation studies carried out by Jordan and Miller (1980) showed that the genotypes with deeper root systems resulted in the recovery of higher grain yield levels in sorghum. Increase in root zone depth has been shown to increase in leaf area growth, photosynthesis, transpiration (Jones and Zur 1984) and finally grain yield (Muchow and Sinclair 1986) under conditions of drought. Considerable genetic variation in rooting characters has been reported in sorghum (Jordan and Miller 1980), soybeans (Raper and Barber 1970) and wheat (Hurd 1974, Blum *et al.*1983); but studies on inheritance of rooting traits are not apparent. The character of deep root system facilitates extraction of sub-soil water which promotes total water use. The potential to manipulate rooting characteristics was demonstrated by Passioura and Richards (cf. Ludlow and Muchow, 1988). In this case selection was practised for narrow seminal roots with low hydraulic conductance to restrict water use prior to anthesis in wheat, resulting in the enhanced water quantum for grain development.

Osmatic adjustment: Another important plant physiological attribute which has bearing for dryland agriculture is the osmatic adjustment, which results from accumulation of solutes within cells, which lowers the osmatic potential and helps in maintaining turgor of shoot system and root system. Osmatic adjustment allows turgor-driven processes such as stomatal opening and expansion to continue. Osmatic adjustment may not have any effect on water use efficiency (Morgan *et al.* 1986; McCree and Richardson, 1987); but contributes to grain yield under drought conditions by increasing or maintaining the harvest index. Osmatic adjustment also reduces the rate of leaf senescence (Wright *et al.* 1983, Blum and Sullivan 1986). Osmatic adjustment maintains stomatal adjustment by which, the stomata remains partially open at progressively lower leaf-water potential as drought stress increases (Ludlow *et al.* 1985, Ludlow1987). Genotypes of sorghum and wheat possessing high osmatic adjustment are found to produce higher root biomass and greater root length density and are reported to extract more water from soil profile (Morgan and Codon 1986; Santamaria 1986). In grain legumes such as pigeon pea, it has been reported (Lawn, 1982) that osmatic adjustment results in enhancement of water extraction which lead to the increase of total biomass and higher harvest index, thereby conferring tolerance against the end-season drought.

Early vigour represents the positive vigour of the seedlings soon after germination which would lead to better initial canopy development. This attribute is particularly of relevance under conditions of dryland agriculture, where seedling establishment and canopy growth are the conspicuous problems causing decline in yield. Turner and Nicolas (1987) reported that vigorous early growth resulted in higher dry-matter production at anthesis, which in turn enhanced grain yields without any decline in harvest index. The authors pointed out that on deep sandy soils, vigorous early growth facilitated better root development and higher moisture use efficiency, which promoted yields without getting restricted due to moisture shortage at the end of the season.

Lawn (1982) also reported that selection for higher initial dry matter production a consequence of early vigour, in turn facilitates high remobilization of pre-anthesis assimilates. Over and above, the superimposition of initial dry matter production together with harvest index would be more appropriate to achieve the desired goals.

Early vigour also facilitates increased radiation interception in cereals which also contributed to higher grain yields. This however, may not be the case in grain legumes where growth and disposition of leaf angle are much different from those of cereals (Ludlow and Muchow 1988).

Leaf area maintenance is another important factor that needs attention under conditions of dryland agriculture, since reduced leaf growth and accelerated leaf senescence are common problems observed under conditions of moisture shortage, the basic feature of dryland agriculture, causing detriment to productivity even after amelioration of moisture deficit. Therefore, sustained leaf area is found to contribute to yield, though it might result in enhanced evapotranspiration. Nevertheless, the sustenance of leaf area is found to enhance yield stability during intermittent moisture stress situations due to better radiation interception. This, however changes to opposite situation at terminal or end season drought as the crop reaches grain maturity. Thus, while sustenance of higher leaf area might be favourable during intermittent drought spells; but may not be useful at the end season drought. Duncan *et al.* (1981) and Rosenow *et al.* (1983), brought out the genetic variation for leaf area maintenance, thereby indicating its genetic control. Delayed leaf senescence and "stay-green" is an important trait under drought stress, indicative of plant water status, and is useful for selection of maize genotypes under drought stress, although relatively less weight is given in the selection index (Banziger *et al.*, 2000)

The stay-green characteristic of maize facilitates a long grain-filling period and a long duration (Choi *et al.*, 1995). Bekavac *et al.* (1998) detected highly significant genetic and phenotypic correlations between stay-green, stalk water content, leaf water content

High temperature tolerance of crop types could be very important in those dry areas where the soil surface temperature might reach around 60 degrees C, which perhaps is more relevant in the case of maize (Peacock 1982; Mc Cown *et al.* 1980, 1982) resulting in loss of yield due to mortality of individual plants in the population. It has been reported that the seedling emergence of cowpea (Onwueme and Adegoroye 1975) and soybean (Emerson and Minor 1979) are adversely affected by high soil temperatures. Improved high temperature tolerance has been reported to result in higher grain yield. Though there are reports of the availability of genetic variability of the trait of high temperature tolerance in grain sorghum (Wilson *et al.* 1982), the inheritance of the trait is not discernible.

Sullivan and Ross (1979) reported a positive correlation between high leaf temperature tolerance and grain yield under hot and dry conditions.

Photoperiodism is a reaction in which a plant responds to the changes in the length of light and dark periods to complete the process of flowering. It also indicates that a particular flower can bloom in a specific period of the year, which may differ in every plant. Rosenow *et al.* (1983) demonstrated value of the photoperiod for plant responses to abiotic and biotic stresses have received increasing attention, since photoperiod control provides a mechanism to achieve matching the flowering time with the average period of the soil moisture availability, as has been done in sorghum (Bunting and Curtis 1970) and cowpeas (Summerfield *et al.*,1974). The reaction of flowering plants to photoperiod can cause alterations in growth and flowering. The major problem with photosensitive crop varieties is their very narrow adaptability (Ludlow and Muchow1988). There is adequate genetic variability for photoperiod sensitivity (Curtis 1968) and its inheritance has been established by Fery (1980).

In the context of dryland agriculture, one can see variation in photo period depending upon the light intensity from season to season. This kind of variation in photoperiod can cause variation in growth and development of crop plants, which again has a bearing on crop productivity under conditions of dryland agriculture. The crop plants that show the variation in response to photoperiod are termed photosensitive.

Occurrence of cloudy days and delays in onset of rainfall too can cause variation in photoperiod, adversely affecting the growth and development and finally yield of crops.

One such example is sesame crop, which is photosensitive resulting in poor yields due to variations in photoperiod due to onset of cloudy weather even under conditions of dryland agriculture. Hence there is a need to incorporate photo-insensitivity in varieties grown on drylands during rainy season which poses marked variation in photoperiod.

Harvest Index: It is the ratio of harvested grain to total shoot dry matter, which is highly heritable under both ideal and stressed environments (Hay, 1995) and very useful in plant breeding as a selection criterion.

Bidinger *et al*., (2000), however, have come up with a novel selection criterion which is termed the '*candidate trait for drought tolerance*'. They advocate '*panicle harvest index*', which is the ratio of grain to total panicle weight. The has been evaluated as a selection criterion for terminal stress tolerance in pearl-millet breeding.

The character of panicle harvest index is currently used as one of the traits for which quantitative trait loci (QTLs) are being identified from a mapping population made from parents that differ in the ability to maintain panicle harvest index under stress. The advantage with this trait is the ease with which it is readily and inexpensively measured in field experiments and can be used as a direct selection criterion. The main potential benefit to identifying QTLs for panicle harvest index would be to allow marker-assisted backcross transfer of improved tolerance of terminal stress to elite lines and varieties, without the requirement for extensive field screening.

Prolongation of Grain Filling

A prolonged vegetative phase of plant no doubt results in high total biomass. If this phenomenon is taken a bit far, it may result in the reduction of the duration of grain filling leading to low harvest index. With a balanced vegetative phase if the duration of reproductive phase of the plant could be enhanced, it may help in higher grain yield. Longer vegetative phase is not always positively correlated with grain yield whereas a longer period of reproductive phase and grain filling most frequently cause increased grain or fruit yield (Gill, 1994). The genetic variability for this trait is available in genetic resources of almost all crop plants. In other words, genotypes with earlier spring growth would facilitate earlier spike initiation and allow a longer period of spike development. The prolongation of grain filling period is of particular importance for the increase of grain dry-matter production. Generally, in many environments, this could be achieved, if flowering starts earlier and the time taken to reach maturity remains the same. This phenomenon also helps in economising the soil moisture particularly under conditions of dryland agriculture.

The duration of the vegetative phase and that of grain filling should be in harmony with the agro-ecological conditions for which the cultivar is developed (Gill, 1994). Of course, the incidence of biotic and abiotic stresses would have a bearing on the duration of above phenomenon. Under the conditions of incidence of end season drought, there would be an eventuality for grain filling to get restricted. If the storage of assimilates for grain development proceed at a very rapid rate during the short period of grain filling, sufficient quantities of photosynthates would have stored for grain formation, as shown by Russel (1984 and 1985) and Tollenaar (1991) in maize. This is an important point which should be taken into consideration for crop breeding under conditions of dryland agriculture.

Developmental adaptability of genotype to drought can be a very potent feature in which the individual plants adjust and acclimatise to the occurrence of intermittent drought through adjustment of their phenology as has been documented in certain grain legumes like pigeon pea.

Under conditions of dryland agriculture, which are highly location specific, it would be necessary to select for stability of performance over years with in the location, considering the temporal heterogeneity of drought patterns under each location.

On the other hand, the crop variety with broad adaptability might exhibit highest average yield over broad range of drought environments, but would be lower yielding than the variety with higher specific adaptability for diverse drought environments of the location.

Nevertheless, in view of complex interactions of drought and plant response, to achieve high yield potential together with broader adaptation seems difficult and hence it is more relevant to capitalize upon breeding of crop varieties with high specific adaptability, since drought problems of dryland agriculture are highly location specific (Ludlow and Muchow 1988).

Crop water productivity (CWP'): Gowda *et al* (2009) suggested what is called '*crop water productivity*' at the farm level through genetic enhancement of crops and natural resources management, particularly in view of the acute drought conditions prevailing in arid and semi-arid regions. In other words, the improved genotypes of dryland crops would become efficient agents of water conservation for the benefit of area of acute water shortage.

According to Gowda *et al* (2009) screening the plant breeding materials under water-limiting environments and imposed stress conditions together with evaluation for their yield stability are the crucial steps of breeding plant types with drought tolerance for dryland situations. Such investigations carried out at ICRISAT have resulted in the development of drought tolerant lines in sorghum, pearl-millet, groundnut, pigeon pea and chick pea which are listed by Gowda *et al* (2009)

Chlorophyll Stability Index: Immanuel *et al.* (2019) considered the effect of water stress on the chlorophyll stability index (CSI) and the relative water content (RWC) of leaves of *Pongamia* in a pot experiment on six-month-old seedlings. The CSI of *Pongamia* was 88.5% and the RWC was 75.4% suggesting adaptive mechanisms which make it possible for the plant to maintain normal photosynthetic rates under drought.

Photosynthetic Efficiency: Photosynthesis is the basic plant physiological process apparently positively linked to the to the dry matter production, which in turn is tuned to the grain yield. The enhancement of photosynthetic rate by the higher irradiance or CO2 levels is expected to lead to increases in yield. Also, there is substantial variation in the rate of photosynthesis per unit leaf area among plant genotypes of many crops.

While there are no many cases of direct selection in the past for photosynthetic rates, it was expected that some indirect selection would have happened, given the central role of the process of photosynthesis (Evans 1983). Though it is logical to expect increases in crop yield following the selection for higher photosynthetic rates, there is no evidence to support it (Gill, 1994). Varieties with higher yield potential have not been found to possess higher light saturation rates of CO2 exchange per unit of leaf area (Gill, 1994). The success achieved so far in plant breeding is attributable to the increase in photosynthetic area rather than to any increase in efficiency of photosynthesis (Borojevic, 1990; Parry *et al*, 2011). The two crops in which higher photosynthetic rates might be associated with higher yield potential are beans and soybeans (Gill, 1994).

Another important factor of photosynthetic mechanisms is the difference between the C3 and C4 plants. The nomenclature viz., C 3 /C4 refers to the length of the Carbon skeleton of the first product of assimilation following photosynthesis. In the case of plants with C4 –dicarboxylic acid pathway of photosynthesis (Ex. maize, sorghum and sugarcane), it varies up to 350 kg dry matter per hectare per day. In the case C3-Calvin pathway of photosynthesis (Ex. temperate grasses, rice, many dicotyledonous crop plants), the potential growth rate is of the order of 200 kg of dry matter per hectare per day.

The major carboxylating enzyme of C4 plants has affinity to CO2, which is about twice as great as that of the carboxylating enzyme of C3 plants. Additionally, the process of respiration occurs during photosynthesis in C3 plants, which results in dependence of assimilation on the oxygen concentration in the air. This process is termed as photorespiration, which is an inherent limitation. It may be noted that photorespiration is absent in C4 plants. Hence, the net assimilation rate of C4 plants at high light intensities is twice as that of C3 plants. The process of selection of C3 plants with the traits of C4 plants has not succeeded so far. Nevertheless, efforts are being made to transfer C4 character into C3 plants through modern techniques of genetic recombination.

The more practical approaches of enhancing photosynthesis according to Borojevic (1990) and Parry *et al* (2011) are through changes in the canopy structure, duration of photosynthetic activity, maintenance of crop health by prolonging the duration of green canopy and other source attributes with chlorophyll and prolongation of grain filling period.

Plant breeding strategies towards developing efficient plant types: An efficient plant type that can yield well under conditions of moisture-cum-temperature stress, should necessarily possess a combination of characters discussed above.

The three strategic approaches mentioned below would lead to breeding crop genotypes with enhanced adaptation to drought (Nigam *et al.*,2003)

1. Developing short duration genotypes that can escape drought, especially end-season drought,
2. Development of genotypes with superior yield performance in drought-prone regions following the adoption of traditional breeding approaches described below.
3. Development of drought resistant genotypes through plant physiological breeding.

Plant breeders have a challenge before them to develop cultivars adapted to changing micro-environments, which is a common problem in dryland zones. In many instances simple and straight recombination breeding involving two parents may be a partial step.

Parental Selection from out of germplasm begins with an understanding of the goal of the breeding program and the plant breeder must consider many characters for improvement and assign priorities to the traits considered for improvement. It is desirable that the parents are genetically diverse in order to facilitate selection of transgressive segregants, i.e., progeny that combines the genes of parental genotypes and with a phenotypic expression level, superior over either parent. The parental lines may be from many different sources including existing cultivars, adapted elite breeding lines, or new introductions from global collections.

- In order to obtain a pure-line cultivar that combines all the genes and meets the breeding objectives, hybridization of homozygous lines with complementary traits is carried out, followed by pedigree selection of recombinant lines from segregating progeny (Culbertson, 1954; Kenaschuk, 1975).
- **Pedigree selection method** begins with the inter-crossing of homozygous lines, followed by selfing of the F1 hybrid plants to produce a large F2 seed population. F2 seed from individual crosses is sown the following growing season and single plants are harvested from F2 plots. Seed harvested from single plants planted in F3 progeny rows. Seed quality traits such as oil quality, oil content, and protein content are determined using near infrared resonance or gas chromatography from samples taken from the bulk seed lots, and data used to further scale down the population of desirable lines. Seed harvested from F3 single plants is used to establish F4 rows and the selection cycle of between sister-lines and within-family selection continues and is repeated in the F5 generation, following which the F5 bulk seed is used to establish replicated small yield plots to facilitate between line selections for seed yield, and the single plant selections are grown as progeny rows. Superior F5 lines are identified on the basis of yield and seed quality attributes in the replicated plots, and sub-lines of these are used to establish the subsequent F6 generation (Culbertson, 1954; Kenaschuk, 1975).
- **The Bulk Population method of selection** may be adopted in the case of small grain crops that cannot be harvested on single plant basis, which is the basic requirement of the pedigree selection method. Selection is delayed until later generations when selection for more complex characters such as seed yield is more effective. In the bulk method, a segregating population is harvested together and a representative sample of the seeds of the harvested diverse population is used to plant the next generation i.e., F3. The F3 plot is similarly harvested and a plot of F4

planted. This is continued until the F6 generation. Selection begins in the F6 generation, which is largely homozygous. Large populations of F6 need to be planted for selection to be effective at this stage (Prasad, 2021). Two methods of selection can be practiced at this stage: either pure-line selection (single plant selections, plant to row progenies, selections among homozygous lines) or mass selection (large number of individual plants selected and bulked and carried forward.

- **Single Seed descent** is proposed as an alternative procedure to Pedigree and Bulk methods. In this method attaining homozygosity and selection are separated in two different phases (Goulden 1939). As a modification of Pedigree method (Brim, 1966), single seed descent method allows accelerated development of recombinant inbred lines (RIL), which are derived from the single seed progenies of large F2 populations.

RIL populations are good materials for genetic studies and genetic mapping of DNA markers of differing traits in RIL population.

- **Backcross method** of breeding is used to incorporate simply inherited characters from one of the first donor parent by repeated back crosses of the FI and subsequent generations to the second recipient parent called recurrent parent, selecting each time for the character in question of the donor parent that is being transferred. Backcross method of breeding was first used in linseed to develop a set of rust differentials by transferring individual rust resistance alleles into the cultivar Bison (Flor, 1955).
- **Recurrent selection** involves selection practiced within consecutive segregating progeny generations after the population has undergone selection for the major traits being transferred (Hull and Fred,1952; Vasal *et al.*, 2010). The procedure involves reselection generation after generation with interbreeding of selects to provide for genetic recombination. Recurrent selection is a proven method in which the selected individual plants are inter-mated among themselves and the succeeding progeny is subjected to selection over certain norms. The plants thus selected were allowed to inter-mate and the steps mentioned above will be continued till the selection advance is achieved for the attributes in question. Thus, it is a **cyclic selection** that is used to improve the frequency of desirable alleles for a character in a breeding population.
- **Biparental mating** permits evaluation of segregating (F2 or later generation) population of an individual cross made between two inbred lines. Each male is mated to different group of females and 'f' crosses

would be obtained. Variance is divided in two fractions i.e., due to males and due to females. Variance due to males provides an estimate of additive genetic variance. Variance due to females provides information on additive and dominance genetic variances. This technique helps in the selection of suitable breeding procedure. The method involves F2, Parent1 and Parent2 generations of a single cross. The method requires three crop seasons for generation of the material and the 4th season for evaluation. It provides information on additive and dominance components of genetic variance. It helps in the choice of breeding method for selection of polygenic characters. The analysis of the data is based on the second order statistics. The method is effective in breaking undesirable linkages of genes. The method can be used for self-pollinated as well as cross-pollinated plant systems

- **Mutation breeding** is worth taking up to induce new variability not available in the genetic resources of cultivated crops, in view of its practical feasibility and reduction of time lag which is characteristic feature of elaborate breeding programmes based on hybridization. Mutagenesis is the process by which the genetic information of an organism is changed in a stable manner. Mutations do occur in nature spontaneously at a low frequency in a vast population, since mutation is an established property of gene (Prasad, 1985). Mutations are the primary source of all genetic variations existing in any organism, including plants. Through mutation breeding the breeder tries to enhance the probability and frequency of occurrence of mutations for needed traits employing physical and / or chemical mutagens (Ahloowalia and Maluszynski 2001). Even in the case of wildspecies of crop plants too, their mutational rectification before their use in regular breeding is an important and productive strategy as has been done in other cases such as induced mutants for superior grain characters of *Triticum carthlicum*, for use in breeding of *T. durum*. (Prasad, 1972). Irradiation of small seeded groundnut variety resulted in, high yielding and disease resistant variety "Georgia Brown". Several large-seeded lines with high variability for disease incidence, pod yield, total sound matured kernels, pod weight, seed weight and seed size distribution were isolated (Branch, 2002).

Gamma rays have so far been the most widely used physical mutagen, since more than 80% of the mutant varieties were developed using such mutants in regular breeding programs(http://www-mvd.iaea.org). According the FAO/IAEA Mutant Variety Database (http://www-mvd.iaea.org) from the total number of registered mutant varieties there are number of 71 varieties are of *Arachis hypogaea*, L.

In addition to radiations like gamma rays, chemical mutagens like Ethyl-methane -sulfonate (EMS), Nitroso-methyl-urea (NMU), Nitroso-guanidine (NG) and Sodium- azide (SA) which are equally effective, could be used (Prasad *et al.*,1985). The M1 generation raised following seed treatment with mutagens must be selfed or grown in isolation to avoid cross pollination. The following diagram (Fig.1) gives schematic procedures to be followed in this context.

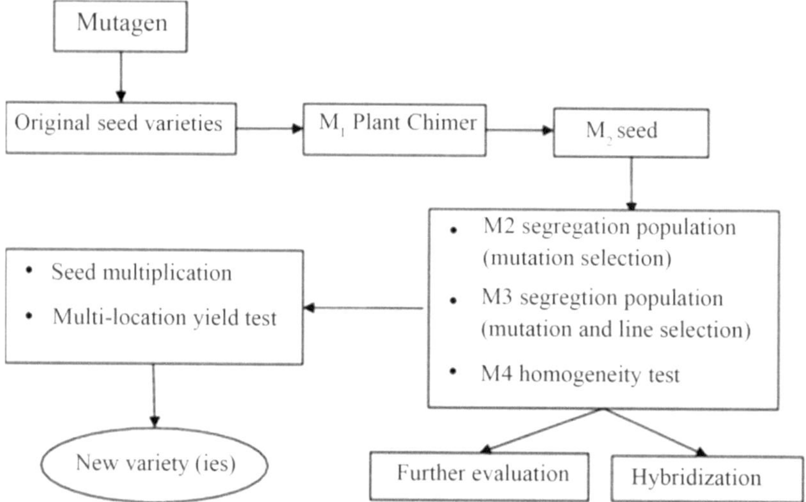

Fig.1. Schematic representation of the procedural steps involved in a regular Mutation Breeding programme. (Oladosu *et al.*, 2016)

• Indirect selection through appropriate plant physiological parameters

Certain surrogate characters have been identified using carbon isotope discrimination in leaf (Δ) and specific leaf area (SLA) which are associated with water use efficiency in groundnut (TE) (Wright *et al.*,1994).

The SLA is an easily measurable trait and as such can be used as a rapid and inexpensive selection criterion for high water use efficiency (TE) (Wright *et al.*, 1994, Nageswara Rao and Wright, 1994). It has also been reported that that the groundnut genotypes having low SLA (high TE) apart from exhibiting high water use efficiency also exhibited more stability in dry-matter production under drought (Nigam *et al.*,2003). A strong relationship between water use efficiency and specific leaf area and between Δ and specific leaf area revealed that genotypes with thicker leaves had greater water use efficiency (Wright *et al.*, 1994)

In this context, it may be mentioned that a portable hand-held SPAD Chlorophyll Meter can be used effectively following necessary protocols for rapid assessment of SLA and specific leaf nitrogen, the surrogate measures of TE (water use efficiency) particularly in groundnut (Nageswara Rao *et al.*, 2001), which can be tried in other low canopy crops.

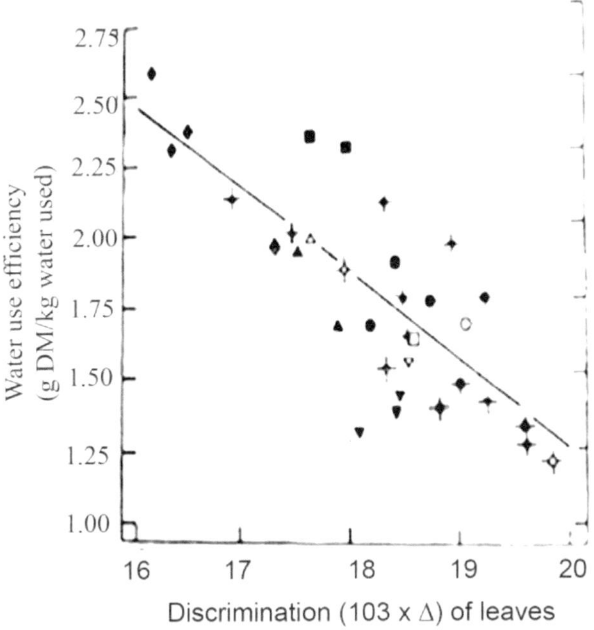

Fig. 2: Water Use Efficiency (TE) Vs. Carbon Isotope Discrimination (Δ) In groundnut cultivars (water use efficiency (TE)=7.347-0.307Δ; r=0.81; P<0.01) (Source: Hubic *et al.*, 1986)

- **Environments for selection**: It is necessary to emphasise that the plant breeding work

  (**a**) starting from study of initial segregating generations to the final selection in the advance generations of the genotypes with desirable character recombinants must be carried out under ideal environmental conditions that are devoid of moisture and nutrient stresses, conducive to the expression of the traits or characters under genetic control. This procedure is recommended in order to facilitate all the genetic recombinants possessing desirable attributes find appropriate environmental milieu for their phenotypic expression. Any initial selection process described above, if carried out under dryland conditions may result in losing of valuable genetic recombinants that can't find adequate phenotypic expression due to environmental stresses characteristic feature of dryland environment together with intra-population competition.

(**b**) The plant populations of selected lines with desirable combinations from the advance generations, however, may be evaluated under typical dryland situations for subsequent selection process.

- **Biotechnology for dryland plant varietal complex:** Wild relatives of crop plants are endowed with important traits that can enhance the adaptability of crop varieties for cultivation under conditions of dryland agriculture. When there exist strong isolation barriers between the cultivated crops and wild relatives, it should be possible take recourse to biotechnological tools to overcome the crossability barriers and achieve gene transfer. The ICRISAT have carried such researches in groundnut, chickpea and pigeon pea for achieving the transfer of traits of value from wild species to the cultivated varieties.

  1. In groundnut breeding (*Arachis hypogeae* L.) the wild species employed are *Arachis glabrata*, *A. cardenasii* and *Arachis pintoi* to transfer multiple resistance to drought, diseases and insect pests (Fisher and Cruz, 1994).

  2. In the case of chick pea (*Cicer arietinum*) the wild relatives of the crop viz., *Cicer stapfianum, Cicer subaphyllum,* and *Cicer pungens* possessing drought tolerance and deep root system (van der Maesen; cf Gowda *et al*., 2009) were crossed with the cultivated chick pea employing biotechnological tools.

  3. Pigeon pea (*Cajanus cajan* L.) is another leguminous crop in which biotechnological approaches gave dividends in crosses of the cultivated pigeon pea with *Cajanus acutifolius* and *Cajanus confertiflorus* to transfer silvery hairs, which confer drought tolerance (van der Maesen, 1986).

  4. An interspecific derivative in chickpea, BG 1103 (renamed as Pusa 1103), has been released for cultivation in northern India because of its high yield and tolerance to *Fusarium* wilt and drought (Abbo *et al*., 2007).

- **Marker-aided introgression of (Qunatitative Trait Loci) QTL associated with drought tolerance:**

Pearl millet Linkage groups (LG) 2, 4 and 6 are reported to harbour several QTLs associated with drought tolerance, flowering, straw and grain yield, panicle numbers, HI, and panicle HI, with some common QTL across stress environments and tester backgrounds (Yadav *et al*., 2002, 2003,).

https://www.nature.com/scitable/topicpage/quantitative-trait-locus-qtl-analysis-53904/

Transgenic approaches for breeding abiotic stress tolerance in plants

Genetic transformation to introduce novel genes for better tolerance to water deficit offers an attractive option (Xu *et al.*, 1996; Sivamani *et al.*, 2000). Therefore, transgenic technology too puts before the scientists an alternative approach to create novel stress tolerant genotype through transgenesis of hitherto inaccessible genes across the genus of species barriers. Novel genes may be accessed from exotic sources of plant, animal and bacteria for introduction into crop plants either physically through biolistics or by using binary vectors based in *Agrobacterium tumefacience.* Further it is also possible to control the timing, tissue specificity and expression level of transferred genes for their optimal function. The transgenic modifications of biosynthetic and metabolic pathways indicate higher stress tolerance may be achieved by introducing multiple mechanisms to water stress tolerance by genetic engineering (Bohnert *et al.*,1995). An efficient tissue and transformation system based on cotyledon explants from mature seeds has been developed for groundnut (Sharma and Lavanya 2002).

This allows the scientist to study the mechanism of drought tolerance by transferring a single gene from distant genetic sources, which otherwise would not be accessed.

The technology also offers to test the efficacy of different genes on gene-by gene basis so as to observe the phenotypic and biochemical alterations that occur before and after specific abiotic stress treatment.

Plants respond to drought stress by displaying complex quantitative traits that involve many genes leading to wide variety of biochemical changes. Processes sensitive to water deficit include protein synthesis, proto-chlorophyll formation, respiration and photo-respiration. The reduced turgor results in stomatal closure followed by reduction in water and gas exchange affecting the process of photosynthesis.

An effective approach to improve the drought tolerance of higher plants is to identify genes, for which transcripts or products are induced or upregulated when plants are subjected to such osmatic stress. Among the approaches used to study stress related gene expression, the one viz., the comparison of mRNA or polypeptide profiles of cells, tissues of plants maintained under normal and water-stressed conditions may be considered. Independently or in combination with the above, differential screening or the variation thereof of cDNA libraries from the above materials is employed to identify the mostly upregulated RNA species. Extensive comparisons with model organisms, halophytic plants and yeast, facilitate a paradigm for many responses to be exhibited by stress sensitive plants. (Bhattacharya and Singh 2003)

One response to plant water deficit is the synthesis of osmolytes, the compatible solutes that can accumulate to high levels without disrupting protein function (Bhattacharya and Singh, 2003). Osmolytes may include amino acids such as proline (Delauney and Verma 1993), sugar alcohols such as pinitol (Vernon *et al* 1993), other sugar viz., glycinebetaine (Rhodes and Hanson, 1993; Hayashi *et al.*, 1997) and poly amines. Recently number of genes involved in the biosynthesis of osmolytes were transferred into model plant species such as tobacco plants and drought resistant phenotypes were observed in the transgenics. (Bhattacharya and Singh, 2003). Over expression of gene encoding for moth bean P5CS in transgenic tobacco plants resulted in the accumulation of proline to the tune of 10mto 18 fold over control plants, thereby exhibiting better growth under dehydration stress (Kavikishore *et al.*, 1995).

Plant types for Intercropping: Intercropping is one of the very important strategies practised under dryland conditions to minimize risk of total crop failure due to inadequate or uncertain moisture availability during crop growth. The traditional intercropping practices in vogue in typical dryland agriculture have been evolved over large number of years with complementary crop components. Thus despite their low productivity levels, the traditional intercropping practices such as sesame+grainlegumes; castor + groundnut; pigeonpea+Groundnut or sorghum/ pearl-millet; cotton + groundnut; safflower+ chick pea are being practised, since the system offers coverage of risk against total crop failure. Agronomic researches have strengthened the practices of intercropping with nutrient management technologies and appropriate plant population densities and row-systems of planting. Nevertheless, the productivity of the system remains rather low since the crop varieties chosen are suited for pure cropping.

Traditionally, intercropping systems have been developed and improved by empirical methods within a specific local context. To support the development of promising intercropping systems, the individual species that are part of an intercrop can be subjected to breeding. Breeding for inter-cropping aims at resource foraging traits of the admixed species to maximize niche complementarity, niche facilitation, and intercrop performance. (Weih *et al.*, 2022).

Plant types for intercropping must possess the attributes of less vegetative growth, complementarity with the main crop and suitable maturity pattern that will not clash with that of main crop component. More studies are needed to synthesize the stabilizing role of genotypes for intercropping on drylands. In this context multifunctional approaches, long-term CA studies and optimal distribution of plant traits in crop combinations are recommended (Palacio *et al.*,2019)

The breeding process can be facilitated by modelling tools that simulate the outcome of the combination of different species' (or genotypes') traits for growth and yield development, reducing the need of extensive field testing (Weih *et al.*, 2022).

Breeding success can be accelerated by adoption of valuable emerging technologies for phenotyping and genotyping. However, proper identification, prioritization of the traits and their combinations to target among the many possible ones remains a major challenge.

In general, trait selection and crop breeding can be performed either on sole crops or intercrops. Yet, selection efficiency for intercropping adaptation under sole cropping conditions is generally moderate or low (Annicchiarico *et al.*, 2019), and elite cultivars selected for sole cropping systems might not be the optimal ones for intercropping. This is because, in a sole crop, desirable traits are often those increasing resource acquisition, whereas, when the same crop is grown in an intercrop, traits that optimize complementarity or facilitation can be more relevant (Costanzo and Bàrberi, 2014) but require complex attributes of above-and below-ground interspecific interactions. Also, traits that are plastic are likely to differ when plants are grown in sole crops or intercrops; and the often observed significant genotype × cropping system interactions indicate that specific breeding for intercropping is needed to exploit the genetic variability of the traits of interest in an intercrop context (Nelson and Robichaux, 2006).

Particularly important, in an intercrop context, are traits related to competitive ability and compatibility, which can be selected for by incorporating the relevant traits into selection indices (Annicchiarico, 2003; Annicchiarico and Filippi, 2007). Still, the relevant adaptive traits can vary with the intercropping systems and over time, reinforcing the need for careful considerations of appropriate trait combinations (Jensen *et al.*, 2015). For example, in general, leaf area, leaf area development, and plant height are all expected to enhance competitive ability. In a specific case of pea crop, the competitive ability was affected mainly by leaf area in early growth stages and plant height later on (Barillot *et al.*, 2014), which needs to be considered when this species is to be grown in an intercrop. While leafless pea types are desired in sole crops to improve population density and standing ability, leafy types of pea might be preferred in intercrops due to a higher growth rate and competitive ability (Semere and Froud-Williams, 2001). Thus, breeding for intercrops requires setting specific objectives for each of the admixed species in relation to the other(s). For example, in cereal-legume mixtures, the legume component is often less competitive due to a lower relative growth rate, so we could admix less competitive cereals or try to improve the competitiveness of the legume.

This can be achieved by selecting for **(i)** higher relative growth rate and plant height in the legume or lower in the cereal or both; **(ii)** greater plasticity of both, with implications for plant competition for light, including higher light absorption capacity under shading (Wang *et al.*, 2005); and **(iii)** early establishment of rhizobium symbiosis in legumes (Hauggaard-Nielsen and Jensen, 2001).

The crop growth models can assist breeding for intercropping, provided that (i) they incorporate the relevant plant features and mechanisms driving interspecific plant–plant interactions; (ii) they are based on model parameters that are closely linked to the traits that breeders would select for; and (iii) model calibration and validation is done with field data measured in intercrops.

Minimalist crop growth models are more likely to incorporate the above elements than comprehensive but parameter-intensive crop growth models. Their lower complexity and reduced parameter requirement facilitate the exploration of mechanisms at play and fulfil the model requirements for calibration of the appropriate crop growth models.

For models to be effective in assisting breeding of intercrops, they need to be designed so that they can be used to predict the best trait combinations for the specific end-use of the intercrop. To this end, models need (i) to incorporate the relevant plant features and mechanisms driving interspecific plant–to-plant interaction in the model; (ii) rely on parameters that are closely linked to the traits that breeders would select for; and (iii) be calibrated and validated with field data that are assessed in intercrops, if possible using advanced field phenotyping technologies to fulfil the parameter requirements of the common crop growth models. Due to their lower complexity and much reduced parameter requirement, minimalist crop growth models are more likely to incorporate the above elements than comprehensive and parameter-rich crop growth models (Weih *et al.*, 2022).

3

Soil and Water Management in Drylands

Soil and water are the basic resources in drylands. Soil supports plant life by providing a medium for their growth and development. It is a non-renewable natural resource and susceptible to rapid degradation through various forms of erosion processes. Worldwide, around 52% of total productive land has been degraded by various kinds of degradation processes and almost 80% of the terrestrial land is affected by water erosion .Further, annually ~10 million hectares (Mha) of cropland becomes an unproductive at the global level due to soil erosion with an average rate of 30 t ha^{-1} $year^{-1}$ soil erosion .In India, out of 328 million hectares of geographical area, 68 million hectares are critically degraded while 107 million hectares are severely eroded. Soil degradation in India is estimated to be occurring on 147 million hectares (Mha) of land, including 94 Mha from water erosion, 16 Mha from acidification, 14 Mha from flooding, 9 Mha from wind erosion, 6 Mha from salinity, and 7 Mha from a combination of factors (Battacharya *et al.*, 2015). This is extremely serious because India supports 18% of theworld's human population and 15% of the world's livestock population, but has only 2.4% of the world's land area

The uppermost weathered and disintegrated layer of the earth's crust is referred to as soil. The soil layer is composed of mineral and organic matter and is capable of sustaining plant life. The soil depth is less in some places and more at other places and may vary from practically nil to several meters. The soil layer is continuously exposed to the actions of atmosphere.

The word erosion has been derived from the Latin word „erodere" which means eating away or to excavate. The word erosion was first used in geology for describing the term hollow created by water. Erosion actually is a two-phase process involving the detachment of individual soil particle from soil mass, transporting it from one place to another (by the action of any one of the agents of erosion, viz; water, wind, ice or gravity) and its deposition. Hence soil erosion is defined as a process of detachment, transportation, and deposition of soil particles (sediment). It is evident that sediment is the end product of soil erosion process. Sediment is, therefore, defined as any fragmented material, which is transported or deposited by water, ice, air or any other natural agent. From this, it is inferred that sedimentation is also the process of detachment, transportation, and deposition of eroded soil particles. Thus, the natural sequence of the sediment cycle is as follows

Soil ⟶ Detachment ⟶ Transportation ⟶ Deposition

Soil erosion is classified into two categories, i.e., accelerated, and geological erosion. Geological erosion is the natural phenomenon, occurs through the constant process of weathering and disintegration of rocks in which the rate of erosion remains lower than the soil formation rate. In contrast, in accelerated erosion, the rate of soil erosion exceeds a certain threshold level and becomes rapid. Anthropogenic activities such as slash-and-burn agriculture, overgrazing, deforestation, mining, and intensive and faulty agriculture practices are accountable for accelerated soil erosion. This higher rate of soil erosion leads to the removal of organic matter and plant nutrients from the fertile topsoil and eventually lowering crop productivity. Hence, the conservation and management of natural resources are essential. Although the soil erosion cannot be eliminated, however it must be reduced to the level that can minimize its adverse impact on productivity and agricultural sustainability.

It also creates the problem in arid and semiarid regions, characterized by an intensive rainstorm and scanty vegetation cover. Water erosion comprises three basic phases, i.e., detachment, transportation, and deposition. Rainfall is one of the major factors which causes the movement and detachment of soil particles. The detached soil particles seal the open-ended and water-conducting soil pores, reduce water infiltration, and cause runoff. The first two phases determine the quantity of soil to be eroded and the third phase determines the distribution of the eroded material along the landscape. If there is no dispersion and transport of soil particles, there will be no deposition. Hence, detachment and transport of soil particles are the primary processes of soil erosion. Splash, sheet, rill and gully erosion are main forms of soil erosion by water.

a) Causes of Soil Erosion

No single unique cause can be held responsible for soil erosion or assumed as the main cause for this problem. There are many underlying factors responsible for this process, some induced by nature and others by human being. The main causes of soil erosion can be enumerated as:

Destruction of Natural Protective Cover by (i) indiscriminate cutting of trees, (ii) overgrazing of the vegetative cover and (iii) forest fires. (2) Improper Use of the Land (i) keeping the land barren subjecting it to the action of rain and wind, (ii) growing of crops that accelerate soil erosion, (iii) removal of organic matter and plant nutrients by injudicious cropping patterns, (iv) cultivation along the land slope, and (v) faulty methos of irrigation

2.Types of Soil Erosion

Soil erosion can broadly be categorized into two types i.e. geologic erosion and accelerated erosion

(i) **Geological Erosion**: Under natural undisturbed conditions an equilibrium is established between the climate of a place and the vegetative cover that protects the soil layer. Vegetative covers like trees and forests retard the transportation of soil material and act as a check against excessive erosion. A certain amount of erosion, however, does take place even under the natural cover. This erosion, called geologic erosion, is a slow process and is compensated by the formation of soil under the natural weathering process. Its effects are not of much consequence so far as agricultural lands are concerned.

(ii) **Accelerated Erosion:** When land is put under cultivation, the natural balance existing between the soil, its vegetation cover and climate is disturbed. Under such condition, the removal of surface soil due to natural agencies takes places at faster rate than it can be built by the soil formation process. Erosion occurring under these condition is referred to as accelerated erosion. Its rates are higher than geological erosion. Accelerated erosion depletes soil fertility in agricultural land. Soil erosion is broadly categorized into different types depending on the agent which triggers the erosion activity. Mentioned below are the four main types of soil erosion

(iii) **Water Erosion**: Water erosion is seen in many parts of the world. In fact, running water is the most common agent of soil erosion. This includes rivers which erode the river basin, rainwater which erodes various landforms, and the sea waves which erode the coastal areas. Water erodes and transports soil particles from higher altitude and deposits them in low lying areas. Water erosion may further be classified, based on different actions of water responsible for erosion, as : (i) raindrop erosion, (ii) sheet erosion, (iii) rill erosion, (iv) gully erosion, (v) stream bank erosion, and (vi) slip erosion.

(iv) **Wind Erosion**: Wind erosion is most often witnessed in dry areas wherein strong winds brush against various landforms, cutting through them and loosening the soil particles, which are lifted and transported towards the direction in which the wind blows. The best example of wind erosion are sand dunes and mushroom rocks structures, typically found in desert

(v) **Glacial Erosion**: Glacial erosion, also referred to as ice erosion, is common in cold regions at high altitudes. When soil comes in contact with large moving glaciers, it sticks to the base of these glaciers. This is eventually transported with the glaciers, and as they start melting it is deposited in the course of the moving chunks of ice

(vi) **Gravitational Erosion**: Although gravitational erosion is not as common a phenomenon as water erosion, it can cause huge damage to natural, as well as man-made structures. It is basically the mass movement of soil due to gravitational force. The best examples of this are landslides and slumps. While landslides and slumps happen within seconds, phenomena such as soil creep take a longer period for occurrence.

Mechanics of Soil Erosion

Soil erosion is initiated by detachment of soil particles due to action of rain. The detached particles are transported by erosion agents from one place to another and finally get settled at some place leading to soil erosion process.

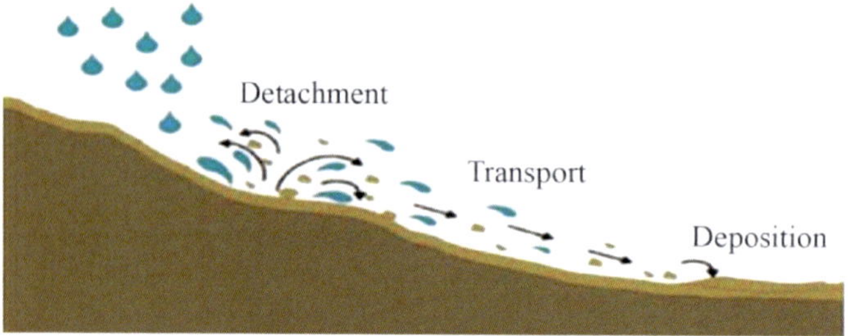

Fig. 1: Process of water erosion by the impact of raindrops (Mishra , and Mal,2020)

2.1. Types of water erosion

The other forms of water erosion are ravine formation, slip, tunnel, stream bank, and coastal erosion The different forms of water erosion are described below:

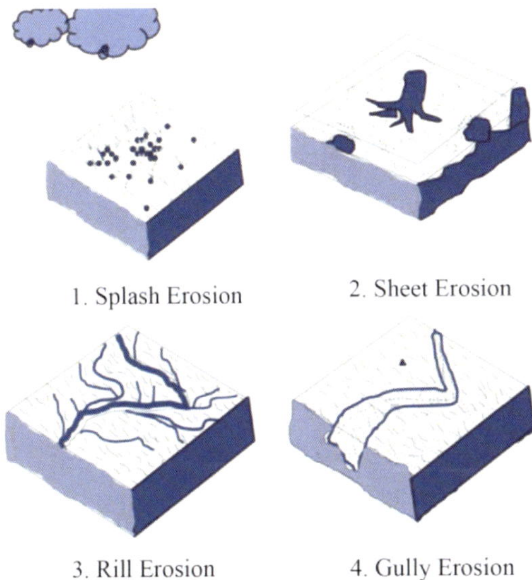

Fig. 2: Kinds of Erosion
(Source: Mishra and Mal.C,2020)

Splash erosion

Splash erosion is the first form of soil erosion by water. Falling raindrops on the soil surface break the soil aggregates and disperse and splash soil particles from their source, known as splash erosion. The process of splash erosion involves raindrop impact on soil particles, a splash of soil particles, and the formation of craters. The raindrops falling on soil surface act like a small bomb which disintegrates soil particles and forms cavities of contrasting shapes and sizes. The depth of craters is equal to the depth of raindrop penetration which is a function of raindrop velocity, size, and shape. In this form, soil particles can move only a few centimeters away from their source.

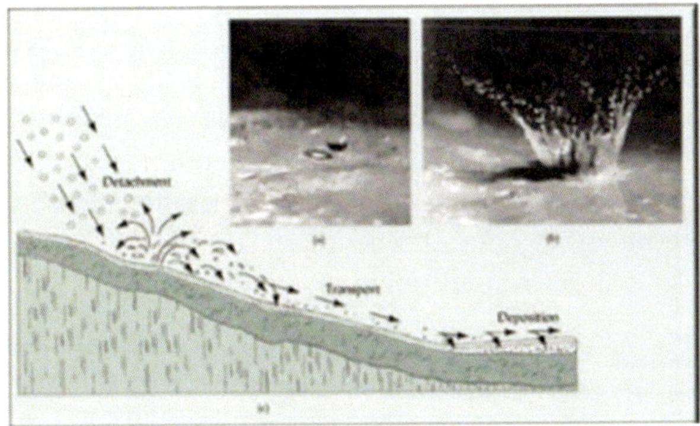

Fig. 3: Splash erosion. (Mishra, B and Mal, B.C,2020)

Sheet erosion

This is the next phase to splash erosion, which promptly initiates sheet erosion. The fertile topsoil surface is removed uniformly as a thin layer from the entire sloping surface area of the field by runoff water. Sheet erosion is a function of particle detachment, rainfall intensity, and land slope. The shallow flow of runoff water causes this type of soil erosion in which small rills are formed. This is the most common and severe form of soil erosion from an agricultural point of view as it removes the nutrient-rich top layer of soil. Out of total soil erosion, nearly 70% is caused by splash and sheet erosion only.

Fig. 4: Sheet erosion.(Mishra, B and Mal, B.C.,2020)

Rill erosion

It describes the flow of runoff water loaded with soil particles and organic matter in finger-like small channels, known as rill erosion. This is the advanced form of sheet erosion for soil loss. Water flow in small channels erodes soil at a faster rate than sheet erosion. Rill erosion is the second most common form of water erosion. These rills can be easily managed by tillage operations but can cause higher soil loss during intensive rainfall. The key factors that cause rill erosion are soil erodibility, land slope, runoff transport capacity, and hydraulic shear of water flow.

Fig. 5: Rill erosion. (Mishra, B and Mal, B..C,2020)

Gully erosion

Gully erosion is the advanced form of rill erosion. When the volume and velocity of concentrated runoff water increase, the rills become deep and broad and forms gullies. The gullies are linear incision channels with 0.3 m width and 0.3 m depth. Concentrated runoff flow is a primary factor for gully formation. Continuous gully erosion results in the removal of the entire soil profile. The extreme form of gully erosion may result in failure of crops, expose plant roots, reduce the groundwater level, and adversely affects landscape stability. It can cut apart the fields and aggravate the non-point source pollution (e.g., sediment, chemicals) to nearby water bodies. Gullies cannot be corrected by usual tillage operations. The dominant factors affecting gully erosion are shear stress of flowing water and critical shear stress of the soil. The further erosion of gullies results in ravines formation. Based on the size, depth, and drainage area, gullies can be classified as:

U-shaped gullies: These types of gullies are usually formed in alluvial soils where the characteristics of both the surface and subsurface soils are similar.

V-shaped gullies: This is the most common shape of gully erosion which occurs in the areas where the subsurface of soil is more resistant than the topsoil surface.

Fig. 6: Gully erosion. (Mishra, B and Mal, B.C, 2020)

Ravine formation

It is referred to as a network of deep and narrow gullies that flows parallel to each other while linking with the river system. Mismanagement and non-judicious use of land result in enlargement of rills and gullies and eventually lead to ravine formation. Abrupt changes in elevation of the riverbed and the adjoining land surface, deep and permeable soil with high erodibility, sparse vegetation, and backflow of river water during the recession period causes severe bank erosion which consequently results in ravine formation.

Tunnel erosion

It is the sub-soil erosion through runoff flow in channels while surface soil remains intact. Tunnel erosion is also known as pipe erosion and commonly occurs in arid and semiarid regions where the soil permeability for water varied with the soil profile. The further widening and deepening of tunnels form large gullies which degrade the productive agricultural lands. Soil with erodible characteristics, having sodic B horizon and stable A horizon are highly prone to tunnel erosion. Runoff flow through natural cracks and animal burrows initiates tunnel formation by infiltrating thorough dispersible subsoil layers. Seepage, lateral flow, and interflow are key indicators of tunnel erosion. It alters the geomorphic and hydrologic characteristics of the affected areas. Management practices for tunnel erosion are ripping, contour farming, vegetation including trees and deep-rooted grasses with proper fertilization and liming, consolidation of surface soil, and diversion of concentrated runoff

Slip erosion or landslip erosion

It is the downward and outward movement of slope forming materials composed of natural rocks and debris from sloppy lands. It is also known as mudslide or mass erosion. This type of erosion mostly occurs in hilly regions having water-saturated soils slips down the hillside or mountain slope. Banks along highways, streams, and ocean fronts are often subject to landslides. The large masses of land slip down which destroy the vegetation and degrade the productivity of lands. The slope can be stabilized through developments of diversion drains, contour trenches, crib structures, geotextiles, kutta—crate structures, and retaining walls.

Stream bank erosion

The scouring of soil material from the stream bed and cutting of stream bank by the action of flowing water is known as stream bank erosion. Streams and rivers change their direction of flow by cutting the bed from one side and depositing the sediment to the other side of the stream. Flash floods enhanced the stream bank erosion which is more destructive. Stream and gully erosion are relatively comparable. Primarily, stream bank erosion predominantly occurs at the lower end water tributaries which have a relatively flat slope and continuous flow of water.

Fig. 7: Stream bank erosion. (Mishra ,B and Mal, B.C,2020)

Sea-shore erosion: It is also called coastal erosion. Sea shore erosion is the wearing away of land and the removal of beach or dune sediments by wave action, tidal currents, wave currents, or drainage . Waves, generated by storms, wind or fast moving motor craft, cause coastal erosion which may take the form of long-term losses of sediment and rocks, or merely the temporary redistribution of coastal sediments. It may be caused by hydraulic action, abrasion, impact and corrosion

Fig. 8: Sea-shore/ coastal erosion.(Mishra, B and Mal, B.C. 2020)

Landslide Erosion: When gravity combines with heavy rain or earthquakes, whole slopes can slump, slip or slide (Fig. 3.7). Slips occur when the soil (topsoil and subsoil) on slopes becomes saturated. Unless held by plant roots to the underlying surface, it slides downhill, exposing the underlying material.

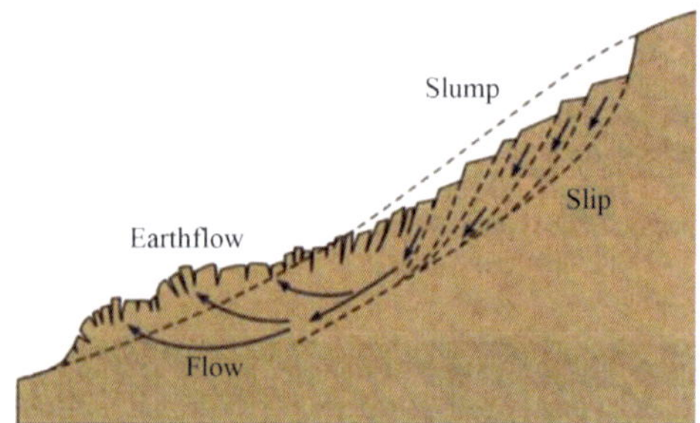

Fig. 9: Cross-section of landslide characteristics. (Mishra, B and Mal, B..c, 2020)

2.1 Forms of Water Erosion

Hydraulic Action

The hydraulic action takes place when water runs over the soil surface compressing the soil, as a result of which the air present in the voids exerts a pressure on the soil particles and this leads to the soil detachment. The pressure exerted by the air voids is called hydraulic pressure. The soil particles so detached from their places, are scoured by the running water. The hydraulic action is more effective when the soil is in loose condition

Abrasion: Soil particles mixed with the running water create an abrasive power in the water which increases the capacity of flowing water to scour more soil particles. Due to this effect, larger soil particles are eroded by the flowing water.

Attrition: This form includes mechanical breakdown of loads running along the moving water due to collision of particles with each other. The broken particles are moved along with the flow velocity, which generate abrasion effect on the bottom and banks of the water course. This effect pronounces the water erosion Solution: the water soluble contents present in the water are transported by the water in solution form.

2. Impact of soil erosion on agriculture

The accelerated soil erosion significantly influences the soil quality, agricultural production, and nutritional quality. Higher soil erosion results in the removal of fertile topsoil along with nutrients which leads to reduced agronomic yield, land degradation, and terrain deformation [Lal, R. 2009 and Lal, R, 2015]. The main causal factors affecting the rate of soil erosion are parent material, soil texture, slope steepness, plant cover, tillage, and climate. According to an estimate of existing soil loss data, the mean annual rate of soil erosion in our country is approximately 16.4 ton ha−1 which results in annual total soil loss of 5334 million tons (m t) and nutrient loss of 8.4 m t throughout the country [Battacharya *et al*, 2015]. However, the mean annual permissible limit of soil loss is 12.0 tons/ ha . Out of total eroded soil around 29% is permanently lost to the sea, while 61% is transported by runoff from one place to another and the remaining 10% is directly deposited in reservoirs. Higher nutrient concentration has been recorded in soil samples collected from runoff loads over the soil of agricultural fields Further, around 45.9 kg C ha−1 and 4.3 kg N ha−1 were recorded in eroded soil during the month of July [Santra, P.*et al*., 2013].

The soil organic matter (SOM) is vital for improving soil bio-physico-chemical properties and contains nearly 95% of N and 25–50% of phosphorus. Higher rate of erosion results in loss of soil and fine organic particles. The soil removed by erosion has 1.5–5 times higher SOM than the soil left behind The availability of SOM also effects the biological activities and soil biodiversity in a particular agro-ecosystem. Moreover, the intensive and erratic rainfall results in higher soil erosion which leads to reduced infiltration and eventually less water availability to the vegetation. Sharda *et al*. (2010) studied the impact of the harshness of water erosion on agricultural productivity and advocated that water erosion reduced the annual crop production by 13.4 Mt in 2008–2009 at the national level . Thus, the soil loss by water and wind severely affects the productive efficiency of all ecosystems The comprehensive impacts of erosion on soil and water resources which are liable to reduce agricultural productivity

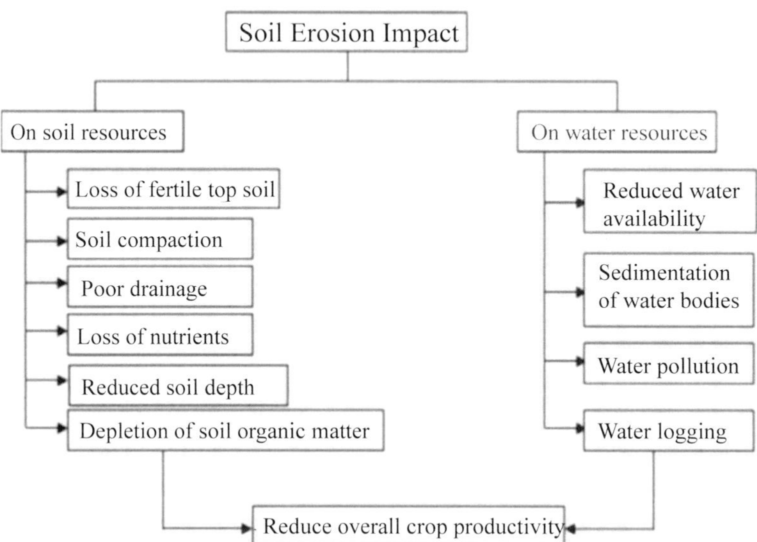

(*Source*: Anita kumawat, *et al* .,2020)

3. Factors Affecting Soil Erosion

Climatic Factor: The climatic factors that influence erosion are rainfall amount, intensity, and frequency. During the periods of frequent or continuous rainfall, high soil moisture or saturated field conditions are developed, a greater percentage of the rainfall is converted into runoff. This in turn results in soil detachment and transport causing erosion at high rate.

Temperature: While frozen soil is highly resistant to erosion, rapid thawing of the soil surface brought about by warm rains can lead to serious erosion. Temperature also influences the type of precipitation. Although falling snow does not cause erosion, heavy snow melts in spring can cause considerable runoff damage. Temperature also influences the amount of organic matter that get collected on the ground surface and get incorporated with the topsoil layer. Areas with warmer climates have thinner organic cover on the soil. Organic matter cover on the surface protects the soil by shielding it from the impact of falling rain and helping in the infiltration of rainfall that would otherwise cause more runoff. Organic matter inside the soil increases permeability of the soil to cause more percolation and reduce runoff

Topographical Factors: Among the topographical factors, slope length, steepness and roughness affect erodibility. Generally, longer slope increases the potential for erosion. The greatest erosion potential is at the base of the slope, where runoff velocity is the greatest and runoff concentrates. Slope steepness, along with surface roughness, and the amount and intensity of rainfall control the speed at which runoff flows down a slope. The steeper the slope, the faster the water will flow. The faster it flows, the more likely it will cause erosion and increase sedimentation. Slope accelerates erosion as it increases the velocity of flowing water. Small differences in slope make big difference in damage. According to the laws of hydraulics, four times increase in slope doubles the velocity of flowing water. This doubled velocity can increase the erosive power four times and the carrying (sediment) capacity by 32 times.

Soil: Physical characteristics of soil have a bearing on erodibility. Soil properties influencing erodibility include texture, structure, and cohesion. Texture refers to the size or combination of sizes of the individual soil particles. Three broad size classifications, ranging from small to large are clay, silt, and sand. Soil having a large amount of silt-sized particles is most susceptible to erosion from both wind and water. Soil with clay or sand-sized particles is less prone to erosion

Structure refers to the degree to which soil particles are clumped together, forming larger clumps and pore spaces. Structure influences both the ability of the soil to absorb water and its physical resistance to erosion. Another property is the cohesion which refers to the binding force between the soil particles and it influences the structure. When moist, the individual soil particles in a cohesive soil cling together to form a doughy consistency. Clay soils are very cohesive, while sand soils are the least cohesive

Vegetation: Vegetation is probably the most important physical factor influencing soil erosion. A good cover of vegetation shields the soil from the impact of raindrops. It also binds the soil together, making it more resistant to runoff. A vegetative cover provides organic matter, slows down runoff, and filters sediment. On a graded slope, the condition of vegetative cover will determine whether erosion will be stopped or only slightly halted. A dense, robust cover of vegetation is one of the best protections against soil erosion.

Biological Factors of Soil Erosion: Biological factors that influence the soil erosion are the activities like faulty cultivation practices, overgrazing by animals etc. These factors may be broadly classified into following three groups:(i) Energy factors, (ii) Resistance factors, and (iii) protection factors

Energy Factors: They include such factors which influence the potential ability of rainfall, runoff and wind to cause erosion. This ability is termed as erosivity. The other factors which directly reduce the power of erosive agents are reduction in length/degree of slope through the construction of terraces and bunds in case of water eroded areas and creation of wind breaks or shelter belts in case of wind eroded area

Resistance Factors: They are also called erodibility factors which depend upon the mechanical and chemical properties of the soil. Those factors which enhance the infiltration of water into the soil reduce runoff and decrease erodibility, while any activity that pulverizes the soil increases erodibility. Thus, cultivation may decrease the erodibility of clay soils but increases that of sandy soil.

Protection Factors: This primarily focuses on the factors related to plant cover. Plant cover protects the soil from erosion by intercepting the rainfall and reducing the velocity of runoff and wind. Degree of protection provided by different plant covers varies considerably. Therefore, it is essential to know the rate of soil erosion under different land uses, degrees of length and slope, and vegetative covers so that appropriate land use can be selected for each piece of land to control the rate of soil erosion. The quantity of soil moved past a point is called soil loss. It is usually expressed in unit of mass or volume per unit time per unit area

4. Soil and water conservation measures

It clearly shows that for deferent regions the problems of soil and water conservations are quite deferent. This information is useful in determining the appropriate soil and water conservation practices for various regions. This classification and related information also assist in utilizing the research and field experience of one place to other places of identical soil, climatic and topographic conditions.

Table 1: Soil and water conservation problems in various soil conservation regions of India

S.No	Soil conservation region	Rainfall (mm)	Important areas	Problems
5.	Desert area	150-200	Western central Rajastan, Contiguous areas of Haryana and Gujarat, Runn of kutch	Shifting sand dunes, wind erosion, extreme moisture Stress and drought ,over grazing, improper land management
6.	Mixed red, black Yellow soils	600-700	Districts of Pali, Bhilwara, Ajmer ,Chittorgarh Udaipur,Rnorthern Majasamand, Jhalawar in Rajasthan and Southern UP(including Bundelkhand area) and Northern MP	Ravine, shortage of moisture recurring drought , problem of drainage, overgrazing, siltation of reservoirs and tanks

7	Black soils	500-700	Southern eastern Rajasthan, Parts of Madhya Pradesh, tracts of Maharashtra, Andhra Pradesh ,Karnataka, and small parts of Tamil Nadu	Sheet erosion, acute water shortage, recurring droughts, ill drained soils, siltation of reservoirs, lack of ground water recharge
8	Black soils(deep and medium deep)	800-1300	Parts of Madhya Pradesh, Andhra Pradesh, and Maharashtra	High soil erosion, gully formation, waterlogging, poor workability of soil, shortage of water during post-rainy season
9	Eastern res soils	1000-1500	Bulk of West Bengal, Bihar, Orissa and Eastern Madhya Pradesh including Chotanagapur and Chhattisgarh area, Part Andhra Pradesh	Problems of sheet erosion gullies ,acute water shortage ,recurring drought, heavy grazing and improper land management, siltation of reservoir and tanks
10	Southern red oils	Around 750 in Kerala upto 2500	Bulk of Kerala, Tamil Nadu hills and Plains, Karnataka ,Andhra Pradesh and part of Maharashtra	Sheet erosion, gullies, acute water shortage, recurring drought, siltation of reservoir and tanks ,lack of ground water recharge
11	East -west coasts	East and West coast about 1000 and rest heavy rainfall	East and Western coast from Orissa to Saurashtra	Problems of coastal salinity. soil erosion ,coastal sand dunes, wind erosion, and flooding of cultivated lands by sea water or rainwater

(*Source*:Pathak *et al*., 2007).

Field based soil and water conservation measures are essential for in-situ conservation of soil and water. The main aim of these practices is to reduce or prevent either water erosion or wind erosion, while achieving the desired moisture for sustainable production. The suitability of any in-situ soil and water management practices depend greatly upon soil, topography, climate, cropping system and farmers' resources. Based on past experiences several field-based soil and water conservation measures have been found promising for the various rainfall zones in India

Table 2: Prioritized land-based soil and water conservation measures for various rainfall zones in India

Seasonal rainfall (mm)			
<500	500-700	750-1000	>1000
• Contour cultivation with conservation furrows	• Contour cultivation with conservation furrows	• BBF (Vertisols)	• BBF (Vertisols)
• Ridging sowing across slopes	• Ridging	• Conservation furrows	• Field bunds
• Mulching	• Sowing across slopes	• Sowing across slopes	• Vegetative bunds
• Scoops	• Scoops	• Tillage	
• Tied ridges	• Tide ridges	• Lock and spill drains	• Graded bunds
• Off-season tillage	• Mulching	• Small basins	• Chos
• Inter row water harvesting system	• Zingg terrace	• Field bunds	• Level terraces
• Small basins	• Off-season tillage	• Vegetative bunds	
• Contour bunds	• BBF	• Graded bunds	
• Field bunds	• Inter row water harvesting system	• Nadi	
• Khadin	• Small basins	• Zingg terrace	
	• Modified contour bunds		
	• Field bunds		
	• Khadin		

(Source:Pathak *et al* 2007)

There are two types of measures for soil and water conservation, that is, mechanical/engineering/structural measures and biological measures. Mechanical measures are permanent and semi-permanent structures that involve terracing, bunding, trenching, check dams, gabion structures, loose/stone boulders, crib wall, etc., while biological measures are vegetative measures which involve forestry, agroforestry, horticulture and agricultural/agronomic practice

4.a. Biological measures (agronomic/agricultural and agroforestry)

Agronomic measures are applicable in the landscape of ≤2% slope. Agronomic measures reduce the impact of raindrops through the covering of soil surface and increasing infiltration rate and water absorption capacity of the soil which results in reduced runoff and soil loss through erosion [FAO., 1984]. These measures are cheaper, sustainable, and may be more effective than structural measures, sometimes.

Important agronomic measures

Contour farming

Contour farming is one of the most used agronomic measures for soil and water conservation in hilly agro-ecosystems and sloppy lands. All the agricultural operations viz. plowing, sowing, inter-culture, etc., are practiced along the contour line. The ridges and furrows formed across the slope build a continual series of small barriers to the flowing water which reduces the velocity of runoff and thus reduces soil erosion and nutrient loss. It conserves soil moisture in low rainfall areas due to increased infiltration rate and time of concentration, while in high rainfall areas, it reduces the soil loss. In both situations, it reduces soil erosion, conserves soil fertility and moisture, and thus improves overall crop productivity. However, the effectiveness of this practice depends upon rainfall intensity, soil type, and topography of a particular locality. . It is suitable in almost all soils having the rainfall upto 1000mm with slopes ranging from 1.5 to 4%.

Choice of crops

The selection of the right crop is crucial for soil and water conservation. The crop should be selected according to the intensity and critical period of rainfall, market demand, climate, and resources of the farmer. The crop with good biomass, canopy cover, and extensive root system protects the soil from the erosive impact of rainfall and create an obstruction to runoff, and thereby reduce soil and nutrient loss. Row or tall-growing crops such as sorghum, maize, pearl millet, etc. are erosion permitting crops which expose the soil and induce the erosion process. Whereas close growing or erosion resisting crops with dense canopy cover and vigorous root system viz. cowpea, green gram, black gram, groundnut, etc. are the most suitable crops for reducing soil erosion .To increase the crop canopy density, the seed rate should be always on the higher side.

Cover crops

The close-growing crops having high canopy density are grown for protection of soil against erosion, known as cover crops. Legume crops have good biomass to protect soil than the row crops. The effectiveness of cover crops depends on crop geometry and development of canopy for interception of raindrops which helps in reducing the exposure of soil surface for erosion. It has been reported that legumes provide better cover and better protection to land against runoff and soil loss as compared to cultivated fallow and sorghum. The most effective cover crops are cowpea, green gram, black gram, groundnut, etc.

Advantages

- Protection of soil from the erosive impact of raindrops, runoff, and wind.
- Act as an obstacle in water flow, reduce flow velocity, and thereby reduce runoff and soil loss.
- Increase soil organic matter by residue incorporation and deep root system.
- Improve nutrients availability to the component crop and succeeding crops through biological nitrogen fixation.
- Improve water quality and water holding capacity of the soil.
- Improve soil properties, suppress weed growth, and increase crop productivity.

Vegetative Barriers

Vegetative barriers or vegetative hedges or live bunds are effective in reducing soil erosion and conserving moisture. In several situations the vegetative barriers are more effective and economical than the mechanical measures viz. bunding. These are recommended in Alfisols, Vertisols, Vertic-inceptisols and associated soils, ranging the rainfall of 400-2000 with more than 2.5% slope. Vegetative barriers can be established either on contour or on moderate slope of 0.4 to 0.8%.

In this system, the vegetative hedges act as barriers to runoff flow, which slow down the runoff velocity resulting in the deposition of eroded sediments and increased rainwater infiltration. It is advisable to establish the vegetative hedges on small bund. This increases its effectiveness particularly during the first few years when the vegetative hedges are not so well established. The key aspect of design of vegetative hedge is the horizontal distance between the hedge rows which mainly depends on rainfall, soil type and land slope. Species of vegetative barrier to be grown, number of hedge rows, plant to plant spacing and method of planting are very important and should be decided based on the main purpose of the vegetative barrier. If the main purpose of the vegetative barrier is to act as a filter to trap the eroded sediments and reduce the velocity of runoff then the grass species such as vetiver, sewan (*Lasiurus sindicus*), sania (*Crotolaria burhia)* and kair (*Capparis aphylla)* could be used.

But if the purpose of vegetative hedges is to stabilize the bund then plants such as Glyricidia or others could be effectively used . The Glyricidia plants grown on bunds not only strengthen the bunds while preventing soil erosion, but also provide N-rich green biomass, fodder, and fuel. The cross section of earthen bund can also be reduced. Study conducted at CRIDA research center indicated that by adding the N-rich green biomass from the Glyricidia plants planted on bund at a spacing of 0.5 m apart for a length of 700 m could provide about 30-45 kg N ha-1 yr-1. In areas with long dry periods, vegetative hedges may have difficulties in surviving. In very low rainfall areas, the establishment and in high rainfall area, the maintenance could be the main problem. Proper care is required to control pests, rodents and diseases for optimum growth and survival of both vegetative hedges and main crops. **Benefits:** *Once properly established the system is self-sustaining and almost maintenance free. *Land under the hedge is used for multipurpose viz. N-rich biomass, fodder and fuel. * Can be successfully used under wide range of rainfall (400-2500 mm) and topography. * Economical and often more effective than other erosion control measures.

Fig. 10: Gliricidia plants on bunds and over view of a watershed with Glyricidia on Graded bunds, ICRISAT, Patancheru, (Pathak *et al.*, 2007)

Crop rotation

Crop rotation is the practice of growing different types of crops in succession on the same field to get maximum profit from the least investment without impairing the soil fertility. Monocropping results in exhaustion of soil nutrients and deplete soil fertility. The inclusion of legume crops in crop rotation reduces soil erosion, restores soil fertility, and conserves soil and water .Further, the incorporation of crop residue improves organic matter content, soil health, and reduces water pollution. A suitable rotation with high canopy cover crops helps in sustaining soil fertility; suppresses weed growth, decreases pests and disease infestation, increases input use efficiency, and system productivity while reducing the soil erosion

Intercropping

Cultivation of two or more crops simultaneously in the same field with definite or alternate row pattern is known as intercropping. It may be classified as row, strip, and relay intercropping as per the crops, soil type, topography, and climatic conditions. Intercropping involves both time-based and spatial dimensions. Erosion permitting and resisting crops should be intercropped with each other. The crops should have different rooting patterns. Intercropping provides better coverage on the soil surface, reduces the direct impact of raindrops, and protects soil from erosion

Advantages

- High total biomass production.
- Efficient utilization of soil and water resources.
- Reduction of marketing risks due to the production of a variety of products at different periods.
- Drought conditions can be mitigated through intercropping.
- Reduce the weed population and epidemic attack of insect pests or diseases.
- It improves soil fertility.

Strip cropping

Growing alternate strips of erosion permitting and erosion resistant crops with a deep root system and high canopy density in the same field is known as strip cropping. This practice reduces the runoff velocity and checks erosion processes and nutrients loss from the field. The erosion resisting crops protects soil from beating action of raindrops, reduces runoff velocity, and thereby increased time of concentration which results in a higher volume of soil moisture and increased crop production . Strip cropping is practiced for controlling the run-off and erosion and thereby maintaining soil fertility.

Types of strip cropping

i. **Contour strip cropping:** The growing of alternate strips of erosion permitting and erosion resisting crops across the slopes on the contour is known as contour strip cropping. It reduces the direct beating action of raindrops on the soil surface, length of the slope, runoff flow and increases rainwater absorption into the soil profile.

ii. **Field strip cropping:** In this practice the field crops are grown in more or less parallel strips across fairly uniform slopes, but not on exact contours. It is useful on regular slopes and with soils of high infiltration rates, where contour strip cropping may not be practical.

iii. **Wind strip cropping:** It consists of the planting of tall-growing row crops (such as maize, pearl millet, and sorghum) and close or short growing crops in alternately arranged straight and long, but relatively narrow, parallel strips laid out right across the direction of the prevailing wind, regardless of the contour.

iv. **Permanent or temporary buffer strip cropping:** It is the growing of permanent strips of grasses or legume or a mixture of grass and legume in highly eroded areas or in areas that do not fit into regular rotation, i.e. steep or highly eroded, slopes in fields under contour strip cropping. These strips are not practiced in normal strip cropping and generally planted permanent or temporary basis.

Mulching

Mulch is any organic or non-organic material that is used to cover the soil surface to protect the soil from being eroded away, reduce evaporation, increase infiltration, regulate soil temperature, improve soil structure, and thereby conserve soil moisture (Pang, H.C. *et al.*, 2010, Kumar *et al.*, 2017 and Jabran, K. 2019). Mulching prevents the formation of hard crust after each rain. The use of blade harrows between rows or inter-culture operations creates "dust mulch" on the soil surface by breaking the continuity of capillary tubes of soil moisture and reduces evaporation losses. Mulching also reduces the weed infestation along with the benefits of moisture conservation and soil fertility improvement. Hence, it can be used in high rainfall regions for decreasing soil and water loss, and in low rainfall regions for soil moisture conservation. Organic mulches improve organic matter and consecutively improving the water holding capacity, macro and micro fauna biodiversity, their activity, and fertility of the soil. Inorganic mulches have a longer life span than organic mulches and can reduce soil erosion, water evaporation losses, suppress weeds but cannot improve soil health. This practice is costly and labor intensive therefore, suitable for cash crops such as fruits and vegetables. Polyethylene mulch is commonly used for the conservation of soil and water resources to increase crop productivity (Sarvede, *et al.*, 2019)

Organic farming

Organic farming is an agricultural production system that devoid the use of synthetic chemical fertilizers or pesticides and includes organic sources for plant nutrient supply viz. FYM, compost, vermicompost, green manure, residue mulching, crop rotation, etc. to maintain a healthy and diverse ecosystem for improving soil properties and ensuring a sustained crop production. It is an environmentally friendly agricultural crop production system.The maintenance of high organic matter content and continuous soil surface cover with cover crops, green manure, and residue mulch reduce the soil erosion in organic farming. It leads to the addition of a large quantity of organic manures which enhances water infiltration through improved bio-physico-chemical properties of soil, and eventually reduces soil erodibility (Mader, P. *et al*., 2002). Organic materials improve soil structure through the development of soil binding agents (e.g., polysaccharides) and stabilizing and strengthening aggregates which reduce the disintegration of soil particles and thus reduced soil erosion. Soil erosion rates from soils under organic farming can be 30–140% lower than those from conventional farming (Blanco, H. and Lal, R.2008).

4.b. Land configurations

Adoption of appropriate land configuration and planting techniques according to crops, cropping systems, soil type, topography, rainfall, etc. help in better crop establishment, intercultural operations, reduce runoff, soil and nutrient loss, conserve water, efficient utilization of resources and result in higher productivity and profitability. Ridge and furrow, raised bed and furrow, broad bed and furrow, and ridging the land between the rows are important land configuration techniques.

Conservation Furrow

The conservation furrow is a simple and low cost in-situ soil and water conservation practice for rain-fed areas with moderate slope. It is recommended for Alfsols and associated soils. This practice is suitable for the regions having rainfall of 400 – 900 mm with slope ranging from 1 to 4%..it can be adopted for groundnut crop in red soils with a reduced gradient along the bed (0.2–0.4%). This practice is highly suitable for soils with severe problems of crusting, sealing and hard setting. Due to these problems the early run-off is quite common on these soils. In this system series of furrows are opened on contour or across the slope at 3-5 m apart. The spacing between the furrows and its size can be chosen based on the rainfall, soils, crops and topography. The furrows can be made either during planting time or during interculture operation using country plough. Two to three passes in the same furrow may be needed to obtain the required furrow size. These furrows harvest the local runoff water and improve the soil moisture in the adjoining crop rows, particularly during the period of water stress. The practice has been found to increase the crop yields by 10-25% and it costs around Rs 250- 350 ha-1. To improve its further effectiveness, it is recommended to use this system along with contour cultivation or cultivation across the slope (Ram Mohan Rao *et al.* 1981). The conservation furrows harvest the local runoff and increase the soil moisture for adjoining crop rows. The Reduced runoff and soil loss. It is Simple and low cost system. It is easy to adopt and can be implemented using traditional farm implements.

Ridge and furrow system

Raising rainy season crops on ridges and *rabi* season crops in furrows reduces the soil crusting and ensures good crop stand over sowing on flat beds. Moreover, inter-row rainwater can be drain out properly during the monsoon period and collected in farm ponds, for life-saving irrigations and profile recharging for the establishment of *rabi* crops. It leads to the increased moisture content in soil profile which reduces moisture stress on plants during the drought period. This method is most suitable for wide-spaced crops viz. cotton, maize, vegetables, etc..

Broad bed and furrow system: This system has been developed by the ICRISAT in India. It is primarily advocated for high rainfall areas (>750 mm) having black cotton soils (Vertisols). Beds of 90–120 cm width are formed, separated by sunken furrows of about 50–60 cm wide and 15 cm depth. The preferred slope along the furrow is between 0.4 and 0.8% on Vertisols. Two to four rows of the crop can be grown on the bed, and the width and crop geometry can be adjusted to suit the cultivation and planting equipment.

Advantages: Increase *in-situ* soil moisture conservation, Safely dispose of excess runoff without causing erosion, Improved soil aeration for plant growth and development, Easier for weeding and mechanical harvesting, It can accommodate a wide range of crop geometry

Fig. 11: Broad Bed Furrow system at ICRIST Center,Patancheru (Source: Pathak *et al.*, 2007)

Agroforestry measures

Agroforestry is a sustainable land management system which includes the cultivation of trees or shrubs with agricultural crops and livestock production simultaneously on the same piece of land (Jhariya, M.K.2015 and Singh,N.R. and Jharya M.K.2016). It is an emerging technology for effective soil and water conservation and comprises a wide range of practices for controlling soil erosion, developing sustainable agricultural production systems, mitigating environmental pollution, and increasing farm economy. The leaf litter addition act as a protective layer against soil erosion improves soil health and moisture retention capacity of the soil and increases crop productivity [Sharma, R. *et al.*,2007, Meena, R.S. *et al.*, 2007 and Gupta, B. *et al.*, 2020)]. It has been reported that different agroforestry practices can reduce up to 10% of soil erosion (Udawatta,R.P. *et al.*, 2015). Agroforestry not only controls soil erosion but also produce tree-based several marketable products.

Types of agroforestry systems

Agri-Silviculture: It is the growing of agricultural crops as a primary component with the secondary component of multipurpose trees (MPTs) on the same managed land unit. The tree species bind soil particles in the root zone and increase water infiltration and reduce runoff.

Agri-Horticulture: Growing of agricultural crops and fruit trees on the same managed land unit is known as agri-horticulture. Fruit tree species like custard apple, wood apple , ber (*Ziziphus mauritiana*), and aonla (*Phyllanthus emblica*) can be successfully planted in agricultural fields and on degraded and low fertile lands with some restoration measures.

Alley Cropping: Growing of agricultural crops in the alley formed between the hedge rows of leguminous nitrogen-fixing tree species. This system is one of the effective measures for soil and water conservation in hilly areas.

Silvi-pasture System: Raising grasses or livestock with MPTs on the same managed land unit is known as silvi-pasture system. This system has the potential to reclaim eroded and degraded lands. Mechanical measures combined with grass species cultivation are more effective for controlling soil erosion processes. The grass species such as *Cenchrus ciliaris* (buffel) grass), *Cenchrus setigerus* (birdwood grass), *Dichanthium annulatum* (marvel grass), *Panicum antidotale* (blue panicgrass), *Panicum maximum* (Guinea grass), *Brachiaria mutica* (para grass) and *Pennisetum purpureum* (elephant grass) are important in ravine restoration.

4. C. Mechanical measures

Mechanical measures or engineering structures are designed to modify the land slope, to convey runoff water safely to the waterways, to reduce sedimentation and runoff velocity, and to improve water quality. These measures are either used alone or integrated with biological measures to improve the performance and sustainability of the control measures. In highly eroded and sloppy landscape biological measures should be supplemented by mechanical structures. Several permanent and temporary mechanical measures are available such as terraces, contour bunding, check dams, gabions, diversion drains, geo-textiles, etc. (Vanwalleghem T, 2016). The mechanical measures are preferred based on the severity of erosion, soil type, topography, and climate (Gauchene, C.K. *et al.*, 2019)

Contour bunding: Contour bunding is used to conserve soil moisture and reduce erosion in the areas having 2–6% slope and mean annual precipitation of <600 mm with permeable soils (Kumar, V. *et al.*, 2018). The vertical interval between two bunds is known as the spacing of bunds. The spacing of bund is dependent on the erosive velocity of runoff, length of the slope, slope steepness, rainfall intensity, type of crops, and conservation practices.

ii. **Graded bunding:** Graded bunds are made to draining out of excess runoff water safely in areas having 6–10% land slope and receiving rainfall of >750 mm with the soils having infiltration rate < 8 mm/h.

iii. **Peripheral bunds:** Peripheral bunds are constructed around the gully head to check the entry of runoff into the gully. It protects the gully head from being eroded away through erosion processes. It creates a favorable condition for the execution of vegetative measures on gully heads, slopes, and beds.

Contour trenching

Trenches are constructed at the contour line to reduce the runoff velocity for soil moisture conservation in the areas having <30% slope. Bunds are formed on the downstream side of trenches for the conservation of rainwater. Trenches are of two types:

Continuous contour trenches: Continuous contour trenches are constructed based on the size of the field in the low rainfall areas with the 10–20 cm trench length and 20–25 cm equalizer width without any discontinuity in trench length (10–20 m).

Staggered contour trenches (STCs) Generally, these trenches are constructed in alternate rows directly beneath one another in a staggered manner in the high rainfall areas, where the risk of overflow is prominent. SCTs are 2–3 m long with 3–5 m spacing between the rows. Planting of tree species is done based on the land slope. It is highly effective in forestalling extension of gully head, soil loss, and arrest the overflow.

Terracing

Terraces are earthen embankments built across the dominant slope partitioning the field in uniform and parallel segments [Blanco, H. and Lal, R.2008)]. Generally, these structures are combined with channels to convey runoff into the main outlet at reduced velocities. It reduces the degree and length of slope and thus reduced runoff velocity, soil erosion and improves water infiltration. It is recommended for the lands having a slope of up to 33%, but can be adopted for lands having up to 50–60% slope, based on socio-economic conditions of a particular region. Where plenty of good-quality stones are available, stone bench terracing is recommended. Sometimes, semi-circular type terraces are built at the downstream side of the plants, known as half-moon terraces. Based on the slope of benches, the bench terraces are classified into the following categories:

Bench terraces sloping outward: These types of terraces are used in low rainfall areas having permeable soils. A shoulder bund is provided for stability of the edge of the terrace and thus has more time for rainwater soaking into the soil.

Bench terraces sloping inward (hill-type terraces): These types of bench terraces are suitable for heavy rainfall areas where a higher portion of rainfall is to be drained as runoff. For this, a suitable drain should be provided at the inward end of each terrace to drain the runoff. These are also known as hill-type terraces.

Bench terraces with level top: These types of terraces are suitable for uniformly distributed medium rainfall areas having deep and highly permeable soils. These are also known as irrigated bench terraces because of their use in irrigated areas.

The raised bed portion acts as an in-situ 'bund' to conserve more moisture and ensures soil stability; the shallow furrows provides good surface drainage to promote aeration in the seedbed and root zone; prevents water logging of crops on the bed's .The BBF design is quite flexible for accommodating crops and cropping systems with widely diverting row spacing requirements Precision operations such as seed and fertilizer placement and mechanical weeding are facilitated by the defined trac zone (furrows), which saves energy, time, cost of operation and inputs.

Thesecan be maintained on the long term (25-30 years). It reduces run off and soil loss and improves soil properties over the years. It facilitates double cropping and increases crop yields.

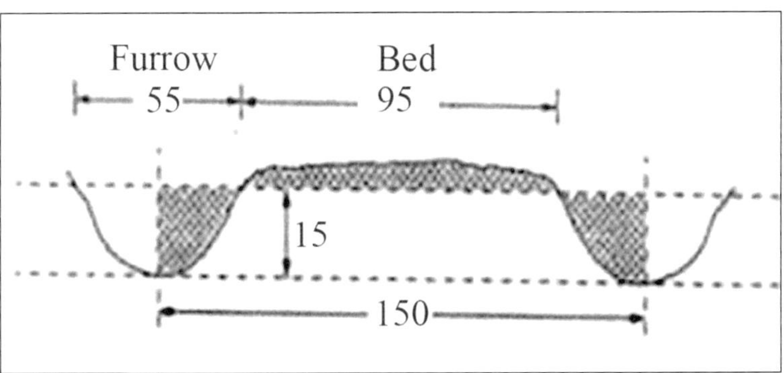

Fig. 12: Diagrammatic representation Broad bed furrow system.

Grassed Waterways

"The grassed waterways and diversions are the means to drain/divert the runoff from the catchment." Grassed waterways are the natural or man-made water courses, lined with erosion resistant grasses, used to dispose the surface runoff water from the bunded/terraced area. The use of grasses in the waterway section acts as a lining material to control the problem of soil erosion caused by turbulent runoff flow through the section. The grassed waterways are constructed along the slope of the area.Apart from disposing the runoff from area, they also act as an outlet for the terraces or graded bunds. Waterways are very important means for removing surplus water from the terraced field, and for erosion control.In addition, they are also used for other purposes such as to handle the natural runoff or to carry the discharge from general fields, contour furrows, diversion channels, or used as emergency spillway in farm ponds or other water storage structures. The grassed waterways should be fully established with grasses before water is turned into them. In other words, the waterways should be ready to hold the water before the bunds, terraces, or diversions etc. are constructed.

5. Community based soil conservation measures

Masonry Check Dam

Masonry check dams are permanent structures effectively used for controlling gully erosion, water harvesting and groundwater recharging. These structures are popular in watershed programs in India. The cost of construction is generally quite high.

Fig. 13: Masonry Check Dam

These structures are preferred at sites where velocity of runoff water flow in gullies/streams is very high and stable structure is needed to withstand the difficult condition. Proper investigations, planning and design are needed before construction of masonry check dams. Masonry check dams are designed based on engineering principles. The basic requirements for designing the masonry check dams are: hydrologic data, information on soils and geology, the nature and properties of the soils in the command area and profile survey and cross-sectional details of the stream or gully. A narrow gorge should be selected for erecting the dam to keep the ratio of earthwork to storage at minimum. Runoff availability for the reservoir should be computed on the basis of rainfall runoff relationship. Depending upon the assumed depth of structure and the corresponding area to be submerged, suitable height of the dam may be selected to provide adequate storage in a given topographic situation (Katyal *et al.*, 1995). The cross-section of dam and other specifications are finalized considering the following criteria: there should be no possibility of the dam being over-topped by flood-water, the seepage line should be well within the toe at the downstream face;the upstream and downstream faces should be stable under the worst conditions, the foundation shear stress should be within safe limit; proper spillway should be constructed to handle the excess runoff and the dam and foundation should be safe against piping and undermining It is long lasting structures with little regular maintenance. This is effective in controlling gully and harvesting water under high runoff flow condition

Khadin System

Khadin is a land-use system developed centuries ago in the Jaisalmer district of western Rajasthan. This system is practiced by single larger farmer or by group of small farmers. It is highly suitable for areas with very low and erratic rainfall conditions. Recommended agro-ecology: Soil : Sandy and other light soils Rainfall : 250-700 mm

In khadin system, preferably an earthen or masonry embankment is made across the major slope to harvest the run-off water and prevent soil erosion for improving crop production. *Khadin* is practiced where rocky catchments and valley plains occur in proximity. The run-off from the catchment is stored in the lower valley floor enclosed by an earthen/stone 'bund' (. Any surplus water passes out through a spillway. The water arrested stands in the *khadin* throughout the monsoon period. It may be fully absorbed by the soil during October to November, leaving the surface moist. If standing water persists longer, it is discharged through the sluice before sowing. Wheat, chickpea, or other crops are then planted. These crops mature without irrigation. The soils in the *khadins* are extremely fertile because of the frequent deposition of %ne sediment, while the water that seeps away removes salts. The *khadin* is, therefore, a land-use system, which prevents soil deterioration (Kolarkar *et al.* 1983). This practice has a distinct advantage under saline groundwater condition, as rainwater is the only source of good quality water in such area.

It improves surface and groundwater availability in the area. The khadin bed is used for growing post-rainy season crops. This requires minimum maintenance (once in 5 years). This system results in assured rainy and post rainy season crops, thereby improving socioeconomic condition of farmer. This system provides source of drinking water for livestock. It reduces flood or peak rate runoff. It conserves the soil and improves rainwater use efficiency

Farm Ponds

Farm ponds are very age old practice of harvesting runoff water in India. These are bodies of water, either constructed by excavating a pit or by constructing an embankment across a watercourse or the combination of both. Farm pond size is decided on the total requirement of water for irrigation, livestock and domestic use. If the expected runoff is low, the capacity of the pond will only include the requirement for livestock and domestic use. Once the capacity of the pond is determined, the next step is to determine the dimensions of the pond. To achieve the overall higher efficiency, the following guidelines should be adopted in the design and construction of farm ponds.

High-storage efficiency (ratio of volume of water storage to excavation): This can be achieved by locating the pond in a gully, depression, or on land having steep slopes. Whenever possible, use the raised inlet system to capture runoff water from the upstream. This design will considerably improve the storage efficiency the structure.

Reduce the seepage losses: can be achieved by selecting the pond site having subsoils with low saturated hydraulic conductivity. It minimizes the evaporation losses: As far as possible, the ponds should be made deeper but with acceptable storage efficiency to reduce water surface exposure and to use smaller land area under the pond

Benefits: Farm ponds are useful for multiple use of stored water. * it is simple to construct by using locally available material. * it is useful for the upstream parts of watershed particularly where groundwater availability is low

Gully Checks with Loose Boulder Wall

Loose boulder gully checks are quite popular in the watershed program for controlling gully erosion and for increasing groundwater recharge. These are very low-cost structures and quite simple in construction. These gully checks are built with loose boulder only, and may be reinforced by wire mesh, steel posts, if required for stability. Often it is found on the land and thus eliminates expenditure for long hauls. The quality, shape, size, and distribution of the boulders used in the construction of gully checks affect the life span of the structures. Obviously, boulders that disintegrate rapidly when exposed to water and atmosphere will have a short structural life. Further, if only small boulders are used in a dam, they may be moved by the impact of the first large water flow. In contrast, a gully checks constructed of large boulders that leave large voids in the structure will offer resistance to in gully checks the flow but may create water jets through the voids. These jets can be highly destructive if directed toward openings in the bank protection work or other unprotected parts of the channel. Large voids in gully checks also prevent the accumulation of sediment above the structures and enhances stabilization of the gully.

Fig. 14: Series of loose boulder wall gully checks at Bundi watershed, Rajasthan (*Source*: Pathak etal., 2007)

It is low-cost and simple in construction made with the locally available materials . These are effective in controlling gully and improving groundwater

Table 3: Recommended conservation measures based on rainfall and soils

Practice	Rainfall situation	Soil type
Deep tillage	All rainfall situations	Deep black soil and moderately deep to deep red soil
Vegetative barriers	>750mm	All soil types
Compartinental bunding	<750mm	Black soils
Furrows	<750mm	Red soil
Ridges and furrows	All rain fall situations	All soil types
Broad bed and	<75(hnm	Latente soil
Furrow	<750mm	Deep black soil
Tied ridging	<750mm	Medium and deep black soil
Scooping	<75(hnm	Shallow and medium black soils
Organic mulching	All rainfall situations	All soil types

(*Source*: Rajani, R. 2014)

5.Experiences with farmers (Maharashtra watersheds)

The Front-Line Demonstrations (FLDs) were conducted in different villages of Akola, Washim, Amravati, Yeotamal and Wardha districts on different in-situ soil and water conservation measures in participatory mode. In all 647 farmers from 71 villages participated in this programme..

In water conservation ditch, the rainwater is stored in-situ without the problems associated with contour/ graded bunds. Evolved at Dr. DPKV, Akola on Rainwater Management.

Economical sub-surface tillage in Vertosols created impact: The yield increment was 45.72% (Soybean) . The rain water use efficiency(RWUE) was 74.19% with sub-surface tillage.

The soil moisture with sub-surface tillage in Vertisols is> 16.72% . The runoff Soil loss is < 100%. The additional cost of operation is Rs` 5000/ ha. There reduction in bulk Density, was from – 1.35 to1.27 .(mg/m)

Cultivation across the slope

Sowing of crops across slope significantly enhanced crop productivity, improved rain water use efficiency and retained higher moisture during dry spell. This method can be easily adopted by farmers using conventional tools

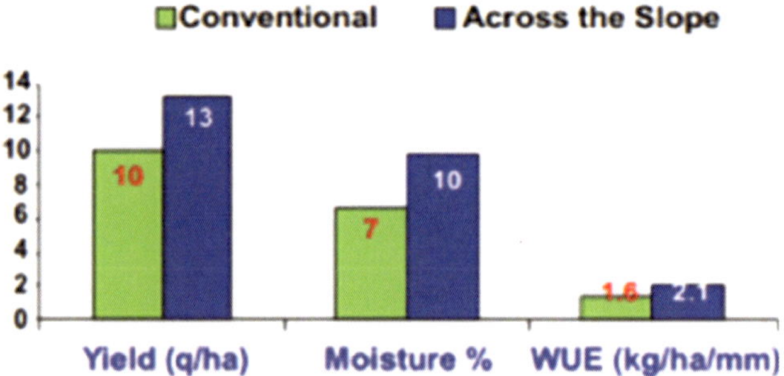

Source: Mishra, P.K.; *et al* (Eds), 2002. In: Soil and Water Conservation Bulletin. Indian Association of Soil and Water Conservationists,Dehradun,Uttarakhand, 96 p. © 2002 Indian Association of Soil and Water Conservationist

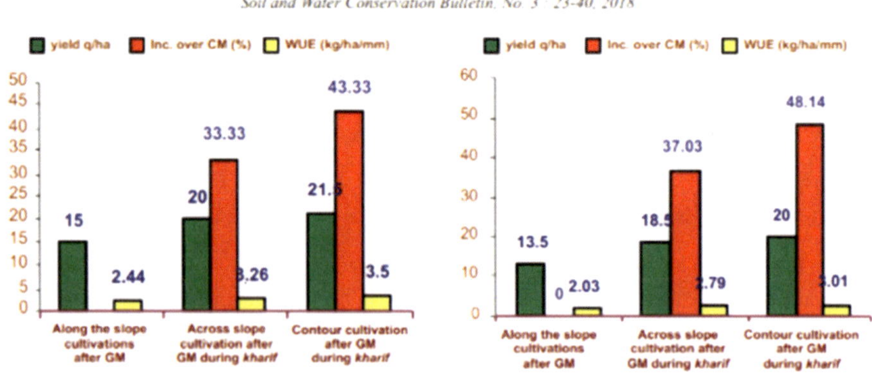

Fig.16 & 17

The crop productivity was increased by 32 to 57%. The rainfall use efficiency (RWUE) was by 31.25%.andsoil moisture was by 44%. The soil erosion was reduced by 50 to 52%. and runoff was reduced by 25 to 30%. Nutrient loss was by 45 to 48%. This method can be easily adopted by farmers using conventional tools to Retained higher moisture during dry spell. Technologies was promoted through 486 demonstrations in 122 villages from 14 talukas of 6 district on 1646 ha area with participation of 300 farmers

Opening of alternate furrows after sowing: Opening of alternate furrows 30 days after sowing retain higher moisture during dry spell, enhances rainwater and crop productivity. The B:C ratio was enhanced by-1.19 to 1.49 – The additional cost for this operation was Rs 4500/ha.

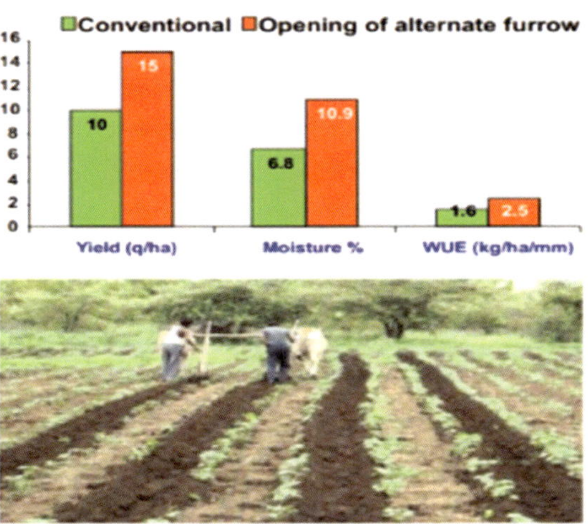

Fig. 18: Opening of furrows.

Opening of furrows 30 days after sowing

Opening of continuous furrows 30 days after sowing retained higher soil moisture during dry spell, enhanced rainwater productivity and increased yield of sorghum and cotton. This can be adopted by farmers using traditional tools. The additional Cost for this operation was Rs 5100 /ha. Technology was promoted through 12 demonstrations in 8 villages from 4 talukas of 3 district on 48 ha area with the participation of 12 farmers.

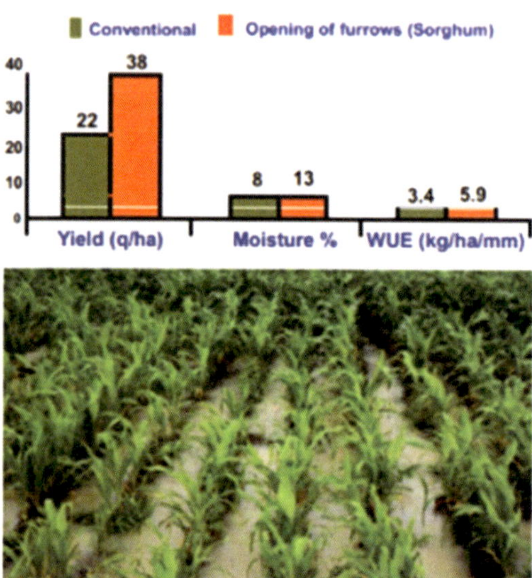

Fig. 19: Opening of tied furrows

Opening of tied furrows after sowing

Opening of tied furrows 30 days after sowing enhanced moisture content, rain water productivity and crop yield. The B.C,ratio was from 1.19 to 1.39 compared to conventional practice. The additional cost for tied ridging was Rs 500/ha. This technology was promoted through 21 demonstrations in 13 villages from 4 talukas of 3 district on 83 ha area with the participation of 16 farmers

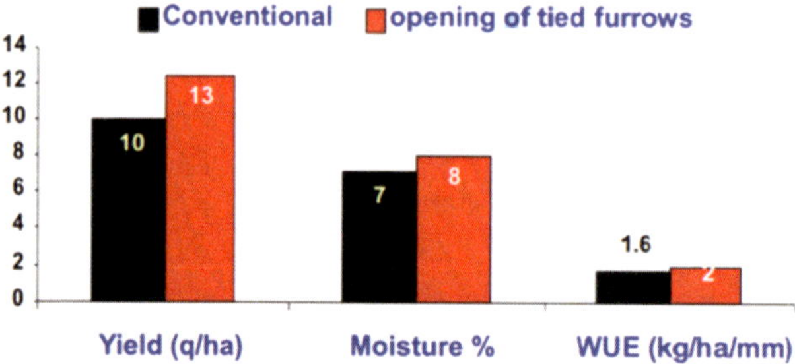

Fig. 20: Histogram on Conventional and opening of tied furrows

Contour cultivation

Contour cultivation enhanced rainwater productivity, moisture availability during dry spell and increase crop yield. It reduces 55% run off, 70% soil loss and 60% nutrient loss. TheB:C ratio for this activity was from 1.19 to 1.80. The cost involved for this operation per ha was Rs 4800/. Contour cultivation technology was promoted through 78demonstrations in 16 villages from 4 talukas of 3 district on 318 ha area with the participation of 75 farmers.

Opening of alternate contour furrows 30 days after sowing retained 93% higher moisture, increased 80% crop productivity and enhanced rainwater productivity. The B:C ratio was from 1.19 to 1.84 The cost of operation is Rs ` 5100/- ha. Technology was promoted through 20 demonstrations in 9 villages from 3 talukas of 3 district on 79 ha area with the participation of 18 farmers.

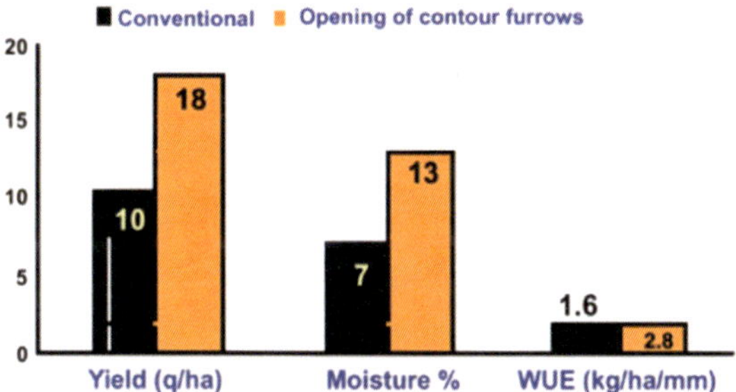

Fig. 21: Histogram on Conventional system and opening of Contour furrows

2. Runoff Harvesting and Recycling

a) Farm pond for protective irrigation

Farm ponds constructed by Department of Agriculture are used for protective irrigation by recycling of runoff through sprinkler and drip irrigation during dry spell. One protective irrigation of 30 mm depth in medium to deep soils and 50 mm depth of irrigation in shallow soils during moisture stress in kharif and rabi season resulted into the significant improvement in yield. The water requirement of the crops at critical stages is given in the following table. During rainy season it is experienced that crops are suffering either in the beginning or later stage, it requires atleast one protective irrigation to sustain till the subsequent rains.

i) **Kharif**

By providing one protective irrigation of 30 mm depth in medium to deep soils during dry spell in kharif the yield levels, water use efficiency and B:C ratio were observed significantly enhanced over the period of four years by 100 farmers as shown below

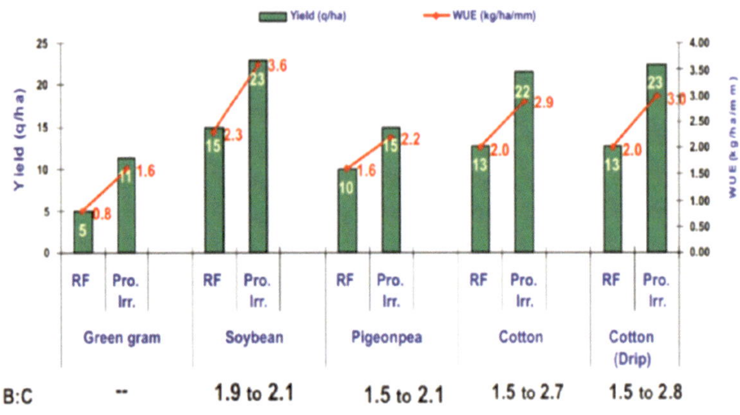

Fig. 22: Effect of the protective irrigation through sprinkler and drip systems during kharif season.

Technology promoted through 48 demonstrations (kharif and rabi) in 32 villages from 5 taluka of 3 districts on 240 ha area with the participation of 45 farmers.

Table 3: Influence of Supplemental irrigation on increment in yield and additional returns in Adilabad, Telangana

Name of the farmer	Crop	No.of SI	Operational cost (Rs/acre)	Increase in yield (q/ ac)	Additional return, Rs/ac
Shri. E. Mallesh	Colton (1 ac)	2	750	1.5	6000
Shri. Ganga Reddy	Cotton (1 ac)	3	1125	1.5	6000
Shri K. Ramarao	Cotton (1 ac)	3	1125	20	8000
Shri S. Bhecm Rao	Cotton (1 ac)	2	750	1.0	4000
Shri. M. Manihu	Sorghum (2 ac)	3	2050	07	q grain + 12q fodder = Rs 20,000/
Shri N Rajanna	Vegetables (0.5 ac)	2	750	-	6800
Shri. K. Ramarao	Vegetables (0.S ac)	3	1125	-	9200

(*Source*:CRIDA,2010)

Table 4: Influence of Supplemental Irrigation on yield and income at Agra

Crop	Yield (kg/ha)		Income from IP (Ks/ha)	Additional income over I'P (Rs/lia)
	With supplemental irrigation (IP)	Without supplemental irrigation (FP)		
Maize	2419	1825	31411	8640
Groundnut	1520	1250	68400	12150
Wheat	3460	2612	44255	17047

(Source: AICRPDA-TSP Report- 2012-15)

The demonstrations on supplemental irrigation in maize, groundnut and wheat benefitted more than 100 farmers in realizing higher maize, wheat and groundnut yield and farmers realized additional net returns up to Rs. 17000/ha (AICRPDA,2018)

Table 5: Influence of supplemental irrigation during rabi in Barley

Treatment	Yield (kg/l			Cost of cultivation (Rs/ha)	NMR (Rs/ha)	B:C ratio	WUE (kg/ ha-mm)
	Grain 2018	Mean (2 yrs)	Stover				
Control (no irrigation)	2062	2164	3110	21975	23267	2.06	9.90
Pre-sowing irrigation	2584	2684	3943	23460	33464	2.43	9.40
f Irrigation @ 30 DAS	2795	2898	4302	23460	38298	2.63	10.00
f Irrigation @45 DAS	2205	2305	3412	23460	25350	2.08	7.90
Irrigation @ 60 DAS	2015	2115	3129	21975	22686	2.03	9.80
	183	-	289	-	-		

At Agra, one protective irrigation at 30 DAS recorded higher barley grain yield (2795 kg/ha), net returns (Rs.38298/ha), B:C ratio (2.63) and WUE (10.0 kg/ha-mm) compared to other treatments except pre sowing irrigation (2584 kg/ha).

Table 6: Effect of supplementary Irrigation in rabi crops at Hissar

Treatment	MEY (kg/ha)		Stalk yield (kg/ha)	Cost of cultivation (kg/ha)	NMR (Rs/ha)	B:C	WUE (kg/ ha-mm)
	Seed	Mean (3 yrs)					
Irrigation							
1,: Supplemental irrigation	2153	2149	4920	24800	70S46	3.84	22.30
12: No supplemental irrigation	2014	1775	4452	24100	64940	3.69	43.60
CD at 5%	74	-	-	-	-	-	
***Rabl* crop**							
C,: Mustard (RH 30)	2670	2434	5723	24450	93413	4.82	37.30
C,: Chickpea (C 23S)	1498	1490	3351	24450	41817	2.71	21.00
CD at 5%	87.	-	-	-	-	-	-

(Source: AICRPDA, 2018)

During rabi, mustard equivalent yield (MEY) was significantly higher (2153 kg/ha) under supplemental irrigation (50 mm) with harvested rainwater in farm pond than without supplemental irrigation (2014 kg/ha). Further, MEY was higher with mustard (2670 kg/ha) than chickpea (1498 kg/ha). The mean yield over 3 years and economics also showed similar trends

Reduction in Evaporation from Pond by using Neem Oil

Background: The water surface in the farm pond of 30 x 30 x 3 m and 20 x 20 x 3 m. with the capacity of 0.741 and 1.971 TMC. Loss of water from farm pond is through seepage and evaporation. About 19% water is lost by evaporation. Seepage can be controlled by using plastics, tiles etc. To reduce the evaporation losses, neem oil should be spread @ 30 to 40 m/m area. Impact: • Reduction in evaporation – upto 15%

Drainage Management

i Drainage management with ridges and furrow system

Lower side of the field suffers from water stagnation causing anaerobic condition. Farmers loose crop on one fifth of area due to poor drainage. System for safe disposal and storage of excess rain water is essential. However, ridges and furrow or BBF system saves the crop from stagnated water. Sorghum crop sown on ridges and soybean on BBF survived as water was stagnated in furrows and safely disposed. Adoption : Adopted by the farmers on about 2700 ha area to improve the drainage and to save the crop during rainy season at least on 20% area at the lower part of their field.

Drainage improved BBF/raisebed at lower part

Drainage improved ridges and furrow lower part

6.0 Ground water recharge in drylands

Rain water harvesting is taken up considering watershed as a unit. Surface spreading techniques are common since space for such systems is available in plenty and quantity of recharged water is also large. Following techniques may be adopted to save water going waste through slopes, rivers, rivulets and nalas.

Gully plug

- Gully plugs are built using local stones, clay and bushes across small gullies and streams running down the hill slopes carrying drainage to tiny catchments during rainy season.
- Gully Plugs help in conservation of soil and moisture.
- The sites for gully plugs may be chosen whenever there is a local break in slope to permit accumulation of adequate water behind the bunds.

Contour bund

- Contour bunds are effective methods to conserve soil moisture in watershed for long duration.
- These are suitable in low rain fall areas where monsoon run off can be impounded by constructing bunds on the sloping ground all along the contour of equal elevation.
- Flowing water is intercepted before it attains the erosive velocity by keeping suitable spacing between bunds.
- Spacing between two contour bunds depends on the slope, the area and the permeability of the soil. Lesser the permeability of soil, the close should be spacing of bunds.
- Contour bunding is suitable on lands with moderate slopes without involving terracing.

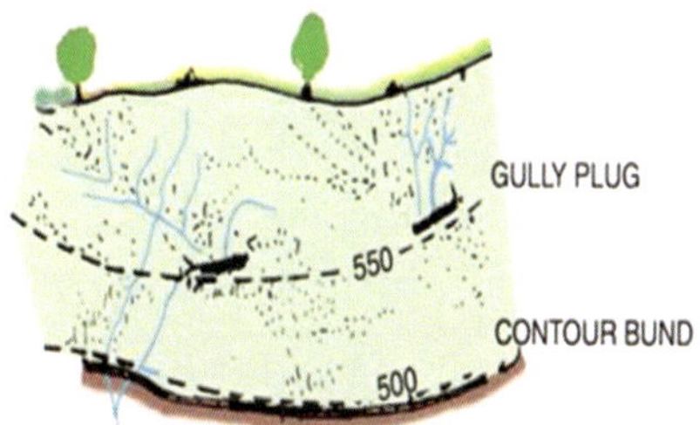

Gabion structure

- This is a kind of check dam commonly constructed across small streams to conserve stream flows with practically no submergence beyond stream course.
- A small bund across the stream is made by putting locally available boulders in a mesh of steel wires and anchored to the stream banks.

- The height of such structures is around 0.5 m and is normally used in the streams with width of less than 10 m.
- The excess water over flows this structure storing some water to serve as source of recharge. The silt content of stream water in due course is deposited in the interstices of the boulders. With the growth of vegetation, the bund becomes quite impermeable and helps in retaining surface water run off for sufficient time after rains to recharge the ground water body.

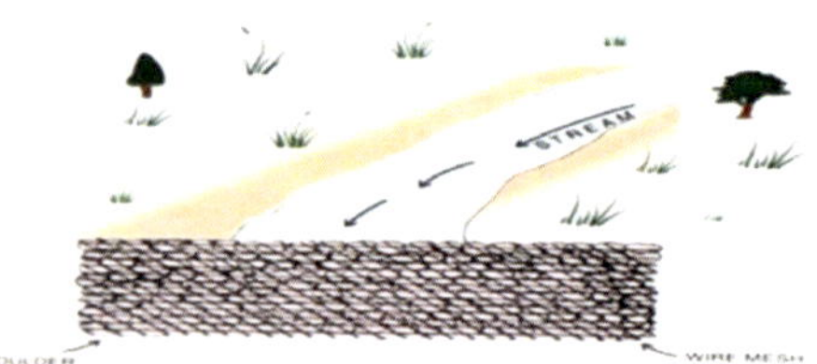

Percolation tank

- Percolation tank is an artificially created surface water body, submerging in its reservoir a highly permeable land, so that surface runoff is made to percolate and recharge the ground water storage.
- Percolation tank should be constructed preferably on second to third order steams, located on highly fractured and weathered rocks, which have lateral continuity downstream.
- The recharge area downstream should have sufficient number of wells and cultivable land to benefit from the augmented ground water.
- The size of percolation tank should be governed by percolation capacity of strata in the tank bed. Normally percolation tanks are designed for storage capacity of 0.1 to 0.5 MCM. It is necessary to design the tank to provide a ponded water column generally between 3 & 4.5 m.
- The percolation tanks are mostly earthen dams with masonry structure only for spillway. The purpose of the percolation tanks is to recharge the ground water storage and hence seepage below the seat of the bed is permissible. For dams upto 4.5 m height, cut off trenches are not necessary and keying and benching between the dam seat and the natural ground is sufficient.

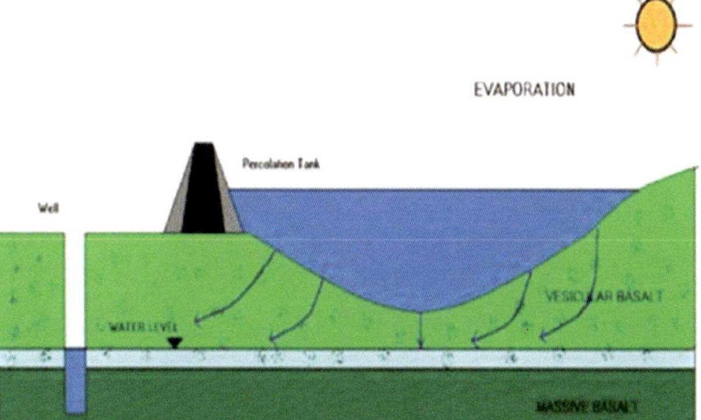

Check dams / cement plugs / nala bunds

- Check dams are constructed across small streams having gentle slope. The site selected should have sufficient thickness of permeable bed or weathered formation to facilitate recharge of stored water within short span of time.
- The water stored in these structures is mostly confined to stream course and the height is normally less than 2 m and excess water is allowed to flow over the wall. In order to avoid scouring from excess run off, water cushions are provided at downstream side.
- To harness the maximum run off in the stream, series of such check dams can be constructed to have recharge on regional scale.
- Clay filled cement bags arranged as a wall are also being successfully used as a barrier across small nalas. At places, shallow trench is excavated across the nala and asbestos sheets are put on two sides. The space between the rows of asbestos sheets across the nala is backfilled with clay. Thus a low cost check dam is created. On the upstream side clay filled cement bags can be stacked in a slope to provide stability to the structure.

Recharge shaft

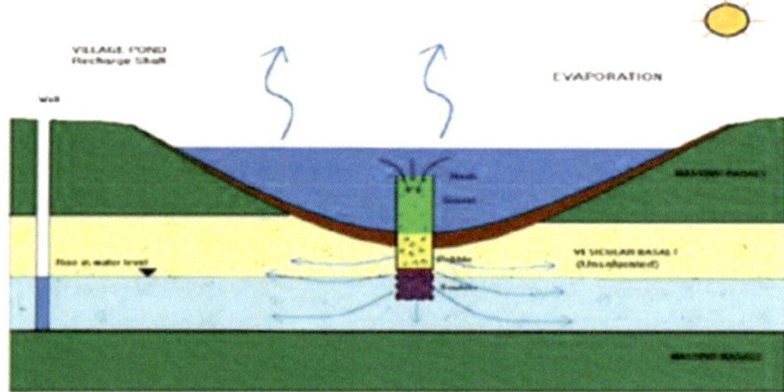

- This is the most efficient and cost effective technique to recharge unconfined aquifer overlain by poorly permeable strata.
- Recharge shaft may be dug manually if the strata is of non-caving nature. The diameter of shaft is normally more than 2 m.
- The shaft should end in more permeable strata below the top impermeable strata. It may not touch water table.
- The unlined shaft should be backfilled, initially with boulders/ cobbles followed by gravel and coarse sand.
- In case of lined shaft the recharge water may be fed through a smaller conductor pipe reaching up to the filter pack.

- These recharge structures are very useful for village ponds where shallow clay layer impedes the infiltration of water to the aquifer.
- It is seen that in rainy season village tanks are fully filled up but water from these tanks does not percolate down due to siltation and tubewell and dugwells located nearby remains dried up. The water from village tanks get evaporated and is not available for the beneficial use.
- By constructing recharge shaft in tanks, surplus water can be recharged to ground water. Recharge shafts of 0.5 to 3 m. diameter and 10 to 15 m. deep are constructed depending upon availability of quantum of water. The top of shaft is kept above the tank bed level preferably at half of full supply level. These are back filled with boulders, gravels and coarse sand.
- In upper portion of 1 or 2 m depth, the brick masonry work is carried out for the stability of the structure.
- Through this technique all the accumulated water in village tank above 50% full supply level would be recharged to ground water. Sufficient water will continue to remain in tank for domestic use after recharge.

Dugwell recharge

Abandoned Dug Well

Abandoned Dug Well fitted with Rain Water Harvesting Mechanism

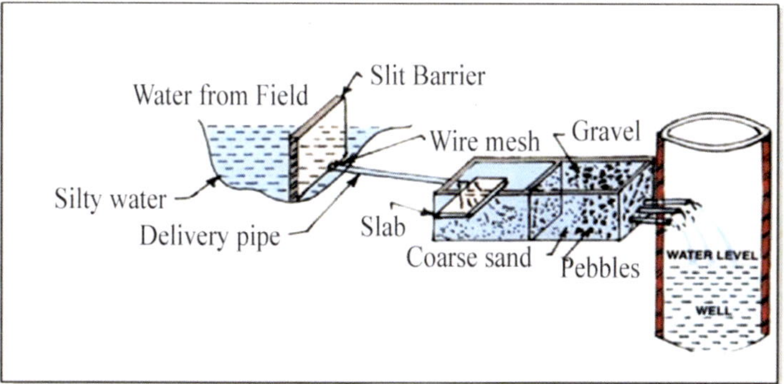

- Existing and abandoned dug wells may be utilized as recharge structure after cleaning and desilting the same.
- The recharge water is guided through a pipe from desilting chamber to the bottom of well or below the water level to avoid scouring of bottom and entrapment of air bubbles in the aquifer.
- Recharge water should be silt free and for removing the silt contents, the runoff water should pass either through a desilting chamber or filter chamber.
- Periodic chlorination should be done for controlling the bacteriological contaminations.

7.0 Indigenous technical knowledge in soil and water conservation

The promotion of appropriate technology with indigenous knowledge base is gaining importance in the natural resource management programme for increasing their adaptability/acceptability and to bring down the dependence on cost intensive technologies. A detail study of Indigenous Technical Knowledge (ITK) on soil and water conservation was taken up through a National Agricultural Technology Project (NATP) entitled "Documentation & Analysis of Indigenous Methods of In-situ Moisture Conservation and Runoff Management" bythe authors at Central Research Institute for Dryland agriculture (CRIDA), Hyderabad as the lead centre.

Some potential ITKs identified for further study, research and development of new projects is presented. A scientific study may change this Indigenous Technical Knowledge to Modern Technical Knowledge (MTK). Prevailing ITKs should invariably be given priority. All the on going projects on resource conservation and management should focus on the viable and appropriate ITKs relating to soil & water for sustainable development and dissemination of the local technology

Table 7: A list of some documented ITKs on soil and water conservation measures under different categories

Name of ITK	Purpose	Researchable Issues
Furrow opening in standing crops	Rainwater conservation	1. Modification of implement with different serrated blades and introducing additional tines 2. Effectiveness in conserving soil moisture

Name of ITK	Purpose	Researchable Issues
Nadi farming system	To collect runolT during kharif for life saving irrigation during drought spell or pre sowing irrigation (Palewa) for rabi crops	1. Documentation and analysis of socio-economic aspect of present nadi system for its sustainability 2. Evaluation of present nadi farming system
Mixed pulses as vegetative barrier	Resource Conservation	1. Proportion of pulses as vegetative barrier 2. Cost effectiveness of the system
Stabilization of gullies using sand bags	Gully control and runoff management	1. Soil conservation efficiency 2. Strengthening of sand bags structure with different vegetative barriers
Application of w hite soil as lining material in farm pond	To wwk as a sealant material for lining dugout farm pond	1. Standardization of application technique and economic feasibility for wider application 2. Study on the seepage losses at different hydraulic heads
Wider row spacing in pearl millet	Rainwater conservation and weed control	1. Plant geometry' and population research in different rainfall situations
Rainwater harvesting in kund/ianka	The harvested water in kund / tanka is used for drinking and establishment of tree	1. Research should be done on the use of stored water for arid horticulture 2. Design of tankas for different geo hydrologic conditions
Crop stubbles and residue management	Improve the organic matter and water holding capacity of soil	1. Quantification of soil and w'ater conserved and yield advantage 2. Better or improved implements for crop residue incorporation 3. Alternate ways of composting and annlieation
Brush wood waste weir	Safe disposal of excess runoff	1. Design and stabilization of structure
Mulching in turmeric	To conserve rainwater	1. Quantification of soil loss, improvement of soil quality and water availability 2. Use of alternative organic material to Sal leaves as mulch
Indigenous stone / brush wood structure across the slope	To check soil loss	1. Shape and size of brush wood structure depending on the runoff and site conditions
Agave sp. as vegetative barrier	To reduce runoff velocity and to increase infiltration opportunity time	2. Different species of Agave can be evaluated 3. Cost benefit analysis.

Name of ITK	Purpose	Researchable Issues
Broad bed and furrow practice	To harvest rain water and dispose of runoff	1. Width of broad bed needs to be evaluated for different crops and rainfall situations 2. Identification of suitable low cost tractor/bullock drawn implement for layout of BBF
Water harvesting and recycling	Rain water harvesting	1. Recharging of water table 2. Cost effectiveness
Standardization of recharging technique	Augmentation of ground water	1. Design of filter and improvement in filtering efficiency with better filtering material. 2. Effect of geology/soil formation on recharge
Set-row cultivation	For harvesting rain water and maintaining soil structure	1. Quantification of rainwater conservation and water use efficiency (WUE) of the crops 2. Improvement in soil health and crop yield over years
Summer / pre monsoon tillage	Conservation tillage-to harvest early showers, facilitate timely seeding and weed control	1. Identification of appropriate tillage implements for soil and water conservation 2. Evaluation of root: shoot ratio and quantification of WUE of crops
Ridge & furrow planting for modulation of overland flow	Conservation of rain water, modulating excess water, control soil loss and boosting productivity	1. Fabrication and development of ridge former accommodating required row spacings and ridge cross- section
Formation of Gurr	Reduction of runoff and soil moisture conservation	1. Effect of bullock and tractor made Gurr on runoff reduction, soil water conservation and crop productivity

(*Source*: Mishra, P.K *et al*., 2002)

Implements used in Dryland Agriculture for various operations

Implement	Purpose	Image
Three row Ridger planter /BBF planter	esoil and water conservation Formation of broad bed furrows besides sowing on ridges,furrows as drainage channels Conservation furrows Approximate cost: Rs. 60000/- Cost of operation: Rs. 800/ha	

Implement	Purpose	Image
Animal Drawn Kap As Ridger	For making ridges and furrows for cultivation in maize, sorghum, cotton, millets, cowpea, groundnut and other arable crops. furrow is formed in a single pass and the ridge in the return pass.	
Bullock Drawn Ridger	During the onward movement, a furrow is formed and ridges get formed in its return pass for making ridges and furrows	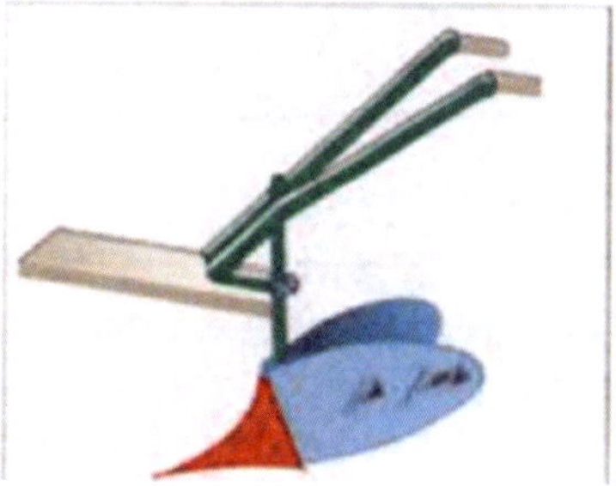
Bullock Drawn Bund Former	3-4Capacity (ha/day) Size (mm) Size ofbunds 750-1050 Adjustable	
Bullock Drawn Weeder	Effective in removing the weeds but also creating a concave structure between the rows for capturing the rain water and also supporting the plants with soil mass. Field coverage : 0.8-0.9 ha/ day Approximate cost: Rs. 2250/- Cost of operation: Rs. 1500/ha	
Improved bakhar	seedbed preparation in black soil:Bullock drawn seedbed preparation, cutting and burying of grass and weeds	

Implement	Purpose	Image
Bullock Drawn Disc Harrow	Disking and soil preparation	
Animal Dra Wn Mouldboard Plough	Ploughing and inversion of soil Primary and secondary tillage operations	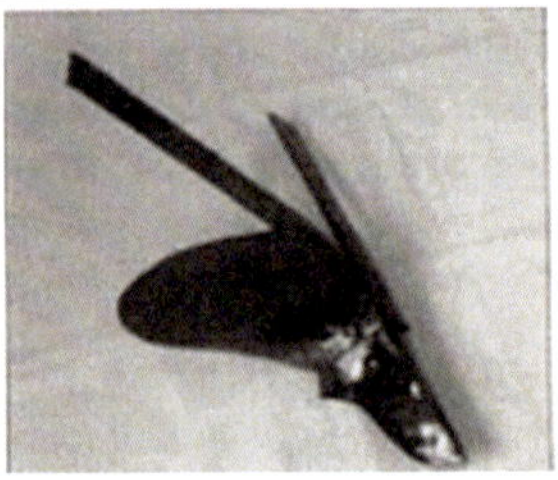
Blade Harrow Local Name: MP Double Dora	The weeding of crops sown in line such as cotton, maize, sorghum and similar crops. capacity/ha/day:0.8 to 1.0 Bullock drawn	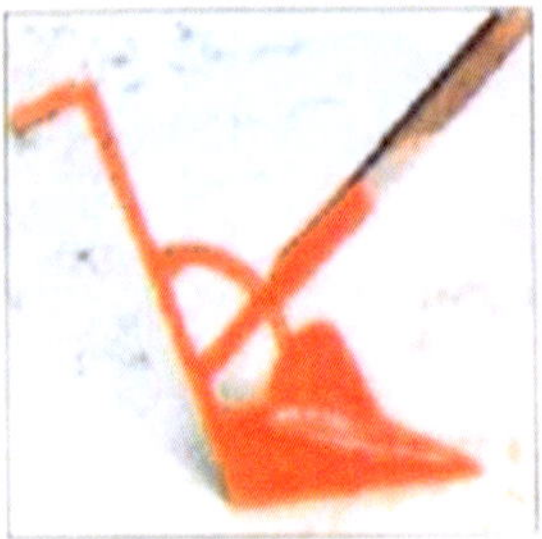
Chiel Plough	Used for deep tillage of compacted soil layers	

4

Crops and Cropping Systems in Dryland Agriculture

Rainfed farming has a distinct place in the Indian agriculture, occupying 67 % of cultivated area contributing 44 % of food grains to our national food basket, supporting 40% of human and 65% of livestock population. The system is characterized by resource poor farmers, poor infrastructure and low investment in technology inputs. Even when full irrigation potential of the country is realized, 50% of the net sown area is continued to be rainfed. The important problems encountered in rainfed areas are unfavorable weather, limited choice of crops and varieties, low cropping intensity and unstable productivity. The research work done by the net work of AICRPDA (All India Co-ordinate Research Project for Dryland areas) centres and SAUs (State Agricultutal Universities) indicated that the production in rainfed environment could be enhanced by 200 % compared to the productivity of farmers' field by adopting improved cropping systems and agro techniques.

But in real situation, a wide yield gap exists between the potentials of the crops and cropping systems in research station and farmers' field. To bridge this yield gap, concerted efforts are to be made to work with the farmers while transferring the technology. For successful technology transfer in rainfed regions there is need to strengthen the infrastructure facilities to supply the critical inputs at the door steps of the farmers on cost basis. Suitable price policy for commodities produced in rainfed situations has to be made by the government for higher production in rainfed regions. Keeping the demands of ever growing population of our country in year, there is need to upscale the productivity of rainfed lands from 1 to 2 tons / ha by adopting efficient cropping systems along with the improved agro techniques.

1.0. Cropping System Strategy in relation to Environment

The traditional cropping systems that are followed currently in arid and semi arid regions are not necessarily efficient in terms of utilization of resources in a given location. These are mostly subsistence oriented and are need-based.

In arid regions, a single cropping system involving a long fallow period (October-June) is the rule rather than an exception. Mixed cropping, as a means of insurance and a form of risk distribution is very common. The deep Vertisols in semi- arid regions of India (about 12 million ha) are left fallow during the rainy season. The post rainy season crops like sorghum; safflower and chickpea are commonly grown either as a sole or intercrop in different combination. This is under utilisation of cropping period, more so in assured and medium to high rainfall areas. In northern central plains the main crop wheat is grown as a sole crop. In some Vertisol regions the common systems are based on cotton. In Alfisols cropping in rainy season is very common. (Singh, R. P. and Subba Reddy, G.1988)

2.Criteria for Selection of Crops and Cropping Systems

In drylands, generally, crop should be of short duration with early vigor, deep root system with ramified roots, dwarf plants with erect leaves and stem, moderate tillering in case of tillering crops and varieties, resistance/tolerance to biotic stresses, lesser period between flowering and maturity resistance/ tolerance to abiotic stresses, low rate of transpiration, less sensitive to photo-period and wider adaptability. Thus, under changing climate conditions

The guiding principles for selection of the crops and varieties for efficient management of resources for watershed /drylands are (i) land use capability concept (ii) water availability concept (iii) crop substitution (iv) quantity and distribution of the rainfall (v) soil depth and performance of the crops. (SubbaReddy and Maruthi 2006)

2.1. Slope of the land and soil depth

The slope of the watershed plays an important role in selection of the suitable crops. Grasses and economic bushes must be planted on the top of the watershed areas to meet fodder requirements of the livestock besides getting some additional income. Arable cropping systems are to be mounted in the middle part of the watershed to impart stability in income, while tree/fruit based systems are to be given priority for the lower portion of the watershed areas.

2.2. Land use capability concept

It is an age old concept but rarely used in dryland agriculture production in India. In subsistence farming, food crops are grown according to household needs of the farmers. Hence, in dryland crop production, it is the moisture storage capacity of the soil and water availability periods that play key role in selection of crops and varieties in soils.

2.3. Water availability period

The cultivars that grow in Vertisols are sometimes longer in duration than water availability period. As a result, the crops invariably undergo moisture stress and resulting in low yields. The water availability periods for different agro climatic zones were worked out as guiding principle in selection of the crops and cropping systems in Vertisols(Table 1). For post rainy season crops, it is the amount of soil moisture store in soil profile at sowing time that dictates the choice of crops. At Hissar Dry farming station (Haryana) have shown that the choice of rainy season crops changes with conserved moisture (Table 1).

Table 1: Effective cropping season at various locations in arid and semi arid tropics of India

		Rainfall (mm)		
Zone	Locarion	Growing seaso (weeks)	Monsoon season (weeks 23-39)	Post monsoon season (weeks 40-48)
Arid	Jodhpur	11	353	8
	Hissar	13	395	19
	Anantapur	13	305	149
	Rajkot	17	572	36
Semi-arid	Hyderabad	22	603	108
	Bangalore	32	400	226
	bijapur	17	381	130
	Sholapur	23	494	101
	Bellary	8	261	133
1. Standard meteorological week 1=1-7 January				

Table 2: Suggested cropping strategies for different mounts of stores moisture at sowing, Hissar, India

Stored moisture (mm)	Suggested crops
300	Wheat, pea,chickpea
200-300	Wheat (Desi) barley, lentil, chickpea
150-200	Chickpea,barley,raya, Brassica Juncea),Sarson Brassica Campestrisncv.Bown sarson), chickpea
75-150	Raya,chickpea, Possible in better catchment areas
50-75	. Taramira (Eruca Sativa)

Results suggest that short duration and low water requiring crops should be preferred under receding moisture situations. In shallow to medium deep Vertisols at Sholapur, there is not great choice among sorghum, chickpea and safflower since water use efficiency (WUE) is almost same (6.6 to 7.6 kg grain/ mm water used) .However, chickpea and safflower prices are higher.

The moisture availability period determines the effective cropping season. Based on the analysis of long term rainfall data in arid and semi-arid areas of India, effective cropping season have been delineated for a number of locations

In arid regions, the effective cropping season is normally 11-17 weeks, which restricts the choice of crops, and limits the farmer to a single crop in a rainy season. In semi-arid regions, the effective cropping season is normally longer (22-33 weeks) with the exception of 8 weeks in Bijapur (Karnataka) region. Rainy season crops are grown in shallow to medium Vertisols at Bijapur, while post rainy season crops are commonly grown in deep Vertisols at Bellary.

2.4 Crop Substitution

Table 3: Alternate crops for crop substitution at different locations

Place	Traditional crop (yield kg/ha)		Alternate crop (Yield (kg/ha)	
Bellary	Cotton	200	Sorghum	2670
Indore	Greengram	1180	Soyabean	3330
Agra	Wheat	1030	Mustard	2040
Hissar	Wheat	320	Eruka Sativa	1610
Bijapur	wheat	940	Safflower	1850

3. Efficient crops and varieties

The crops and cultivars in dryland areas are not necessarily the most stable and efficient in terms of moisture use. Many of traditional cultivars of sorghum, pearl millet, pigeonpea ,groundnut,castor,cotton and other crops are not adapted to the rainfall pattern where they are grown. For example, the crop duration is longer than effective cropping season. They experience drought stress at the most critical stages of their life cycle, which leads to low and uneconomic yields. In order to achieve yield stability, it is necessary to grow crops and crops and cultivars with water requirement pattern that match the effective growing season. The food needs of the farmer, storability and marketability of the produce, the price at harvest and susceptibility to pests and diseases also governs the choice(Singh, and Subba Reddy, 1988).

In dryland production systems, generally, the crop should be of short duration with early vigour, deep root system with ramified roots, dwarf plants with erect leaves and stem, moderate tillering in the case of tillering crops and varieties, resistance/tolerance to biotic stresses, lesser period between flowering and maturity so that the grain filling is least affected by adverse weather, resistance/ tolerance to abiotic stresses, low rate of transpiration, less sensitive to photo-period and wider adaptability. Thus, under changing climate conditions, introduction of high yielding, drought resistant/tolerant varieties hold the promise for getting higher yields In general ,dryland crops are sown early with the onset of monsoon to realize higher yields during kharif season. And any delay in monsoon beyond normal period affects sowing of many crops of longer duration or with narrow sowing window. The crops with wider sowing windows can be taken up till the cut-off date without major yield loss. Under delayed monsoon, choice of alternate crops or cultivars depends on the farming situation, soil, rainfall and cropping pattern in the location and extent of delay in the onset of monsoon. For example, pulses and oilseeds are preferred over cereals with respect to remaining shorter growing period during kharif sowing. Cluster bean, moth bean and horse gram are better choice for low rainfall areas as compared to other kharif season pulses. For cultivation on conserved soil moisture during rabi season, chickpea and lentil are preferred over peas and French bean. Similarly, among oilseeds, groundnut, castor sesame and niger perform well under rainfed/dryland conditions during kharif season.

Among the kharif cereals, coarse cereals (millets, ragi and sorghum) are better choice over maize. Among the millets, setaria is most suited for late sown condition without any serious effect on productivity. . The resilient crops and varieties to cope with delayed monsoon in various rainfall and soil zones are given in Table 4.and 5

Table 4: Suitable varieties of dryland crops under delayed sowing in Kharif season

Location/ MCSR (mm)	Crop	Suggested contingency crops and cultivate		
		Delay by 2 weeks	Delay by 4 weeks	Delay by 6 weeks
Anantapur (Alfisols)/ 352 (kharif). 144 (rabi)	Groundnut	-	Kadiri- 9. Prasuna. Narayani	Kadiri- 9, Prasuna. Narayani
	Pigconpca	Palnadu. LRG-30, PRG-158	LRG-41. Palnadu. LRG-41. Lakshmi. Lukshmi. LRG-30. PRG-158	1CTP 8203. ICMV-221.ICMH-451
	Sorghum	PSV-15, 19	CSH-9,14. CSV-12	CSH-9. 12. 13, 14, CSV-12. NTJ-1.3
	Castor	Kninthi. Jyothi. GCH- 4.6	Kninthi. Jyothi. GCH-4, 6	Krunllii, Jyothi. GCH-4,6
	Cotton	Narsimha, MClJ-5, LRA-5166. NHH-44. NDLHH-240	Narsimha. MClJ-5. IRA-5166. NHH-44. NDLHH-240	Narsimha, MClJ-5. LRA-5166

Location/ MCSR (mm)	Crop	Suggested contingency crops and cultivate		
		Delay by 2 weeks	Delay by 4 weeks	Delay by 6 weeks
Bangaluru (Alfisols).' 517 24l(ra6<)	Fingcrmillct	MR-1,6,1.-5	MR-1.6.1.-5. HR-911	Fingcrmillct: GPU-28 Little millet: CO-2, PRC-3 Foxtail millet: RS-118. K-22I-I
	Maize	Deccan-103. NAC-6004, Ganga-11	NAC-6OO2. 6004	DHM- 2. Ganga-11. Deccan-103
	Groundnut	JL-24. K-134. GPBD-4. K-134	TMV-2. JL-24	JL-24. K-134
	Pigconpca Cowpca	BRG-2 TVX-944-2E, KBC-I,	BRG-2 TVX-944.1T-38956-1	K-134, VRA-2
Bijapur	Pigconpca	C.S-1,	BJ-221, GS-1, BJ-221, Maruthi.	Durga, GC-11,39, WRG-I. Maruthi, TS- 3 R. ICPC-87
(Vcrtisols)/ 388 (k/iarif). 134 (rabi)	Sunflower	Maruthi, Asha, TS-3, ICPC-87	Asha, TS-3. ICPC-87	
		KBSH-I. 44. DSH-I. RSFH-I, MSFH-17, SH-41	KBSH-I. 44. DSH-l. RSFH-1. MSFH-17. SH-41	KBSH-1. 44. DSH-1. RSFH-I. MSFH-17. SH-41
Akola (Vcrtisols)/ 688 (kharif). 82.3 (rabi)	Soybean	Samrudhi. JS-335. JS-93-05	JS-335. JS-93-05	-
	Pigconpca	BDN 708. AKT-8811	AKT-8811. Vipula, PKV-Tara, BSMR-736	AKT-8811, Vipula. PKV-Tara. BSMR-736
Parhhani (Vcrtisols V 800.5 (*kharif*). 110.5 (*rabi*)	Sorghum	CSH-9.14	CSH-9, II, 14, 16. 401, 809	PVK-CSH-9, 11, 14, If PVK-401,809
	Pigconpca	Vaisali	BSMR-736 853. BDN-708.711	BSMR-736, 853 BDN-708,711

MCSR: Mean crop seasonal rainfall; Source: Annual Reports - AICRPDA-NICR A 2011 to 2015

Table 5: Performance of improved varieties dryland crops with higher sustainability

AICRPDA Centre	Crop	Variety/Hybrid	SYI	Agro-climate/soil type
Arjia	Maize	PF.HM 2	0.4	Southern zone of Rajasthan (Semiarid Vertic Inceptisols)
Bangalore	Finger millet	MR-1	0.77	Southern dry zone of Karnataka (Semi arid Alftsols)
Indore	Soybean	NRCS 37	0.50	Mahva Plateau zone of Madhya Pradesh (Semiarid Vertisols/Vertic Intcrerades)
Ananta pur	Groundnut	Nurayani	0.49	Scarcity zone of Andhra Pradesh (Arid to semiarid Alfisols/Aridisols)
Akola	Soybean	JS335	0.55	Western Vidarblu zone of Maharashtr (Semiarid Vertisols)

(*Source*: AICRPDA Annual Reports; SYI- Sustainable yield index)

On-Farm experience

Among the various options for addressing the climate variability, improved cultivars play an important role in stabilizing productivity in rainfed environments. In regions receiving low rainfall (<750 mm), where the cropping season is restrictedto 10-14 weeks and the delay in the onset of the monsoon and the early withdrawalare the most commonly occurring contingencies limiting the crop production. Short duration varieties which can complete life cycle within the growing period of a region can escape drought during early withdrawal of monsoon and can also be grown under delayed onset of monsoon conditions and thus an adaptation strategy. For example, Rajkot district in Gujarat state receives 660 mm of rainfall annually with irregular distribution and mid-season droughts in July and August resulting in lower yields of traditional spreading-type groundnut varieties which are of 120 days duration. Bunch type high yielding groundnut cultivars were introduced (GG-5, TG-38, GG-20) which are short in duration (95-100 days) and performed better under late planted conditions with 10-15% higher yield than that of traditional varieties of groundnut. Similarly, in black soil (Vertisols) regions of Maharashtra, Madhya Pradesh and Kota district of Rajasthan, soybean is an important crop distributed in about 9 m ha with signifi cant area under JS-335 variety which is of 110-115 days duration. Due to delay in onset of monsoon in recent years, planting of soybean is extended up to fi rst fortnight of July and the crop at maturity stage is vulnerable to late season dry spells. Hence, a short duration variety JS-93-05 (95-100 days) in Maharashtra, and JS-95-60 (85-95 days) in Madhya Pradesh and Rajasthan were introduced which gave 15- 20% higher yield and benefi t: cost (B:C) ratio (2.5) compared to the farmers' practice

4. Cropping systems

Traditionally, double cropping including relay cropping is practiced in rainfed regions with sufficient rains (usually >750 mm) and good soil moisture holding capacity (>150mm). However, some more areas could bring under double cropping through use of available dryland technologies viz., rainwater management, choices of crops, short duration varieties and agronomic practices. Out of the two crops, one could be short durations (usually legumes) and another, medium duration (usually cereals) for optimum use of available growing season. For example, a second crop could successfully grown in high rainfall regions of Odisha, Eastern Uttar Pradesh and Madhya Pradesh by replacing medium to long duration (>120 days) rice variety with short duration (Table.8)

Table 6: Selection of crops and cropping systems in Dryland Agriculture

Rainfall (mm)	Soil type	Effective growing season in weeks	Suggested cropping systems
350 – 600	Alfisols and shallow Vertisols	20	Single rainy season croppimg
350-600	Deep Aridisols and Entisols	20	Single cropping either in kharif or rabi
350-600	Deep Vertisols	20	Single post rainy season cropping
600-750	Alfisols, Vertisols and Entisols	20-30	Intercropping
750-900	Entisols, Deep Vertisols, Deep Alfisols and Inceptisols	30	Double cropping with monitoring
>900	Entisols, Deep Vertisols, Deep Alfisols and Inceptisols	More than 30	Double cropping assured
Depth	Available moisture (cm)	Soil moisture depth availability period in days	Cropping system
Shallow	10	90	Sole crop sorghum or maize or soybean
Medium	15	150	Sole soybean ,Sorghum + Pigeonpea
Deep	20	180	Maize, Safflower, Soybean, Chickpea and Maize-Chickpea

4.1 Intercropping systems

Growing more than one crop of different durations on the same piece of land in a definite row proportion. A combination of two or three crops of cereal, pulse, oilseed and cotton crops having short and long duration, deep and short rooted are preferred for risk minimization, high productivity, enhancing soil fertility etc. Strip cropping consisting cultivation of more or less uniform and alternate strips of erosion permitting and erosion resisting crops. The erosion permitting strips slow down the runoff and reduce the

wind velocity which otherwise caused erosion. Row crops must be limited in width to avoid excessive run-off and erosion while the erosion resistant crops must be in wide enough strip to afford adequate protection and capacity to filter dement from runoff water. Adoption of wide row spacing ensures availability wider cross section of effective root zone, which may provide moisture for long period and frequent deep intercultivation reduces evaporation considerably by creating dust mulch.

In mixed intercropping, two or more crops are grown together without any definite row proportion. Sometimes it is also referred to as mixed cropping (Voncossal, *et al*., 2019). In pasture-based cropping system, grass-legume intercropping is an ideal example of mixed intercropping [Gulwa. etal., 2017]. The mixed intercropping is commonly observed to fulfil the requirement of food and forage where the land resource is a limiting factor (Undie, Uwah, Attoe 2012). Furthermore, a review work clearly described perennial polycultures as an agroecological strategy in cropping system with enough potential for the sustainable intensification of agricultural systems spatially and temporally.

The row intercropping is raising of one or more crops sown in regular rows, and growing intercrops in a row or without row at the same time. The row intercropping is a usual practice targeting maximum and judicious use of resources and optimization of productivity

Relay intercropping is raising two or more crops at a time during a portion of the growing period of each. In this system, the second crop is seeded when the first crop completes a major part of its life cycle and reaches reproductive stage or close to maturity but before harvest. The areas with limitation of time and soil moisture are more appropriate for relay cropping Blade *et al*., 2011) before harvesting of the preceding crop, the next crop is sown and both the crops remain in the field for some period of their cycle. However, the succeeding crop yields less compared to normal sowing in sequential cropping and more seeds of the succeeding crop are required to obtain a good stand..

Table 7: Promising intercropping systems in different zones

Soil zone/ Agroclimatic zonc/Stalc a.	Intercropping system
a. Vertisols and Vertic Inceptisols	
Malwa plateau. Madhya Pradesh	Soybean - pigeon pea (4:2) Sorghum + pigeon pea (2:2)
Western Vidarblu zone of Maharashtr	Cotton + greengrarn (1:1)
Southern Rajasthan	Maize + blaekgrum (2:2) Groundnut + sesame (6:2)
Northern Dry zone, Karnataka	Groundnut + caslorbean (3:1) Pearl millet - pigeonpea (4:2)
Northern Saurushtru. Gujarat	Groundnut + caslorbean (3:1) Groundnut + pigeonpea (3:1)
Southern Tamil Nadu	Cotton + blaekgramgrccngram (2:1)
b. Inceptisols and related soil zone	
Western plateau, Jharkhand	Pigeon pea + rice (2:3) Maize + cowpca (2:2)
Alfixolv Oxisols zone	
Eastern Glial zone. Odisha	Maize + pigeon pea (2:2) Fingennillet + pigeon pea (4:2)
Alfisols zone	
Southern dry zone. Karnataka	Groundnut + pigeon pea (8:2) Fingennillet + pigeon pea (10:2)
Southern zone. Telangana	Sorghum + pigeon pea (2:1)
Scarcity zone, Andhra Pradesh	Groundnut + pigeon pea (7:1)
Aridbob zone	
Northern zone. Gujarat	Castor bean cowpca (1:2) Pearl millet + cluster (2:1)

Sustainability of Cropping Systems

In rainfed agriculture sustainability of yield becomes more important. Than "sample mean "as the magnitude of yield is rainfall dependant. While considering rainfall yield index it is observed, rice + radish (4:2) and rice + black gram (4:2) in Oxisols of Phulbani, Pigeonpea + okra (1:1), Maize + Okra in dry sub humid inceptisols of Ranchi, Sorghum + Cowpea (F) under delayed sowing conditions of Arjia, Pearl millet + pigeonpea in Vertisols of Solapur, Sunflower + pigeonpea in Vertisols at Indore and greengram + castor in semi arid entisols of Danthiwada recorded high sustainability over other systems

On-Farm evaluation

Among cropping systems, intercropping systems of foxtail millet (Setaria)+pigeonpea(5:1)andcastor+pigeonpea(1:1)atKurnool,cotton+pigeon pea in Srikakulam, soybean pigeon pea (4:2 pearlmillet+pigeonpea (3:3), cotton+green gram (1:1) in NICRA village of Aurangabad district, soybean+

pigeon pea (3:1), groundnut+ soybean (4:1) and rabi sorghum+safflower (3:3) Nandurbar in Maharashtra, imparted stable income in different NICRA villages

Adoption of maize + blackgram (2:2) intercropping system under delayed onset of monsoon during 2012-14, gave on an average 27.5% higher maize grain equivalent yield (1842 kg/ha) with net returns of Rs. 19424/ha, B:C ratio of 2.53 and RWUE of 3.93 kg/ha-mm, as compared to farmers' practice (mixed cropping of maize and blackgram) (1444 kg/ha)in southern zone of Rajastan

Demonstrations on maize + blackgram (2:2) intercropping system during 2015-18 recorded 29% higher maize equivalent yield (2092 kg/ha), net returns (Rs. 19714/ha) and B:C ratio (2.17) compared to mixed cropping of maize and blackgram (1628 kg/ha). Similarly, during 2015-18, groundnut + sesame (6:2) intercropping system gave 38.8% higher groundnut pod equivalent yield (836 kg/ha) with net returns of Rs. 21555/ha, RWUE of 1.80 kg/ha-mm and B:C ratio of 1.82 compared to mixed cropping of both crops (602 kg/ha). At Shektha village, Aurangabad, Maharashtra farmers realized the advantage of higher net incomes with intercropping of the main crops such as cotton, pearl millet, pigeon pea and soybean ranging from Rs.9216 to Rs.12330/ha. closely followed by pigeon pea + groundnut (2:4) intercropping system (Rs.156937/ha) compared to pearlmillet + groundnut (2:4) intercropping system (Rs. 59276/ha) in Northern zone of Karnataka(AICRPDA 2018 NICRA2015 to 2019)

Management of Intercropping

A seed-bed is prepared before sowing by physical manipulation of soil and suitable tillage is required for different crops [Lal, R, 2016]. In intercropping when two or more crops with dissimilar morphological characters are sown together, uniform bed preparation may not be ideal for different crops and it greatly depends on crops (Bybee-Finley K.A. and Rayan, M.R. 2018). For example, deep-rooted crops need deep tillage, while cereals require shallow tillage. The crops with small seeds (like mustard, sesame and jute) require pulverized soil and fine beds. Some crops are sown on ridges (cotton, maize) whereas others such as green gram, black gram and mustard prefer flat-beds. In additive series, the bed is prepared as per the requirement of the base crop. In maize + green gram/black gram intercropping, generally, crops are sown on a flat-bed [Manasa, etal.,2018]

The varieties of crops chosen in intercropping should have some desired characteristics as the highest level of complementarity and least competition occur [Fukai; and Trenbath, 2016). The crop varieties are required to be photoperiod insensitive as these can be cultivated at any time of the year [Saxena, M. *et al.*, 2018]. The short duration sorghum hybrids like CSH 17

(103 days) and CSH 23 (105 days) are suitable for intercropping long-duration pigeon pea varieties like GAN 1 and WRP 1 (both are of 160–165 days). The varieties selected for intercropping should have some morphological and physiological characteristics like thin leaves, less branching and be tolerant to shading.

In the intercropping system, modification or alteration is done in planting geometry, spacing and thus plant stand In a pearl millet + green gram intercropping system, paired row planting resulted in more yield than uniform row planting, and intercropping of paired row maize + pigeon pea performed well in the southern dry zone of Karnataka [Kumar *et al.*, 2018]. In replacement series after sowing of a crop with some uniform rows, replacement is done by another crop of some rows and row proportion is determined mainly by the farmers or as per the recommendations. Sometimes closer spacing within rows is also followed to accommodate more number of plants and generally a greater number of plants is accommodated in intercropping. Other factors related to optimal and uniform plant stands include seed treatment, bed preparation, sowing at the proper depth and so on, are maintained as per the standard procedure. Further, gap filling is an important operation in drylands, which can also be taken care of to obtain the desired plant population.

The nutrient removal by the crop is greater because of more dry matter production or biological yield in intercropping. In a cereal–legume combination of intercropping, legumes use less N from the soil and it may be either from inherent soil fertility or in the form of applied fertilizer. On the other hand, cereals are more N demanding and use a major portion of applied N. Nevertheless, legumes initially use P for better nodulation, but after nodulation, the root exudate of legumes and other rhizospheric micro-organisms make P available to both legumes and the companion cereals. In N-deficient soils, legume fixes a considerable quantity of N, but when sufficient N is supplied as fertilizer, biological N fixation is reduced. Moreover, sufficient supply of N fertilizer promotes the growth of cereals because cereals are more aggressive in nature and the growth of legumes is suppressed. Considering the above, it is advisable to apply N as basal and topdressing to cereal rows and P and K to the whole plot. In the molybdenum-deficient soils, the micronutrient application should be done as basal or by foliar spray, because sufficient molybdenum enhances nodulation as well as biological N-fixation by legumes [Alam, etal.,2019]. The mutual benefit or complementarity observed in the cereal–legume combination is the result of below-ground chemical and biological processes which can assure the availability of some micro-nutrients like iron and zinc

Studies carried out with cereal based cropping system showed that 25-50 % fertiliser NPK dose of kharif crops can be curtailed to the use of FYM,

Sesbania green manure and crop residues under different situations. Beneficial effect of integration of chemical fertilisers with green manuring or FYM on total productivity of systems involving cereals, oilseeds and cotton increased by 7-45 % over farmers practice in different agro- ecological zones. In Pearl millet + Pigeonpea system, pearl millet responded up to 10 kg N and 15 kg P_2O_5 in arid environment, In case of pearl millet + castor significant response were observed up to 80 kg /ha. In crop rotation of finger millet and groundnut, the yield of groundnut was maximum when only FYM of 10 t/ ha annually was added and decreased with addition of fertilisers. Thus the fertiliser management in cereal based intercropping systems showed that there is need to apply basal application of the nutrients of cereal crop for both the component in the system. While nitrogen as a part of the top dressing should be applied to the cereal component only. In pulse and oilseed based cropping system, application of nutrients on respective row proportionate bases gave higher yield and monetary advantages as compared to the recommended dose of nutrients of individual sole crops

Under dryland moisture-stress conditions , the extra need for water for crop mixture in intercropping will be an additional burden to resource-poor farmers. However, simple In -situ moisture conservation technologies like Conservation furrows, use of green leaf materials like Glyricidia raised on farm bunds, green matter from alley cropping systems as mulch cum manure can be adopted in dryland intercropping systems.

In intercropping systems, chemical herbicide application is difficult once crops have emerged particularly when a combination of dicotyledonous and monocotyledonous plants are chosen in combination .. However, as the greater portion of the land area is covered by the crops in intercropping, there will be fewer weeds. Fast-growing crops like mung and black gram under intercropping cover maximum land area and suppress weed growth. The weed suppression ability in the intercropping system depends on some factors like selection of crop, the genotypes used, plant population, the ratio of crops considered in the intercropping and spatial arrangement, fertility and soil moisture. Mostly hand weeding is practiced in intercropping

Experimental results showed that intercropping of maize + soybean and maize + cowpea significantly reduced weed growth than sole cropping and the pre-emergence application of alachlor 2 kg ha−1 or metolachlor 1.0 kg ha−1 controlled weeds successfully.

The insect-pest population is regulated by the intercropping system itself. In marginal farming mixed cropping is chosen because of the low incidence of insect pests [Maitra etal., 2021]. The crop mixture attracts beneficial insects which have the potential to maintain the harmful pest population below the threshold level.

In intercropping, the maximum ground area is covered; hence there will be a minimum chance of run-off, soil erosion and nutrient loss [Nayawade, 2019]. In an agroforestry system, Gliricidia alley cropping can reduce run-off by 28.2% and soil loss by 49.3–51.1% over no alley cropping system. Furthermore, Gliricidia alley can conserve soil organic carbon, N, P and K by 63.4, 5.0, 0.3 and 2.4 kg ha−1 . Similarly, the Leucaena-based alley cropping system is also effective in terms of checking run-off of water, conserving

soil and preventing nutrient loss. Leucaena alley with a miniature trench can reduce run-off by 18.3–18.7% and soil loss by 37.2–43.0%. The alley cropping of Leucaena can conserve organic C, N, P and K by 57.7, 4.6, 0.3 and 2.2 kg ha−1 [Madhu, etal., 2019]. The intercropping combination of finger millet + black gram recorded the lowest runoff (10.2%) and losses of soil and nitrogen, phosphorus and potassium through erosion over sole when sown in contour because black gram covered enough ground area in intercropping with finger millet [Das and Sudhushir 2010)

The soil fertility status was also improved in finger millet + pulses intercropping which was due to contribution of leaf fall and biological nitrogen fixation by legumes [Das, A. and Sudhushir,2010]. The cereal–legume combination of intercropping is known to enable long term immobilization of N . Nutrient balance studies indicated that cereal–legume intercropping enhanced N fertility of the soil. Among legumes, groundnut in combination with maize added N to the soil because of above- and below-ground architecture of groundnut and more soil coverage [Choudhary and Choudhary.2016)

The nutrient removal by the crop is greater because of more dry matter production or biological yield in intercropping. In a cereal–legume combination of intercropping, legumes use less N from the soil and it may be either from inherent soil fertility or in the form of applied fertilizer. On the other hand, cereals are more N demanding and use a major portion of applied N. Nevertheless, legumes initially use P for better nodulation, but after nodulation, the root exudate of legumes and other rhizospheric micro-organisms make P available to both legumes and the companion cereals. In N-deficient soils, legume fixes a considerable quantity of N, but when sufficient N is supplied as fertilizer, biological N fixation is reduced. Moreover, sufficient supply of N fertilizer promotes the growth of cereals because cereals are more aggressive in nature and the growth of legumes is suppressed. Considering the above, it is advisable to apply N as basal and topdressing to cereal rows and P and K to the whole plot. In the molybdenum-deficient soils, the micronutrient application should be done as basal or by foliar spray, because sufficient molybdenum enhances nodulation as well as biological N-fixation by legumes. The mutual benefit or complementarity observed in the cereal–legume combination is the result of below-ground chemical and biological processes which can assure the availability of some micro-nutrients like iron and zinc.

In intercropping systems, chemical herbicide application is difficult once crops have emerged particularly when a combination of dicotyledonous and monocotyledonous plants are chosen in combination. However, as the greater portion of the land area is covered by the crops in intercropping, there will be fewer weeds. Fast-growing crops like mung and black gram under intercropping cover maximum land area and suppress weed growth. The weed suppression ability in the intercropping system depends on some factors like selection of crop, the genotypes used, plant population, the ratio of crops considered in the intercropping and spatial arrangement, fertility and soil moisture. Mostly hand weeding is practiced in intercropping. Weed control by the application of chemical herbicides is difficult as most of the herbicides are crop-specific. The more complex the intercropping system, the less likelihood of a finding of herbicides. Earlier Reddy (2005) mentioned that Isoproturon (1.0 kg ai ha−1) was effective in intercropping wheat + chickpea and wheat + mustard. Likewise, Alachlor (1.5 kg ai ha−1) was beneficial in maize + cowpea and sorghum + black gram intercropping systems. Further, Butachlor (1.25) resulted in successful control of weeds in maize + mung intercropping system

4.2. Indices for Measuring the Efficiency of Intercropping

In intercropping systems, most of the competition studies have examined growing two crop species and also in a ‘replacement series’. The component crops involved in the system may be related to each other in the manners mentioned below:

(i) **Competitive:** In this relationship, the output of one crop would be increased through the decline in the production of the other. This is also known as ‘compensation’. Willey *et al.*, 2014 referred to the two species as ‘dominant’ and ‘dominated’ species.

(ii) **Complementary**: This is another type of relationship in which an increase in output of one crop helps to bring about an increase in output of the other species. This is termed as ‘mutual cooperation’ [85] and is not very common.

(iii) **Supplementary**: In this case, the output of one crop may be increased without having any influence on the output of the other. This situation commonly occurs when the maturity of two crop species differ widely.

(iv) **Mutual Inhibition:** Mutual inhibition happens when the actual productivity of each component of crops harvested is less than the expected yield. The competitive and supplementary relationship is very common in different intercropping systems.

Land Equivalent Ratio (LER) Willey and Osiru [1978) gave the idea of the LER and it is described as the proportionate land area required under a pure stand of crop species to yield the same product as obtained under an intercropping at the same management level. The LER of intercropped plots are estimated for each component crops separately by adding the estimated total of two varieties; the LER of the sole crop is taken as unity (1). In a replacement series of intercropping with a combination of two crops at the ratio of 50:50, the LER can be calculated by the following expression.

(1). In a replacement series of intercropping with a combination of two crops at the ratio of 50:50, the LER can be calculated by the following expression. Land Equivalent Ratio (LER) Willey Osiru *et al.*, 2014 gave the idea of the LER and it is described as the proportionate land area required under a pure stand of crop species to yield the same product as obtained under an intercropping at the same management level. The LER of intercropped plots are estimated for each component crops separately by adding the estimated total of two varieties; the LER of the sole crop is taken as unity (1). In a replacement series of intercropping with a combination of two crops at the ratio of 50:50, the LER can be calculated by the following expression.

$$LER = \frac{Yba}{Yaa \times Zab} + \frac{Yba}{Ybb \times Zba} = La + Lb \tag{1}$$

where, Yab is the yield of "a" crop grown in association with "b" crop and Yba is the yield of "b" crop grown in association with "a" crop. Yaa and Ybb represent the yields of "a" and "b" crops grown in a pure stand, respectively. The modified formula for any other situation is:

$$LER = \frac{Yba}{Yaa \times Zab} + \frac{Yba}{Ybb \times Zba} = La + Lb$$

The LER denotes the benefits of an intercropping system to utilize the resources as against their pure stands. The LER value greater than unity (1.0) indicates the advantages of the intercropping system and less than one (1.0) is considered as a poor performance of the intercrops. The LER value of some intercropping system with major crops is presented below(Table 9)

Table 8: Land equivalent ratio (LER) in intercropping systems

Intercropping System	Ratio	LER	Country
Sorghum + Scsbania	2:1	1.06	Syria
Wheat + Faba bean	1:1	5.24	UK
Sorghum + Cowpea	2:1	1 08	Nigeria
Wheat * Mustard	1:1	1.46	Bangladesh
Wheat + Fenugreek	1:3	1.4	Pakistan
Wheat + Maisc	1:1	1.19	China
Sorghum + Soyabean	1:1	1.40	Nigeria
Pearlmillet + Soybean	"	2.77	Nigeria
Sorghum + Ground nut	1:1	2.10	Ethiopia
Maize + Soybean	1:1	1.54	Nigeria
Maize + Groundnut	2:2	1.42	Ghana
Maize + Potato	1:2	1.58	Ethiopia
Maize + Garden pea	1:2	1.56	Bangladesh
Maize + Groundnut	2:2	1.82	India
Maize + Soybean	2:2	1.90	China
Wheat + lentil	2:2	1.34	India
Potato + Dolichos	1:2.4	1.24	Kenya
Potato + Vetch	1:2	1.75	Kenya
Pearlmillet + Green gram	1:1	2.03	India

(*Source*; Maitra *et al*.,2021)

Area Time Equivalent Ratio (ATER)

The LER emphasizes on the only land area without considering the time factor for which the crop occupies the field. As time factor is not a part in the LER, researchers needed another expression considering the field occupancy by the crops in an intercropping to correct this constraint of the LER. Hiebsch [108] developed the concept of Area Time Equivalent Ratio (ATER) in which the duration of crops (starting from seeding to harvest) was considered. The ATER is calculated by the following formula

$$\text{ATER} = \frac{(\text{RYc x tc}) + (\text{RYp x tp})}{\text{T}} \quad (3)$$

where, RY = Relative yields of crop species "c" and "p" = Yield of intercrop ha-1/ Yield of sole crop ha t = duration (in days) for species "c" and "p" and T = duration (in days) for the intercropping system. However, the LER generally overemphasizes and the ATER undervalues the land-use efficiency . Researchers revealed the advantageous ATER values in different intercropping system.

Table 9: Area time equivalent ratio (ATER) in maize-legume intercropping systems

Intercropping System	Proportion	ATFR	Country
Cotton + Cowpea	-	1.13	Pakistan
Lupine + Wheat	75% + 100%	131	Ethiopia
Mai/e + Soybean	2:6	132	India
Make + Black cowpea	2:2	151	India
Pearlmillet + Green gram	2:1	125	India
Wheat + Paba bean	.	128	Pakistan
Potatot + Dolichos	1:2.4	1.13	Kenya

(Source Maitra, S2021)

Competitive Ratio (CR)

In an intercropping system, competitive ratio (CR) denotes the competitive ability of the component species The CR expresses the number of times by which one component crop is more competitive than other (Willey, R.W.and Rao, M.R. 1980) actually represents the proportion of individual LERs of the crops considered in intercropping and also takes into account the ratio of the crops sown in a mixed stand.

The CR can be calculated by the following formulae.

CRa = (LERa/LERb) × Zba/Zab) (6)

CRb = (LERb/LERa) × (Zab/Zba)

where, CRa and CRb are indicative of the competitive ratios of the crop species "a" and "b" and LERa and LERb are the LERs of the crop species "a" and "b" respectively. Zab is the sown ratio of species "a" in mixture with "b" and Zba is the sown proportion of the species "b" in mixture with "a". If the value of CR is 1), there is a negative impact.

In this condition, the competition between intercrops in mixture is too high, and they are not recommended to grow as intercrops.

Table 10: Competitive ratio (CR) of finger millet + legume intercropping (4:1) systems

Intercropping Systems	Competitive Ratio (CR)	
	Finger Millet	Legumes
Finger millet + Red gram (4:1)	0.28	3.59
Finger millet + Green gram (4:1)	0.71	1.41
Finger millet + Groundnut (4:1)	0.58	1.73
Finger millet + Soybean (4:1)	0.68	1.48

the values of CR of legumes appeared as >1, representing that legumes were more competitive than finger millet. Among the legumes, green-gram was found to be the least aggressive on affecting the growth of finger millet and thus it provided a balanced competition with finger millet.

4.2. Double Cropping

In areas receiving more than 800 mm annual rainfall and soil moisture storage of 200 mm/M, double cropping is a distinct possibility. Such options were recommended for rainfed areas in Orissa, Bihar, Madhya Pradesh and eastern Uttar Pradesh. With some adjustment of sowing dates (early planting and harvesting of *kharif* crops) double cropping is also possible in Vidharbha and Malwa plateau as well. Late onset of monsoon and the consequent delay in planting of *kharif* crop is the principal constraint in wider adoption of recommendations related to double cropping. The fluctuating process of the component crops and the overall profitability of the system in terms of additional costs and benefits also play a crucial role in farmers' choice of cropping systems

Table 11: Promising double cropping systems in drylands

Production System/ Centre/	AESR	Rainfall (mm)	Soil Type	Predominant double cropping systems
Maize Based Production Systems				
Hoshiarpur	9.1	1032	Entisols/ Inceptisols	Sunhemp-Wheat, Pearl millet (F)-Wheat, Maize - wheat/mustard, Maize-wheat + chickpea
Rakh Dhiansar	14.2	1100	Inceptisols	Maize - toria + Wheat, Maize - wheat/mustard, Blackgram-wheat
Arjia	4.2	862	Vertisols	Greengram – safflower
Oilseed Based Production Systems				
Indore	5.2	964	Vertisols	Soybean - Wheat, Soybean - Safflower/chickpea, Soybean – mustard
Rewa	10.3	1048	Vertisols	Rice-chickpea/lentil, Rice – wheat
Cotton Based Cropping Systems				
Akola	6.3	825	Vertisols	Greengram-safflower, Sorghum – safflower
Nutritive Cereal Based Cropping Systems				
Jhansi	12.4	936	Alfisols	Sunhemp - rabi sorghum, Kharif sorghum - cowpea - barley + chickpea (1:1)
Agra	4.1	538	Entisols	Pearl millet- chickpea/Barley, Greengram- mustard

Source: Subba Redy and G. Maruthi,V (2006)

5.0. Management of dryland crops against weather Aberrations

The common behavior of weather aberrations are: : 1. The commencement of rains may be quite early or considerably delayed. 2. Dry spell immediately after sowing. 3. There may be prolonged breaks during the southwest monsoon season during which most of the dry land crops are grown and 4. Rains may terminate earlier than normal cessation date or may continue beyond the normal rainy season

Contingency cropping is growing of a suitable crop in place of normally sown highly profitable crop of the region due to aberrant weather conditions. Four important aberrations in the rainfall behaviour have been more commonly observed are.

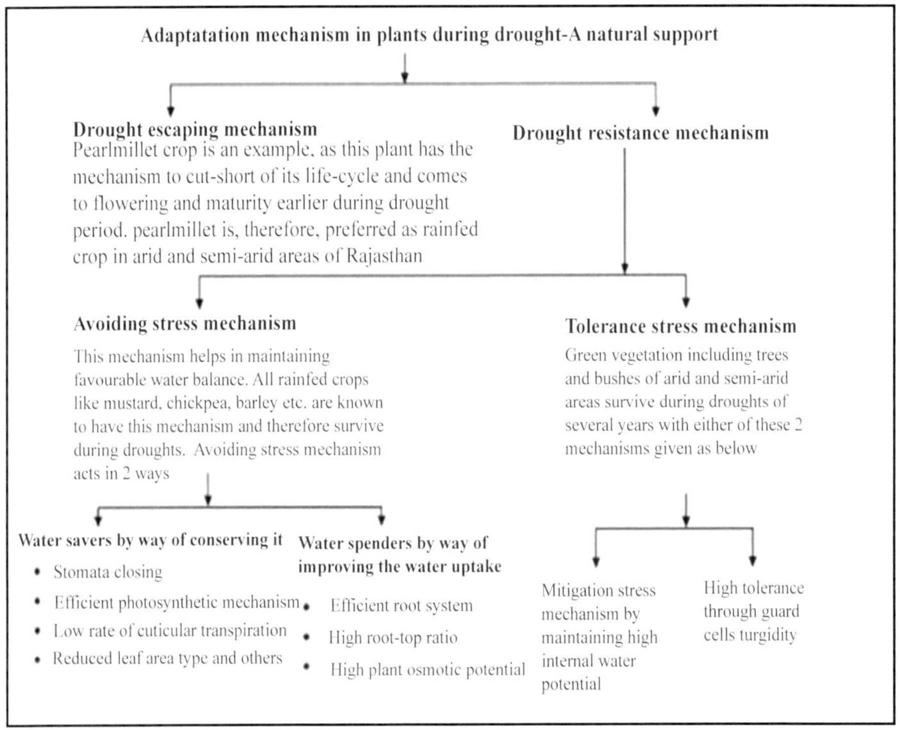

5.1. Mitigation of Agricultural Drought

Resilient crops and cropping systems

Intercropping has a potential option to minimize risk in crop production, ensure reasonable returns at least from one of the intercrop during the years of low rainfall and also maximizes returns during the years of favorable rainfall and utilizes the natural resources such as light, moisture effectively and contributes to the soil fertility build up. Intercropping systems with the predominant crops of the region were assessed in various drought-prone regions of the country. Intercropping of foxtail millet with pigeon pea (5:1 ratio) sown under delayed onset of monsoon conditions showed that the system was more profitable with highest B:C ratio (5.1 compared to 1.5 to 2.6 in sole cropping) in all the 3 years and can also tolerate dry spell by up to 25 days without impacting yields. Intercropping of soybean + pigcon pea (4.2), pearl millet + pıgeon pea (3:3), pigeon pea + green gram (1:2) and cotton + green gram (1:1) performed significantly better than respective sole crops at Aurangabad, Maharashtra which received an average rainfall of 645 mm. Other intercropping systems which performed better over sole crops in Maharashtra were soybean + pigeonpea (4:2) and cotton + green gram (1:1). In regions, receiving rainfall of more than 750 mm, intercropping systems of groundnut + pigeon pea (4:1) at Gumla, pigeon pea + black gram (2:4) and pigeon pea + pearl millet (3:3) at Koderma, outperformed the sole crops. Among rabi crops, intercropping of sorghum + chickpea (6:3) performed well compared to farmers' practice of sole cropping of either crops at Belgaum (NICRA, 2014). A separate chapter on Intercropping is presented with more details.

Table 12: Promising intercropping systems for various drought prone regions receiving rainfall of less than 750 mm

Resilient intercropping system	Equivalent yield of intercropping system (t/ha)	Meld of sole crops (t/ha)	District/ location
Soybean + pigeonpea (4:2)	1.6	1.2	Aurangabad
Pigeonpea-*- pcarlmillct (3:3)	2.0	1.6	Aurangabad
Rubi Sorghum + safflower (6:3)	2.9	2.8	Aurangabad
Rabi Sorghum + chickpea (1:4)	2.9	2.5	Bclgaum
Safflower + chickpea (1:1)	2.7	2.3	Bclgaum
Maize + blackgram (2:2)	2.5	1.7	Arjia
Groundnut + sesame (6:2)	1.7	1.5	Arjia
Chickpea + mustard (4:2)	14	10	Agra
Pigeonpea + sunflower (1:2)	5.5	4.2	Solapur
Groundnut * pigeonpea (4:1)	1.7	1.5	Gumla
Maize + pigeonpea (6:2)	3.2	2.8	Chatra
Pigeonpea + blackgram (2:4)	1.7	0.8	Koderma
Pigeonpea + sorghum (1:5)	1.0	0.9	Gumla
Soybean + pigeonpea (6:2)	2.2	1.8	Amravati

(Prasad *et al.*, 2018)

5.2 Contingency measures for mitigating impact of drought in standing crops

A multi-pronged strategy is essential to meet the fodder requirement of the region:

- Production of quality seeds (breeder and foundation seeds) of selected promising varieties/hybrids with participation of farmers.

- Organizing fodder production programme by promoting appropriate and region-specific varieties.
- Post-harvest management techniques like fodder block making units, chaff cutter for fodder processing and silage making.

A. During June and July

- Normal onset of monsoon followed Early season drought by prolonged dry spell of 15 days or more immediately after sowing of kharif crops leads to poor germination. Sesame and millets sown at a shallow depth of 1 to 3 cm are the most affected followed by pulses, sunflower and sorghum which are sown at a depth of 3 to 5 cm. Germination failures are more common in lighter soils than in heavy soils.
- If the moisture stress occurs at a very early stage i.e., within a week to 10 days after sowing, it is recommended to resow with subsequent rains for better plant stand.
- Thinning of plants is advocated in small seeded crops which are closely planted, to reduce crop stand to conserve soil moisture
- Undertake interculture to break soil crust, remove weeds and create soil mulch for conserving soil moisture. Use organic mulches such as weed biomass, subabul lopping, tree leaves, straw and other available crop residue or organic manures to conserve soil moisture
- Avoid top dressing of fertilizers till sufficient moisture is available in soil.
- Make conservation furrows at 10 to 15 m intervals or adopt ridge and furrow across the slope for effective conservation of soil moisture as well as rainwater in bold seeded, dibbled and wide spaced crops (>30 cm) such as cotton, maize, pigeon pea and oilseed crops. In medium deep black soils of north interior Karnataka and Maharashtra, take up compartmental bunding (bunds of 15 cm height formed on all the four sides to form a check basin of 6 m x 5 m size) for better retention of soil moisture.
- In cotton sown in shallow and medium deep black soils in Maharashtra, Andhra Pradesh and Karnataka, pot watering may be taken up along with gap filling when the crop stand is less than 75%. In case of non-availability of hybrid cotton seed, gap filling may be done with pigeonpea in Maharashtra to maintain adequate plant stand. When germination is less than 30%, re-sowing may be taken up with a closer plant to plant

spacing of 45 cm. Raising of cotton seedlings in polythene bags for transplanting when sufficient moisture is available after receipt of rains can be practiced to compensate loss in plant stand with seedlings of similar age.

- In northern dry zone of Karnataka, when the normal sown crops completely wither, sow alternate crops like sunflower, setaria, dolichos, horsegram soon after receipt of rains. 286 Contingency Crop Plans • In this zone after planting of rabi sorghum and safflower if the soil moisture is inadequate thin out the plant population.
- In Vidarbha, in sorghum, groundnut and chickpea spray 2% urea during prolonged dryspells.
- In Vidarbha region, incase of failure of kharif crops, take up sowing of photoinsensitive crops such as pearlmillet, sunflower, sesame and pigeonpea once adequate rains are received.
- Provide micro-irrigation with drip for wide spaced crops such as cotton, maize, chillies and vegetables and sprinklers for groundnut, maize and vegetables wherever ground/ surface water is available

B. Mid-season drought during mid-July to mid-September

- Take up repeated interculture to remove weeds and create soil mulch to conserve soil moisture.
- During 40-45 DAS, if there is a severe moisture stress, ratooning or thinning may be done in kharif sorghum and pearlmillet. Thinning by removing every third row in rabi sorghum in black soils of Maharashtra and Karnataka is advantageous.
- Avoid top dressing of fertilizers until receipt of rains.
- Make conservation furrows for moisture conservation.
- Foliar spray of 2% KNO3 or 2% urea solution or 1% water soluble fertilizers like 19-19-19, 20-20-20, 21-21-21 to supplement nutrition during dry spells.
- Prepare shallow furrow while hoeing by tying ropes to prongs, which will provide soil support to plants and conserve soil moisture.
- Open alternate furrows in row crops such as soybean or furrows for every 6-8 rows of pigeonpea with Balaram plough in medium to deep soils of Maharashtra. Open conservation furrows at 10-15 m interval in shallow to medium deep red and black soils.

- Adopt surface mulching with crop residue or tree loppings of Glyricidia wherever possible.If farm waste is not available, use blade to form a thin layer of soil mulch to avoid cracks.
- Apply 10 Kg N/ha in sorghum and oilseed crops soon after receipt of rains.
- Provide supplemental irrigation (10 mm depth) with harvested rain water in ponds by adopting micro-irrigation (sprinklers) wherever possible.
- Apply 2 sprays of Planofix 2 ml/ 9 lit of water at 45 and 55 DAS to prevent shedding of squares and small sized bolls in cotton grown in medium deep soils in north interior Karnataka. Practice topping to reduce transpirational losses or restrict excess growth by spraying growth regulator (NAA 4 ml/15 l of water) in cotton cultivated in medium to deep black soils.
- In medium to deep black soils and rabi cropping areas of Karnataka and Maharashtra, close the soil cracks by deep intercultivation in rabi sorghum and safflower.
- In castor and pigeonpea, if the drought affected plants to recoup with the revival of the rains, spray 2 to 3% urea after the foliage is wetted with the rains.
- In deep black soils of Vidarbha, the land may be tilled properly in case kharif crop fails sow rabi crops like safflower, pigeonpea in September.
- In Madhya Maharashtra provide protective irrigation from the harvested water in farm pond or any other source, if available for eg: 5 cm to rabi sorghum during pre-boot stage (65-70 DAS).

C. Late season or terminal drought during mid-season drought

Give life saving or supplemental irrigation, if available, from harvested pond water or other sources.

- Harvest at physiological maturity with some realizable yield or harvest for fodder and prepare for rabi sowing in double cropped areas
- In rabi sorghum and chickpea mixed/ intercropping in Bijapur region, uproot rabi sorghum.
- Ratoon maize or pearl millet or adopt relay crops as chickpea, safflower, rabi sorghum and sunflower with minimum tillage after soybean in medium to deep black soils in Maharshtra or take up contingency crops (horsegram / cowpea) or dual purpose forage crops on receipt of showers under receding soil moisture conditions

Contingency plan for fodder production in drylands

Early season drought: Short to medium duration cultivated fodder crops like sorghum (Pusa Chari Hybrid-106 (HC-106), CSH 14, CSH 23 (SPH-1290), CSV 17 etc) or Bajra (CO 8, TNSC 1,APFB 2,Avika Bajra Chari (AVKB 19) etc., or Maize (African tall, APFM 8 etc.,) which are ready for cutting by 50-60 days and can be sown immediately after the rains under rainfed conditions in arable lands during kharif season.

If normal rain occurs during later part of the year, rabi crops like Berseem (Wardan, UPB 110 etc varieties) or Lucerne (CO 1, LLC 3, RL 88 etc.) can be grown as second crop with the available moisture during winter.

In waste lands, fodder varieties like Bundel Anjan 3, CO1 (Neela Kalu Kattai), Stylosanthes scabra etc. can be sown for fodder production.

Mid season drought

Suitable fodder crops of short to long duration may be sown in kharif under rainfed conditions. Mid season drought affects the growth of the fodder crop. Once rains are received in later part of the season the crop revives and immediate fertilization helps in speedy recovery. If sufficient moisture is available, rabi crops like berseem (Wardan, UPB 110 etc varieties), lucerne (CO 1, LLC 3, RL 88 etc.,) can be grown during winter. In waste lands fodder varieties like Bundel Anjan 3, CO1 (Neela Kalu Kattai), Stylosanthes scabra etc., can be sown for fodder production

Other strategies

Alley cropping is a system in which food/fodder crops are grown in alleys formed by hedge rows of trees or shrubs (*Leucaena leucocephala, Gliricidia, Calliandra, Sesbania* etc.) The main objective of alley cropping is to get green and palatable fodder from hedgerows in the dry season and produce reasonable quantum of grain and stover in the alleys during the rainy/cropping season. This calls for cutting back (lopping) of hedge rows during the dry season. A welcome feature of alley cropping is its ability to produce green fodder even in years of severe drought. At Rajkot in 1985, rainfall received during kharif season was only 30% of the normal. There was total failure of 3 legume crops tried in the system. In sole crop plots, production was limited to 0.5 – 1.7 t/ha of green fodder. However, in alley cropped plots, Leucaena hedge rows produced over 5t/ha of green fodder. Similar was the experience at the Anantapur Centre in 1984 cropping season rainfall was only 144 mm as against the normal of 495 mm. All crops (groundnut, pigeonpea and sorghum) failed, and even stover production was severely affected. However, the Leucaena hedgerows produced 2 t/ha of dry leaf material. Thus, alley cropping systems can allay a part of the risk faced by small farmers in India.

Fodder production systems at homesteads: Azolla, a blue green alga which has more than 25 % crude protein and a doubling time of 5-7 days can be grown in pits in backyards depending on the number of milch animals owned by the farmer. Azolla yield is much more than the perennial fodder varieties like APBN-1/CO-3 etc and is around 1000 MT per ha at the rate of 300 gm/sq.m/day even after taking into account of unused space between two beds. Intensiv e fodder production systems

Growing of two or more annual fodder crops as sole crops in mixed strands of legume (Stylo or cowpea or hedge Lucerne, etc) and cereal fodder crops like sorghum, ragi in rainy season followed by berseem or Lucerne etc., in rabi season in order to increase nutritious forage production round the year. Fodder crops like *Stylo hamata* and *Cenchrus ciliaris* can be sown in the inter spaces between the tree rows in orchards or plantations as hortipastoral and silvopastoral systems for fodder production

Revival of Common Property Resources (CPRs)

Majority of the total feed requirements of ruminants are met by the CPRs. There is no control over the number of animals allowed to be grazed, causing severe damage on the re-growth of number of favourable herbaceous species in grazing lands. Thus causing severe impact not only on herbage availability from CPRs but also quality of herbage affecting the productivity of animals adversely; hence there should be some restriction on number and species of animals to be grazed in any CPR as a social regulation. CPRs need to be reseeded with high producing legume and non-legume fodder varieties at every 2-3 years intervals as a community activity. Further, grazing restriction till the fodder grows to a proper stage and rotational grazing as community decision would improve the carrying capacity of CPRs

Table 13: Some Fodder Cops and their management

Crop	Variety	Seed rateha	Spacing	Fertilizers (NPK)/ha	Sowing	Harvest
Sorghum	CSV-23	Seed				
(12 kg)	45X10	40:20:0	June	September		
Pigeonpca	PRO-154	Seed (10 kg)	90X20	20:50:0	June	January
Guinea gra«	Riversdalc	Rooted slip				
(Slip* 66.000 iws.)	50X50 cm	50:50:40	June	1 st cut at 75 days after establishment and subsequent 60 days		
Desman thus	Vcltmasal	Seed (20kg)	50X50 cm	25:40:20	June	1 st cut at 75 days after establishmen: and subsequent 60 days
Kinkier Cluster be an	Bundcl Guar-3 Seed (40kg)	30X10 cm	25:40:20	June	August	
Fodder Cow pea	APFC-I0-1	Seed (40kg)	30X10 cm	25:40:20	August	October
Fodder horsegram	CRHG-4	Seed (20kg)	30X10 cm			
25:40:20						

#overlapping cropping systems by integrating drought resistant, easy to grow perennial forage legumes or grasses like Hedge Lucerne and Guinea grass that ensure availability of fodder even during the summer, into proven and existing promising cropping systems such as sorghum+pigeon pea (2:1). Integrating multiple annual fodder species such as cluster bean, cowpea and horse gram into the cropping system is also a possible strategy for relieving fodder scarcity.

Utilization of waste land: There are large stretches of degraded wastelands which are not only lying idle and are underutilized but are also accelerating soil erosion, surface run off of rain water and hosting a wide range of pests and diseases. Development of these lands for forage production will not only ensure enhanced production

Trees are often referred to as 'green blood of mankind'. It is most suited to marginal dryland preferably where the fodder shortage is high. Enough care should be taken to select compatible tree species with forage crops. Promising tree and grass species for silvipastoral system

Table 14:. Promising Tree and grass species for silvipastoral system

Grasses	Legumes	Fodder trees
Borthichloa pertusa (Pitted beardgrass)	Cenchrus ciliaris (Buffel-grass)	Acacia arabica (Wattles)
Cenchrus setigeras (Rirdwood grass)	Chrysopogon fulvus (Guria grass)	Cynodon plectostachyus (Stargrass)
Dtchaiilhium unnululutn (Marvel grass)	Hetcropogon conlortus (Speargrass)	Panicuin antidotale (Panicgrjss)
Penniselum pcdicellatum (Fountain grass)	Sehinta nervosum (Rat's tail grass)	Clitoria lematea (Butterfly pea)
Dolichos lablab (Kidney bean)	Stylosanthes hamata (Pencil flower)	Stylosanthes humilis (Pencil flower)
Mucuna pruricns (Velvet bean)	Acacia nilotica (Babul)	Acacia tortilis (Umbrella thorn)
Ailanthus cxcclsa (Tree of heaven)	Alhi/ia amara (Siris)	Alhi/ia Icbbcck (Woman's tongue)
A/adirachta indica (Neem)	Cholophospcrmum mopanc (Mopanc)	Ficus hcngalcnsix (Fig)
Gliricidia sepium (Quick stand)	Hardwickia binata (Indian Blackwood)	Leucaena leucocephala (Subabul)
Pongamia pinnata (Pongam)	Prosopis cineraria (Khcjri)	Making utilization of available residues

Trees or shrubs must be amenable to lopping management besides being multi-purpose (including nitrogen fixing) and fast growing (i) *Leucaena leucocephala*, (ii) *Sesbania sesban*, (iii) *Cassia siamea,* (iv) *Gliricidia maculata*, and (v) *Calliandra* spp

Crop residues can be a best option to share a good percentage of fodder needs in rainfed area provided they are well processed and stored to increase the quality. Sorghum and maize are the major crops grown in rainfed areas that are well suited for fodder needs. Feed availability is also affected by its non-feed uses. For instance, the paddy straw, otherwise a fodder for livestock, is used as packaging and thatching material, and as filler in particle boards

Fodder bank

A community fodder bank is nothing but, a group of farmers coming together to raise multiple fodder crops consisting of trees, grasses and legumes largely in non-arable or wasteland soil in order to meet the fodder requirement especially in lean periods. These fodder banks also help in the preservation and storage of surplus fodder, availability of nutritious fodder during the period of fodder scarcity and enhance nutritive value of crop residue and other cellulosic waste for animal feeding by conventional and nonconventional fodder

There are many unexploited species still available in the nature that could be used as fodder. Several species of the genus Prosopis are such an example. eg: *Prosopis tamarugo* and *Prosopis juliflora* are reported to grow at salinities almost equivalent to sea water salinity, Several species of Cactus or Naghphani are found growing naturally in rainfed areas of the country. These shrubs are mostly thorny and farmers use them as biofence on the field boundaries to protect agricultural crops from wildlife Both species are known to have highest water use efficiency per unit dry matter production, hence need exploitation for rehabilitation of rangelands / wastelands in India to create alternate food and fodder resources and drought proofing option in drought prone areas of the country.

Management of Livestock for climatic vulnerabilities

Dairy Animals

1. Drought

Feed and fodder Resources

- Harvest and use biomass of dried up crops (soybean, wheat, green gram, black gram, sorghum, bajra, maize, chick pea) material as fodder
- Use of unconventional and locally available cheap feed ingredients especially soya meal waste for feeding of livestock during drought

- Harvest all the top fodder available (Subabul, Glyricidia, Pipol, Prosopis etc) and feed the live stock during drought
- Concentrate ingredients such as grains, brans, chunnies & oilseed cakes, low grade grains etc. unfit for human consumption should be procured from Govt. godowns for feeding as supplement for high productive animals during drought
- Promotion of horse gram as contingent crop and harvesting it at vegetative stage as fodder
- All the hay should be enriched with 2% Urea molasses solution or 1% common salt solution and fed to LS.
- Continuous supplementation of minerals to prevent infertility.Encourage mixing available kitchen waste with dry fodder while feeding to the milch animals

Drinking water

- Adequate supply of drinking water.
- Restrict wallowing of animals in water bodies/resources

  Add alum in stagnated water bodies

Health and disease management

- Carryout deworming to all animals entering into relief camps
- Identification and quarantine of ck animals
- Constitution of Rapid Action Veterinary Force
- Performing ring vaccination (8 km radius) in case of any outbreak
- Restricting movement of livestock in case of anyepidemic
- Tick control measures be undertaken to prevent tick borne diseases in animals
- Rescue of sick and injured animals and their treatment
- Organize with community, daily lifting of dung from relief camps

Heat and cold wave

- Allow the animals early in the morning or late in the evening for grazing during heat waves

- Feed green fodder/silage / concentrates during day time and roughages / hay during night time in case of heat waves
- Put on the foggers / sprinkerlers during hcat weaves
- In severe cases, vitamin 'C' and electrolytes should be added in H2O during heat waves.
- Apply / sprinkle lime powder in the animal shed during cold waves to neutralize ammonia accumulation

2. Poultry

Drought

1. Feed and fodder sources
2. Supplementation only for productive birds with house hold grain
3. Supplementation of shell grit (calcium) for laying birds
4. Culling of weak birds

Drinking water

Mixing of Vit. A,D,E, K and B-complexincluding vit C in drinking water (5ml in one litre water).

Heat Wave

Shelter Management

- In severe cases, foggers/water sprinklers/wetting of hanged gunny bags should be arranged
- Don't allow for scavenging during mid day

Health and disease management

- Supplementation of house hold grain
- Provide cool and clean drinking water with electrolytes and vit. C in hot summer, add anti-stress probiotics in drinking water or feed

5

Challenges in Crop Production on Drylands

Dryland agriculture limits the crop growth to a part of the year due to lack of sufficient moisture (Peterson *et al.*, 2006). According to the Fourth five-year plan of India, dry lands are defined as areas which receive rainfall ranging from 375 mm to 1125 mm and with very limited irrigation facilities. Dry lands are economically fragile regions which are highly vulnerable to environmental stress and shocks. Degraded soils with low water holding capacities along with multiple nutrient deficiencies and depleting ground water table contribute to low crop yields and further leading to land degradation.

In a country like India, 44 per cent of the total food production is supported by dryland agriculture, thereby playing a critical role in nation's food security. With the increasing population, food production has to be increased. A real need to second green revolution has been envisioned, which can be achieved by improving the dryland agriculture. Geographically area under dryland agriculture in India includes the north western desert regions of Rajasthan, the plateau region of central India, the alluvial plains of Ganga Yamuna river basin, the central highlands of Gujarat, Maharashtra and Madhya Pradesh, the rain shadow regions of Deccan in Maharashtra, the Deccan Plateau of Andhra Pradesh and the Tamil Nadu highlands (Singh *et al.*, 2004)

1. Dryland Vs Rainfed situation

Indian agriculture is predominantly a rainfed agriculture under which both dry farming and dry land agriculture is included. Earlier, in the decades prior to 1970s, dry faring zones was considered as those that sustain agriculture with rainfall of less than 500 mm annually. The modern revised concept, however, indicates that dry land areas are those where the balance of moisture is always on the deficit side resulting in annual evapotranspiration exceeding precipitation. In dry land agriculture, there is no consideration of amount of rainfall. It may appear quite strange to a layman that even those areas which receive 1100 mm or more rainfall annually fall in the category of dry land agriculture under this concept. To be more specific, the average annual rainfall of Varanasi is around 1100 mm and the annual potential evapotranspiration is 1500 mm. thus the average moisture deficit so created comes to 400 mm. this deficit in moisture is bound to affect the crop production under dry land situation ultimately resulting into total or partial failure of the crops. Accordingly, the production is either low or extremely uncertain and unstable which are the real problems of drylands in India. The success of crop production in these areas depends on the amount and distribution of rainfall, as these influences the stored soil moisture and moisture used by crops.

The amount of water used by the crop and stored in the soil is governed by the water balance equation: ET = P-(R+S). When the balance of the equation shifts towards right, precipitation (P) is higher than ET, so that there may be water logging or it may even lead to run off (R) and flooding. On the other hand, if the balance shifts to the left, ET becomes higher than the precipitation, resulting in drought of various degrees of severity. As per the meteorological report, severe drought is experienced in a vast area of the country once in 50 years and partial drought in five years, while floods are expected every year in one part of the country or the other, especially during rainy season. In fact, the balance of the equation is controlled by the weather, season, crops and cropping pattern.

Types of Dryland Agriculture

Depending on the amount of rainfall received, dryland agriculture has been grouped into three categories:

(i) **Dry farming:** it is production of crops and animals without irrigation in areas where annual rainfall is less than 75 cm. Crop failures are more frequent under dry farming condition owing to prolonged dry spells during crop period. The growing season is less than 200 days. It is generally practiced in arid regions of the country.

(ii) **Dryland farming:** cultivation of crops and management of animal husbandry in areas receiving rainfall above 75 cm is known as dryland farming. Dry spell during crop duration occurs, but crop failures are less frequent. Semi-arid regions are included under this category.

(iii) **Rainfed farming:** It is practice of crop cultivation and animal husbandry without irrigation in areas receiving 115 cm rainfall, mostly in sub-humid and humid areas. Here chances of crop failure and water stress are very less.

Table.1: Rainfall patterns in relation to different systems of rain-grown farming

Sr.No.	Particular	Dry Farming	Dryland Farming	Kainfcd farming
1	Rainfall (mm)	<750	750- 1150	>1150
2	Moisture availability	Acute shortage	Shortage	Enough
3	Crop growing season	<75 days	75-120 days	> 120 days
4	Growing region	Arid	Semi-arid	Humid
5	Cropping systems	Single crop	Single crop/ intercropping	Inter/ Multicropping
6	Dry spells	Most common	Less frequent	No occurrence
7	Crop failure	More frequent	Less frequent Rare	rare
8	Constraints	Wind erosion	Wind erosion/ water erosion	W'ater erosion
9	Measures	Required Moisture conservation practices	Moisture conservation Practices & drainage for vertisols	Proper drainage required

(*Source*: Mevada *et al*., 2021)

2. SWOT Analysis in Dryland Agriculture

SWOT analysis is a strategic planning method used to evaluate the Strengths, Weaknesses, Opportunities, and Threats involved in any venture.

Strengths: These are the characteristics of an enterprise that give it an advantage over the others

Weaknesses: These are the characteristics that place it at a disadvantage relative to others.

Opportunities: These are the external characteristics that can be exploited to get maximum advantage for the enterprise.

Threats: These are also due to the external elements in the environment that could cause trouble for the enterprise

Strengths

- Wide area of dryland is still available,
- It has wide bio diversity,
- The willingness of the farmers is very high in improving the productivity and profitability of dryland,
- It receives average of the rainfall from low to high and
- It had topography from flat to steep.

Weakness

- Low productivity due to poor nutrient availability.
- Low investment capacity of farmers
- Low farmers' income
- High risk due to climate variability (climate change)
- Weak management capabilities of farmers, technical and appropriate technologies at farmers' level
- Limited facilities and infrastructure support for dryland development
- Weak access to capital, market networks and partnership systems to improves the farmers bargaining position and
- Low value addition and Food processing*high post-harvest losses

Opportunities

- High Diversification of commodities
- Limited Research Institutions and universities that concentrate in development of dryland technologies and resources
- Increased needs of public demand for dryland Agricultural commodities
- Availability of Agribusiness technologies and
- Wider market share of agricultural markets

Threats

- Fragmentation of land holdings (Size of holdings)
- High climatic risks
- Land degradation
- Reduced interest of younger generation (Declining Interest in Agriculture):
- Fluctuating market prices and
- Lack of profitability

3. Constraints in crop production on drylands

Rainfed drylands encounter various constraints which restrict the productivity of these ecosystems. Drylands ecosystems are highly fragile, which can be attributed to periodic recurrence of drought and enhanced over-exploitation of minimum resources. They are highly prone to climatic anomalies and eventually encounter problems viz., soil erosion, low nutrient status, loss of fertile surface soil and low levels of organic carbon consequently resulting in reduced productivity levels. (Vijayan and Roshni 2016)

3.1. Climatic constraints

Climate is one of the most important determinants of agricultural production. Rainfed agriculture is characterized by its harsh and variable climatic condition, which is a major determinant of crop yields. Low and erratic rainfall, high temperature, high solar incidence, high wind velocity, low humidity, weather aberrations, drought, flood, frost, gale, cyclones and burning winds with high potential evapo-transpiration rate make the unfavorable condition or the crop cultivation. Severe moisture stress can cause partial or complete crop failure Rainfall aberrations during south-west monsoon continue to be a major factor contributing to instability in kharif crops production. The rainfall distribution is becoming more skewed with less number of rainy days, with high intensity causing more soil erosion.. Distribution of rainfall is more important than total rainfall to have longer growing period in rainfed agriculture. Delay in the onset of monsoons severely affects the crop due to delay in the sowing . Further, loss of soil moisture and its duration of availability in the soil, crop growth and duration of the crop is directly governed by the temperature. It highly influences germination, growth and development, physical and chemical processes within the plants. Solar radiation governs all physical processes taking place in the soil, plant and environment and controls distribution of temperature. Extreme events of rainfall, flood, heat and cold wave cause significant impact on the performance of crop growth and yields. Therefore, climate is the most important factor that significantly influences the growth and yield of crops in general and the rainfed condition in particular.

3.2. Edaphic constraints

Soils of the rainfed / dryland areas are pre-dominantly poor and marginal lands with poor organic matter content. Singh *et al.*2002 reported that 2.13 Pg carbon (C) stock in the 0-100 cm depth of which 1.23 Pg as soil organic C and 0.90 Pg as soil inorganic C in arid and semi-arid regions of Rajasthan. Major soils of the rainfed areas are mapped in Alfisols (30%), Vertisols (35%), and Entisols (10%). All these soils have a wide diversity in texture, structure and organic matter status, and soil depth. Variations found in the different soils induce differences in infiltration rate, water holding capacity, bulk density, aggregation, aeration, permeability and moisture, and nutrient release capacity. Moreover, rainfed areas have uneven topography and poor quality of groundwater, which contains high soil pH, EC, and SAR. The abundance of sand makes the soil prone to erosion, as a result of low organic matter content. Many soil factors like soil moisture, soil temperature, soil air, nutrients, and soil reactions are significantly governed by the soil type and affect the plant growth and yield

4.3. Technological constraints

Adoption of modern agricultural technologies by farmers is an important factor for improving the agricultural productivity in any kind of production systems. Limitations in adoption of improved production technologies and low adoption of technologies in the rainfed regions cause low crop production. Of course, inadequate awareness about the technologies, economic viability and lack of easy access to technologies and credit also play vital role in the low adoption of technologies. Despite good progress made so far, the adoption and diffusion of key rainfed technologies are still low at the national level, resulting in large yield gaps crop yileds between research stations and farmer's fields. In a study Singh et.al. (1999) found that none of the farmers used recommended modern production technologies for barley and chickpea crops in arid region of Rajasthan, indicating 100% adoption gap.

4.4 Socio-economic constraints

About 80% of farm holdings are managed by marginal and small categories of farmers. Majority of these farmers are resource-poor coupled with poor socio-economic conditions. Low risk-bearing capacity, low use of improved inputs of seed, fertilizer and pesticides , poor mechanization, low credit availability, low level of knowledge, lack of inspiration and difficulty in the use the improved production technologies, dominate in the rainfed situation. Low input -low risk -low yield concept is prevailing in the rainfed condition.

5. Resources for dryland agriculture

5.1 Climate

The, rainfed farming systems are practiced in regions of strong climate contrasts. For example, southern Tamil Nadu experiences typical tropical temperatures with north-east monsoon being the main source of rainfall, whereas Punjab and Haryana in north-western India experience continental climates with extremes of temperatures varying from 45-50 degrees C and near freezing temperatures in winter. The rainfed farming in north-western India is practiced under south-west monsoon rains. Thus, the climate of rainfed-dryland farming ranges from arid, semiarid to sub-humid, with mean annual rainfall varying between 412 and 1378 mm. Length of the growing season ranges from 60-90 days in the arid regions to 180-210 days in the sub-humid regions. In terms of the distribution of rainfall, 15million hectare (Mha) of the area receives an annual rainfall of <500 mm, 15 Mha of 500-750 mm, 42 Mha of 750-1150 mm, and 25 Mha of >1150 mm. The country is divided into 20 agroecological regions on the basis of topography, climate, soils, and effective growing seasons This classification characterizes the suitability and potential of each sub-region for a given land use and cropping and farming system. It could be seen that there is a decrease of arid region in Gujarat and Haryana. There is also an increase in semi-arid region in Madhya Pradesh, Tamil Nadu, and Uttar Pradesh with a shift of the climate from dry sub-humid to semi-arid. The moist sub-humid regions in Chhattisgarh, Odisha, Jharkhand, MadhyaPradesh, and Maharashtra are now mostly dry sub-humid in nature.

5.2 Soils

Soils of India are grouped under eight orders: Entisols, Inceptisols, Vertisols, Aridisols, Mollisols. Ultisols , Alfisols Oxisols and mixed types Alfisols and Vertisols are dominant soils of the peninsular India and Aridisols occur in extremely dry climates along with some Entisols and Inceptisols. In terms of land use and management, alluvial (Inceptisols) soils are the most dominant followed by red (Alfisols), black (Vertisols), desert (Entisols, Aridisols), and lateritic (Plinthic horizon, soils). The dominant soil orders in rainfed production systems of India are Inceptisols followed by Entisols, Alfisols, Vertisols, Mixed soils, Aridisols, Mollisols, Ultisols, and Oxisols .

With the exception of Vertisols, most soils in the rainfed areas are generally coarse-textured. Thus, their capacity to retain water and nutrients is low, and crops grown on these soils are prone to drought stress and nutrient deficiencies. With low soil organic matter (SOM) concentration, aggregate stability is low and erosion is a serious problem. Water infiltration rates are low in Vertisols due to clayey texture and in Alfisols due to formation of surface crusts. With the exception of some Vertisols, which are rich in bases, inherent fertility of rainfed soils is generally low.

Table 2: Major soil groups and their moisture storage capacities in rain-dependent areas of India

Broad soil group	Subgroup (based on soil depth!	Moisture storage capacity (mm)
Vertisols and related soils	Shallow to medium (up to 45 cm)	135-145/45 cm
	Medium to deep (45-90 cm) Deep (>90 cm)	145 - 270/90 cm 300/m
Alfisols and related soils	Shallow to medium (up to 45 cm) Deep (>90 cm)	40 - 70/45 cm (sandy loam) 70-100/45 cm (loam) 180-200/90 cm
Aridisols	Medium to deep (up to 90 cm)	80 - 90/90 cm
Inceptisols	Deep	90 - 100/m (loamy sand) 110- I40/m (sandy loam) 140 - I80/m (sandy loam)
F.ntisols	Deep	110- 140/m (sandy loam) 140 -180/m (loam)

Source: (Srinivasarao *et al.*, 2017)

Biodiversity

The rainfed areas are more diverse and heterogeneous. These areas in India harbor a great deal of biodiversity, influenced by both climate and latitude. Variation in topography, geology, soil type, seasonal patterns of rainfall, fires, herbivore pressure, and human management, along with water scarcity, account for a wide diversity of species in the drylands. In the drylands of India, for example, farmers still maintain many of their traditions of nurturing biodiversity of wild and cultivated food crops and medicinal plants. In addition, the traditional reverence of Indian farmers for multipurpose trees (e.g., *Prosopis cineraria* in croplands of Western Rajasthan) enhances agrobiodiversity. Similarly, farmers maintain different varieties of maize, sorghum, pulses, and minor millets. Ancient civilizations in drylands, like everywhere else, cared for medicinal plants because of the experiences of generations of tried and tested curative methods and products. In India, for example, Shankar *et al.* (2008) reported that 4671 plant species are used in folk medicine. Although not unique to drylands, it is a remarkable fact that the use of medicinal plants is a living tradition of rural people in drylands. However, the loss of biodiversity in Indian dry lands is alarmingly high for diverse reasons including degradation and loss of habitats, reliance on high yielding varieties (HYVs) and hybrids, shift from intercropping/ mixed cropping to monocropping, cutting of trees and shrubs, urbanization, mining, and industrialization. For example, the Indira Gandhi Canal in Rajasthan has adversely impacted dryland biodiversity, with wet conditions affecting the habitat of local species. The change in land use, population pressures, and irrigated farms favor species that demand more water, and this transition has led to the loss of several indigenous shrubs and grasses species

6. Rainfed Production Systems

Under the National Agricultural Technology Project (NATP), the concept of production systems was introduced by ICAR. For prioritization of rainfed districts of India, data on crop, livestock and socio-economic parameters for all districts with less than 30 percent irrigated area (considered as rainfed) were analyzed. Districts with highest area under a given crop/cropping system but with stagnant, declining or low productivity were considered as high priority districts. The underlying implication of this approach was that any improvement in the productivity of crops and livestock in such districts will have a greater impact at the state and national level due to the involvement of large area/ number of farmers

In this approach, the rainfed agro-ecosystem was sub-divided into 5 homogenous production systems, viz.(i). Rainfed rice-based system (ii). Nutritious (coarse) cereals-based system (iii). Oilseeds based system (iv). Pulses based system and (v). Cotton based system

The rainfed rice production system is mostly prevalent in eastern and north eastern parts of India. Coarse cereals are staple food of poor people and principal source of fodder for livestock is mainly confined to western and central parts of the country and the semi-arid hot high lands of Deccan plateau. Oilseed-based production system which is mostly rainfed wherein crops are grown both during kharif and rabi seasons under sole, inter and sequence cropping systems. The groundnut is mostly cultivated in western plains, central high lands, semi-arid Deccan plateau and Eastern Ghats while soybean is mainly confined to Madhya Pradesh and Uttar Pradesh in the central high lands, Malwa-Gujarat plains and Kathiawar peninsula and is now spreading to areas like Vidarbha region in Maharashtra. In the case of pulses-based production system, ninety per cent of pulses are grown under rainfed conditions as intercrops or in sequence cropping systems all over the country. Pigeon pea and chickpea are the two most important pulse crops and grown during kharif and rabi seasons, respectively. Cotton-based production system has sixty per cent of the cropped area under rainfed condition mostly in the Deccan plateau and hot semi-arid peninsular parts of India. All the above efforts made earlier for characterization of rainfed areas by-passed many key parameters essential for drawing meaningful conclusions and need- based interventions. The agriculture and allied developmental efforts need to match with the available natural resources and livelihood. The integration of natural resources and livelihood parameters is required to characterize the rainfed areas of the country.

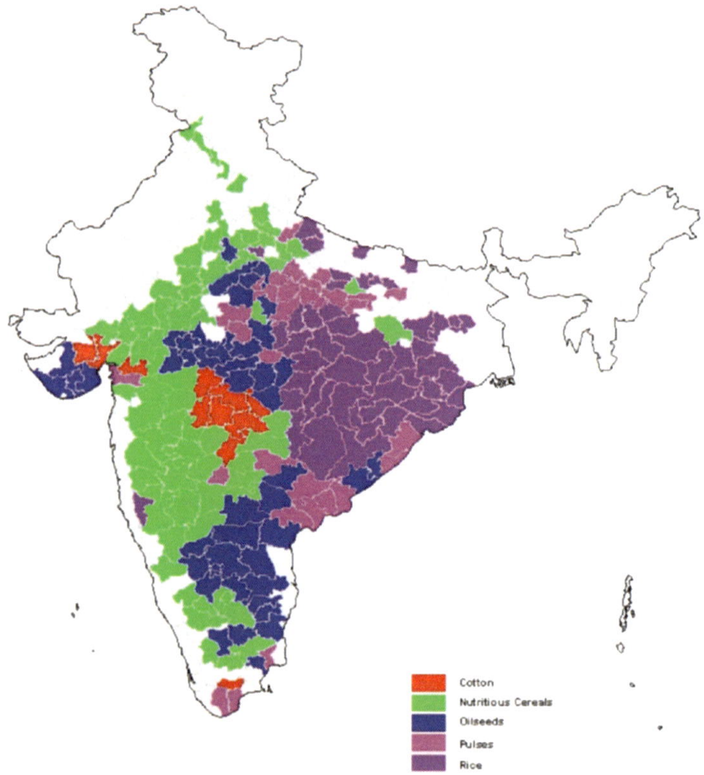

Fig. 1: Rainfed production systems in India

7. Crop Production Technologies

Numerous technologies have been generated over the years through the network of research Institutions in the country dedicated to address the problems of dryland agriculture. These technologies have been refined on farmers' fields through operational Research project sites(ORPS), Institute Village Linkage programmes and farm science centers.

7.1. Rainwater Management

Inter terrace land treatments

Adoption of inter terrace land treatments is highly variable depending on the intensity of rainfall, slope, soil type, and prevailing crop and cropping systems (Singh etal.,2000). Compartmental bunding in shallow Vertosols resulted in a 25%higher grain yield in sorghum. Whereas graded border strips in deep Alifisols enhanced the yield of finger millet by 42%. In better rainfall areas, significant benefits due to inter terrace land treatments were recorded for crops such as upland rice, maize, and soybean.(Table2).In view of capital cost of earthen bunds and problem related with their maintenance ,research efforts were focused on vegetative barriers and live bunds in1980. Although, Vetiver grass ([*Vetiveria zizanoides(L)*Nash] received the most attention, number of plant species such as buffel or *anjan* grass (*Cenchrus ciliaris L.),Cymbopogan flehiosus,Pennisetum hollennackeri,*and subabul(*Leucaena leucocephala* Lam.*)* were found effective in conserving soil moisture and controlling erosion. These crops also provide fodder and green manure. Conservation furrows at 3.6m horizontal interval supplemented with glyricidia *[Gliricidia maculate (Jack.)Walp]* mulch at 5t/ha increased the sorghum grain yield to4003kg/ha from 1821 kg/ha under control, reduced the runoff by 73%,decreased the soil loss by about 50%. Apart from this , glyricidia application can supply considerable amount of N as on dry weight basis since its leaves contain 3 to4%N. About 80% of the dryland farmers follow simple soil and water conservation measures like sowing across the slope, formation of conservation furrows, and key line cultivation (ICAR 2001)

Table 3: Recommended Inter terrace land treatments for soil and water conservation in drylands

Annual rainfall(mm)	Soil type	Land treatment	Yield advantage
>500	Aridisols	Inter-plot rainwater harvesting of 1:1 cropped area to uncropped land	50%pearl millet
	Alfisols (Shallow)	Conservation furrows at 3.6m interval	10% groundnut
500-1000	Alfisols (Shallow	Sowing across the slope	10% sorghum
	Vertisols (Shallow)	Compartmental bunding	25% sorghum
	Vertisols	Contour bunding	35% sorghum
	Alfisols	Graded border strips	42% finger millet
>1000	Entisols	Inter-plot rainwater harvesting	17%Maize
	Vertisols	Raised bed and sunken system	21% rice
			34% soybean

Source: Singh, H.P. *et al.*,2004

Water harvesting

The advantage of harvesting rainwater in dugout ponds and using it for critical irrigation of rainy season crops and pre-emergence irrigation in post rainy season crops has been reviewed by Singh (1986) and Singh and Khan

(1999). On an average,24x10(6) ha m out of 400x10(6) ha.m of precipitation received annually in dryland areas of India, is estimated to be available as harvestable rainwater through on farm facilities. Drylands receiving 500 to 1000 mm annual rainfall has estimated to have harvestable rainwater potential of 5.54x10(6) ham (Katyal,1997).

In medium to high rainfall area, there will be excess run-off due to intense storms in spite of following *in-situ* moisture conservation practices. The excess water needs to be harvested in surface ponds for recycling through supplemental irrigation or to recharge for ground water for later use in post rainy season. Once the surplus water is harvested in surface ponds or ground water recharge, its efficient use is important for increasing the crop productivity in a sustainable manner. Efficient use of water involves both timing of irrigation to the crop and efficient water application methods. Application of irrigation water can be done either by surface irrigation border, basin and furrow or by pressurized irrigation systems sprinkler and drip application of water is carried out firstly by the conveyance of water from the source to the field and secondly by application of water in the field. In Verisols, there is no need to line the open field channels, while on Alfisols close conduits plastic, rubber are becoming popular with the farmers growing high value crops like vegetables and horticultural crops. BBF system in Vertisols and ridge and furrow system in Alfisols are found to be a suitable methods, while sprinkler and drip systems can play the important role to improve the water productivity, especially in high value crops like fruits and vegetables. However, currently the use of these improved irrigation methods is limited primarily due to the high initial cost. Favorable government policies and availability of credit is essential for popularizing this irrigation system methods. (Table 4)

Table 4: Effect of supplemental irrigation on dryland crop yields (Katyal, 1997)

Crop	No. of replications	Yield kg/ha)		Water use efficiency
		With irrigation	Without irrigation	Kg/ha/mm
Wheat	18	3180	1760	28.2
Barley	4	2270	1580	13.8
Sorghum	16	1870	1110	15.1
Pearl millet	4	1900	1660	5.8
Finger millet	4	2320	1620	14.0

7.2. Choice of crops, varieties, and cropping systems

Many traditional crops grown in drylands are of long duration and do not match periods of water availability. In order to match crop water needs with rainfall and water availability periods, extensive trails were conducted over the past five decades. A large number of varieties of millets, pulses and oil seed were evaluated for their yields compared with local varieties used by the farmers (AICRPDA,2000to2016). Yield improvements of the order of 15 to 50% were recorded when traditional varieties were replaced by high yielding varieties. (Table4). Also a benefit of 15 to 25% in yield was demonstrated by the crop substitution strategy, which means by replacing the traditional crop with more appropriate crop that fits into the rainfall and moisture availability pattern. In rainy season, the quantum and distribution of rainfall dictates the effective growing season and cropping system for a given region. In regions, receiving 350 to 600mm rainfall with 20 weeks effective growing season, only single cropping is possible in all soil types except in the case of deep vertisols.

In deep vertisols, a single post rainy crop could be grown in areas receiving 350 to 600 mm rainfall and having 20 weeks effective growing season. Inter-cropping (giving150% cropping intensity) is possible in regions having 20 to 30 weeks effective growing season and 650-750 mm rainfall. In areas receiving more than 750 mm of rainfall and having an effective growing season more than 30 weeks, double cropping (200% cropping intensity) is assured. (Singh and Subba Reddy,1986).

Table 5: Relative potential of traditional Vs efficient crops (AICRPDA, 2000-2016)

Location	Traditional crop	Yield (kg/ha)	Efficient crop	Yield (kg/ha)
Bellary	Cotton	200	Sorghum	2670
Indore	Green gram	1180	Soybean(yellow)	3360
Agra	Wheat	1030	Rape seed/Mustard	2040
Hissar	Wheat	320	Sativa	1610
Udaipur	Maize	1800	Sorghum	2900
Rewa	Soyabean(Black)	400	Soybean(Yellow)	1200

Intercropping systems

Intercropping is traditionally practiced in dryland areas of India since component crops gives higher yields, enhance land use efficiency per unit time, improves the soil fertility and use resources more efficiently.

The productivity of intercropping depends mainly on complementarity of component crops. At least one of the component crops succeeds normally in producing economic yields even in during severe droughts. In high rainfall areas, there are greater chances of success of both component crops and the returns are higher than for sole crop. The ratio of rows of principal crop to component crop (row ratios) in intercropping systems were optimized to minimize the competition and realizing optimum biological productivity. Most inter cropping systems are additive series, where the pant population of a main crop is maintained nearly equal to the sole crop.

Double cropping

In areas receiving more than 800mm of annual rainfall and soil moisture storage of 200 mm per meter of soil depth, double cropping is feasible. Such options are recommended to the zones in Orissa, Bihar, Madhya Pradesh and eastern Uttara Pradesh. With some adjustment of sowing dates such as early planting and harvesting of rainy season crops, double cropping is possible in Vidarbha and Malwa plateau as well.

7.3 Contingency Crop Planning for Weather Aberrations

In dryland agriculture, drought is common phenomenon due to either a late onset or early withdrawal of monsoon or prevalence of dry spells within cropping season. Several technologies have been developed and tested that enhance the crop production during the weather aberrations. Generally short duration crops and varieties have replaced long duration varieties. (Table 5.) When rainfall is inadequate after planting, mid-season corrections such as inter-culture, thinning for reduction of plant population density; plowing in green organic matter for recycling; and additional application of small doses of nitrogen (N)after alleviation of drought have been shown to impart production stability. For late on set of monsoons followed by normal rainfall, selection of short duration varieties of pulse and oil seed crops and fodder crops are important strategies to overcome the drought situation. Under the situations of late onset and low rainfall, leguminous crops and oil seeds showed promising results. To avoid risks from weather aberrations, intercropping is advisable to ensure the success of at least one crop, even under severe drought thereby eliminating the risk of complete crop. (Table5).

Table:5. Contingency crop plans for different rainfed regions in India

S.N.	Agro-Eco Sub-Region (AESR)/Major Production System (MPS)/Area Domain (AD)	Crop plan as per the on-set of monsoon		
		15-31 July	1-15 August	16-31 August
1.	AESR: 9.1 MRS: Maize-Rice AD: Submontaneous districts of Punjab, J&K, HP and Western UP	Short duration maize, moong, mash as grain crops. Bajra, guar, sorghum and maize as fodder crops. Vegetable cowpea and clusterbean	Moong and mash bajra, guar and maize as fodders. Sunhemp or sesbonia as green manure. Vegetable type clusterbean and cowpea	Bajra as fodder crop
2.	AESR: 4.1 NIPS: Pearl millet/Rapeseed and Mustard AD. Agra, Mathura, Aligarh, Bulandshahar, Meerut, Etah, Manipun and western part of Muzaffamagar	Pearl millet, clusterbean, green gram, short duration pigeonpea, vegetable type cowpea and clusterbean	Transplanted pearl millet, clusterbean, green gram and cowpea. Cowpea and clusterbean (vegetable type)	Clusterbean, cowpea
3.	AESR: 4.4 MPS: Fodder sorghum/pulses AD: Jhansi, Banda, Hamirpur, lalitpur, Morena, Gwalior	Bajra, guar, cowpea, lablab bean, pigeonpea and black gram as grain crops	Bajra, guar and cowpea as grain and fodder. Pigeonpea and black gram as grain crops	Bajra, guar and cowpea as fodder
4.	AESR: 2.3 MPS:Peari millet - Rapeseed/ Mustard AD:Hisar, Bhiwani, Sirsa, Mahendergarh, Gurgaon & part of Rohtak district	Short duration bajra (HHB-67), moong, urdbean (T-9), Cowpea (Charodi), guar (HG-365) and also vegetable type clusterbean and guar	Transplanting of HHB-67 variety of bajra as grain crop or direct sowing as fodder crop	Moisture may be conserved for tori a sowing

S.N.	Agro-Eco Sub-Region (AESR)/Major Production System (MPS)/Area Domain (AD)	Crop plan as per the on-set of monsoon		
		15-31 July	1-15 August	16-31 August
5.	AESR: 4.2/21 MPS Maue/ Peart millet AD.Bhilwara, Tonk, Dungarpur, Ajmer, Chittaurgarh, Rajasamand, Jalore, Sikar. Jodhpur, Churu	Sesame (RT-46), green gram (K -851, RMG-62), sorghum and cowpea as fodder, snapmelon and moteera as vegetable crops	Sesame (RT-125), green gram (RMG-62), sorghum as fodder crop	Sorghum as fodder. tona (TL-15), taramira T-27
6.	AESR: 14.2/14 3 MPS Maize AD Jammu, Punch, Riasi, Muzafarbad, Udhampur,	Bajra, cowpea, moong (direct sown), bajra (transplanting)	Bajra + cowpea/guar (fodder) jowar + cowpea/guar (fodder), maize cowpea/guar(fodder)	Fodder as shown in 15 August and/or field preparation for September sowing to torla, qobhl sarson
7.	AESR. 9.2 MRS Rice/Pearf millet ftp Varanasi, Wliraapur, launpui, Ghazipur, Sitapm, parts of Shahiahanpur. Lucknow, Bara bank; Rai Baraeli, Sullanpui	Short duration upland rice varieties (NDR-97, NDR 118, Barani deep, Cauverv, Akashi, Mutmuri). In light texture soils green gram (T-44, Pant moong 1), black gram (T-9, Pant Urd 19. 35, Narendra Urd-1), pigeonpea (Bahar and Narendra Arhar 1), sesame (T-4, T-12, T-13), Vegetable type cowpea, labiab bean and guar	Hybrid bajra (NHB-3 NUB 4, Bj 104), green gram, black gram, pigeonpea, sesame, niger (GA-10, Ootacamund), short duration upland rice varieties. Vegetable type cowpea, guar and labiab bean	Green gram, Bahar variety of pigeonpea sowing at 30 cm row spacing using 25 kg/ha seed. Niger varieties GA 10 and Ootacamund
8.	AESR 6.1 MRS Rabi sorghum AD Solapur, Bidar, Osmanabad, AhmeOnagar, parts of Satara, Latur and Sangli	Sunflower, pigeonpea, horsegram, setaria, castor, pearl millet sunflower + pigeonpea (2:1) Pearl millet + horsegram (2:1) Pigeonpea + clusterbean (1:2) Castor + clusterbean (1:2)	Sunflower, pigeonpea, castor. Sunflower + pigeonpea (2:1)	Sunflower, pigeonpea, castor sunflower * pigeonpea (2:1) Sorghum for fodder

S.N.	Agro-Eco Sub-Region (AESR)/Major Production System (MPS)/Area Domain (AD)	Crop plan as per the on-set of monsoon		
		15-31 July	1-15 August	16-31 August
9.	AESR: 6.3 WHS. Cotton/sorghum AD Akola, Wardha, parts of Aimavatl, Vaotmal, Parbhani, Duldana and Khandesh and w,ti of Adilabad of AP	Pigeonpea, pearl millet, maize, sunflower	Pigeonpea. pearl millet, rnaue. sunflower, castor	Pigeonpea, castor, reserve the land for rabi safflower
10.	AESR: 8.2 MPS: Finger millet AD: Bangalore, Kolar and Tumkur	Sowing ot long duration varieties (Indaf 8, L-5,MRI) transplanting of nursery of above varieties, finger millet + red gram (8:1) and finger millet c field bean little millet and foxtail, groundnut, Red gram + groundnut Sunflower hybrids, castor, soybean, chillies	Sowing of medium duration varieties (GPU 28, HR Oil, PR 202) or transplanting sowing of short duration varieties (GPU-26) as nursery, sunflower hybrids (K8SH-1, KBSH-42 or varieties Morden) and Niger, cowpea (KBC 1, KBC-2) and soybean (KBSH-2), Transplanting of chillies, maize, sorghum, bajra as fodder crop	Transplanting of short duration varlties (GPU 28, HR 011 and PR-202) Cowpea (KBC-1, KBC 2, Lolita), horsegram (K8H-1/ PHG-9) Transplanting of chillies 4 protective irrigation available Maize, sorghum, bajra, sorghum, bajra, as fodder crops
11	Type krna h			
12	AESR: 3.0 MPS: Groundnut AD. AnantaPur- Kurnool, Chittoor districts of AP	Groundnut (Vemana, TMV-2) + redgram (Palnadu)	Groundnut (TMV-2, ICGV-91114)	Pearl millet (ICTP-8203, ICMV-221); green gram (MGG-295, MGG-40, PDM-54); dual purpose sorghum (M-35-1-1, NTJ-1, 2, 3, 4); horsegram (AK-21, Marukulthi and local)

S.N.	Agro-Eco Sub-Region (AESR)/Major Production System (MPS)/Area Domain (AD)	Crop plan as per the on-set of monsoon		
		15-31 July	1-15 August	16-31 August
13	AESR: 4.2 MRS: Pearl millet AD: Khera, Gandhinagar, Mehsana, Sabarkanta, parts of Ahmedabad Panchmahal, Banaskantha & Vadodara dist.	Cluster bean, castor, fodder sorghum	Thinning of already planted crops. sorghum, Castor and fodder sorghum	Castor, fodder sorghum, fodder sorghum+ Karingado
14	AESR: 2.4 MPS: Pearl millet/groundnut AD: Rajkot, Sundergarh, Jamnagar, parts of Junagarh, Bhavnagar and Amreli	Erect groundnut (GG-2, 5, 7); sesame (G-l, G-2); hybrid bajra (GHB-235, 316, 558); green gram (KB-51, GM-4); black gram (T-9); pigeonpea (ICPL-87, GT-101)	Black gram (T-9); forage maize/ sorghum (GFS-5), Castor (Gauch-1); sesame (Purua-1)	Forage maize/ sorghum (Gundri, GFS-5), sesame (Purua-1)
15	AESR: 18.4 MPS: Rice Uplands and medium “"ds of Balasore, Cuttack, un and Ganjam	Black gram (Pant-30); green gram (PDM-54/K-851); sesame (Uma or local), early pigeonpea (UPAS-120/ICPI -87); short duration radish, okra, cowpea (SEB-1,2) & clusterbean as vegetables	Niger, black gram radish, beans and cowpea as vegetables, early pigeonpea (ICPL-87/ UPAS-120)	Horsegram, sesame, niger, cowpea

Souce: ICAR, 20061

Cropping strategy for aberrant weather situations and stability of production:

Drought is a recurrent phenomenon resulting from deficit in soil. The effect is pronounced due to delays in onset of monsoon, prolonged breaks in monsoon and deficit rainfall, poor inflows into water bodies or poor recharge leading to reduction in cropped area, cropping intensity, dip in crop and livestock productivity and sometimes total crop failures adversely impacting rural livelihoods and economy.

1. Delay in the onset-of monsoon modifies sowing operations in relation to seed rate and changes in spacing, and intercropping and inter terrace land management practices like ridges and furrows, dust mulching, broad bed furrow method of planting etc.

Early season drought after sowing may cause poor germination and poor crop establishment, in which case re-sowing, gap filling, thinning may be necessary depending upon the crop and soil type. In addition, moisture conservation measures, nutrient management and life-saving irrigation wherever possible are suggested. In order to void above problems, it is advisable to carry out sowing operation only after saturation of soil profile with moisture, which results in robust seedlings without any loss of plant population (Prasad and SudhakarBau (1997). In the above case early season drought is observed to be beneficial in strengthening the root system and promoting better root growth (Prasad and Sudhakar Babau1997)

In the case of mid-season drought, occurring at the peak vegetative phase of the crop, the interventions with regard to crop and soil nutrient management and moisture conservation measures are to be undertaken with no loss of time in view of the stage of crop growth and kind of drought faced. These include inter-cultivation for weed management, mulching and other *in situ* moisture conservation measures, protective irrigation, split or postponement of fertilizer application, foliar sprays of nutrients and (or) anti-transpirants etc, as the need be.

In the case of terminal drought occurring in September to October coinciding with the gran development and maturity stages of the crop in question, due to dry spells or early withdrawal of the south west monsoon, crop management measures such as life-saving irrigation are to be undertaken as per the feasibility. In order to avoid any adverse effect on the crop grain recovery etc., harvesting at physiological maturity or early harvest of grain crops and use as fodder for livestock are suggested. In the unlikely case of total crop failure, early rabi crop planting with suitable crops/varieties is suggested

Examples of the above include the following

- Ratoon maize or pearl millet with the adoption of relay crops as chickpea, safflower, rabi sorghum and
- Sunflower with minimum tillage after soybean in medium to deep black soils in Maharashtra or take up contingency crops (horse gram / cowpea) or dual purpose forage crops on receipt of showers

Management of receding soil moisture conditions

In the recent past, continuous high rainfall in a short span leading to water logging and heavy rainfall coupled with high speed winds in a short span are being experienced at various growth stages of annual and perennial crops leading to serious crop losses, outbreak of pests and diseases and sometimes total crop failure. Suggested contingency measures include re-sowing, providing surface drainage, application of hormones / nutrient sprays to prevent flower drop or promote quick flowering / fruiting and plant protection measures against pest / disease outbreaks with need based prophylactic / curative interventions. At crop maturity stage suggested measures include prevention of seed germination and harvesting of produce. Post harvest measures include shifting of produce to safer place for drying and maintaining the quality of grain / fodder and protection against pest / disease damage in storage.

In the regions where the normal maximum temperature is more than 40 degrees C, if the day temperature exceeds 30 C for 5 days it is defined as heat wave.

Similarly, in regions where the normal temperature is less than 40 degrees C, if the day temperature reaches 50 degrees C for 5 days, it is considered to be experiencing heat wave. Eastern Uttar Pradesh, Punjab, East Madhya Pradesh, Saurashtra and Kutch in Gujarat are highly heat prone areas and heat waves were experienced in recent years during 1998, 2002, 2003, 2004 and 2007. Generally affected crops due to heat wave are wheat, mustard, rapeseed, linseed, vegetables. Ameliorative measures include conserving soil moisture through appropriate tillage practices, frequent and light irrigations, use of efficient water harvesting techniques, choosing heat tolerant crop varieties (Golden Halna (K0424), heat tolerant variety of wheat developed at CSAUA and T, Kanpur), Provide shelter to young fruit plants and Shelter management for livestock and poultry to alleviate heat stress.

Table 7: Most vulnerable areas of India affected by heat waves

March to July with peak temperatures in April. May. 2"1 fortnight of October	South India: Khaminum and Ramagundam (Telanguna). Kalburgi and Bangalore (Karnataka)
	Eastern India: Bankura and Kolkata (West Bengal) and Bhubaneswar. Titlagarh and Jharsuguda (Odisha)
	North India: Punjab. Allahabad and Lucknow (UP), Ciaya (Bihar), Delhi
	West India: Vidarbhu and Maralhwudu (Maharashtra). Churu (Rajasthan). Ahmadabad (Gujarat)
	Central India: Jashpur(Chhattisgarh), llarda (Madhya Pradesh)

(*Source*: Bal and Minhas, 2017)

In regions where normal minimum temperature remains 100 C or above, if the minimum temperature remains 50 C lower than normal continuously for 3 days or more it is considered as cold wave. Similarly, in regions where normal minimum temperature is less than 100 C, if the minimum temperature remains 30 C lower than normal it is considered as cold wave. The adverse impacts observed are on growth, flowering, fruiting, delay in ripening and mortality of young and aged orchard plants. Poor growth rate is observed and disease outbreaks are experienced in case of fisheries. Increase of minimum and maximum temperatures during March and April 2022, resulted in dry winds, high evapotranspiration and moisture stress. Several districts of Punjab were affected with heat wave events due to increase in temperature which has resulted in yellowing and shriveling of wheat grain, forced maturity, resulting in reduction in yields up to 25%. Increased whitefly infestation, poor vegetative growth and poor pod setting was observed in green gram causing reduction in yields up to 20%. Retarded growth and fall army worm attack was observed in maize, led to reduction of yields up to 18% in maize in Faridkot, Bathinda and Gurdaspur districts of Punjab. Various technologies are available to minimise the yield loss in wheat due to heat wave and some of them are transferred to farmers' fields. Several heat tolerant wheat varieties, PBW 803, DBW 187 and DBW 222 can tolerate high temperatures and can produce normal yields compared to local variety HD-3086 Technologies such as residue management of rice by various machines enable timely sowing of wheat. Direct seeding of rice can result in early maturity by 10 days which can enable timely sowing of wheat. Spray of KNO3 @ 0.5% at boot leaf and anthesis stages can minimise the yield loss. Providing additional irrigation through effective methods during heat stress period can alleviate the stress with optimal water use.(Bal etal,2022)

Suggested measures include Proper selection of fruit species / varieties which are cold tolerant, use of windbreaks or shelter belts, frequent smoking in the orchards and covering young fruit plants with thatches or plastic shelter.

8.4 Cultural Practices and Crop Management

8.4.1. Timeliness

Sowing dryland crops with onset of monsoon rainfall can significantly improve the crop yields. Timely sowing helps in achieving optimum utilization of seasonal rainfall, reduces the incidence of pest/diseases, and is an escape mechanism from terminal drought. Yield losses from delayed sowing by 9 to14 days in sorghum and upland rice are as high as 43 to 137 and 36kg/ha /day respectively. A 15days delay in sowing of sorghum led to reduction in grain yield of 850kg/ha. Similarly, sowing of castor (*Ricinus communis* L) during the second fortnight of July reduced the bean yield of 850 to 250kg/ha. Lack of efficient implements and adequate draft power are the major constraints to timeliness in farming operations.

8.4.2. Tillage

Tillage has a marked influence on the conservation of soil and rainwater. Tillage makes the soil surface more permeable and thus, supports water intake. Deep tillage(25-30cm) helps in soil pulverization, increased rainwater infiltration, and better root growth thereby increasing yield (Thyagraj etal.,1999; AICRPDA,2000). .Off-season or pre-monsoon tillage has a significant impact on weed control and rainwater infiltration .Grain yields of sorghum and barley (Hordeum vulgareL.) were 2600 and 1570kg/ha with off-season tillage compared to 1870kg/ha without off-season tillage. (AICRPDA1986).

Studies on reduced till farming indicated that conventional tillage using recommended fertilizer and wedding, with and without off season tillage, resulted in higher grain yields of barley, rice, entil *(Lens culinaris* Medik.), wheat, soybean, groundnut, finger millet [*Eleusine coracona* (L) Gaertn] and pearl millet (AICRPDA,1999). However, excessive tillage reduces organic carbon(C) and actuates soil erosion.

8.4.3 Mulching

Mulching and crop residues incorporation contribute to the conservation of the soil and water. Mulching reduces runoff from cropped fields, prevents evaporation from the soil surface and controls weeds. Incorporation of sorghum stubbles at 5t/ha to cover 69% soil surface resulted in a0.24t/ha soil loss and 25mm runoff compared to a 1.58t/ha soil loss and 83 mm runoff when the treatment was not applied (AICRPDA). Mulching also reduced soil temperature and resulted in 25% greater moisture storage in 0 to 30cm soil profile. Frequent cultivation between crop rows create dust mulch and break the soil crust which results in reduced capillary water movement and reduced evaporation losses. Cultivation during the vegetative stage enhanced the productivity of castor, sunflower and Pigeonpea [*Cajanus cajan* (L) Millsp.] by 15 to 20% compared to no cultivation (Subba Reddy *et al.*,1996). Organic wastes and crop residues such as sorghum and maize stubbles, dry grass, wheat straw, and pigeon pea stalk can also be used as surface mulch. In vertisols, spreading followed by incorporation of the crop residue at 5t/ha enhanced the productivity of post rainy season sorghum and sunflower (IAEA, 2003)

In Alfisols, incorporation of corn residue at 4t/ha increased crop yield in succeeding crop by about 80% (Gajanan *et al.*,1999). Vertical mulching by embedding sorghum or maize stalks in rows at regular intervals across the slope is usual practice in reducing erosion in Vertisols. Sorghum yield under vertical mulching at 5m interval was about 25% higher than no mulching (Kibuku *et al.*, 2022)

Of late, work has been initiated on mulch cum green manure technique at different locations in the country. Using *Gliricidia* spp. Branches/loppings at 5t/ha in sorghum+pigeonpea-castor inter cropping rotation reduced the run-off by 56%and soil loss by 72% and increased the bean yield from 328 to 984kg/ha (ICAR-ACIAR, 2001)

9. Integrated Nutrient Management

The productivity of rain fed crops is constrained with low fertility. The soils in these regions have low organic matter and nutrient reserves. Extensive surveys in farmers' fields revealed that deficiencies of sulphur, boron and zinc are wide spread in most cases, 80 to 100% of the farmers' fields. To enhance and sustain rain fed agricultural productivity and food security there is need to adopt integrated Nutrient Management (INM) strategy. The INM strategy includes maintenance or adjustment of soil fertility and plant nutrient supply to sustain the desired level of productivity using all available sources of nutrients. viz.: soil organic matter, soil reserves, biological nitrogen fixation (BNF), organic manures, composts, non-toxic organic wastes, mineral fertilizers, and nutrients.

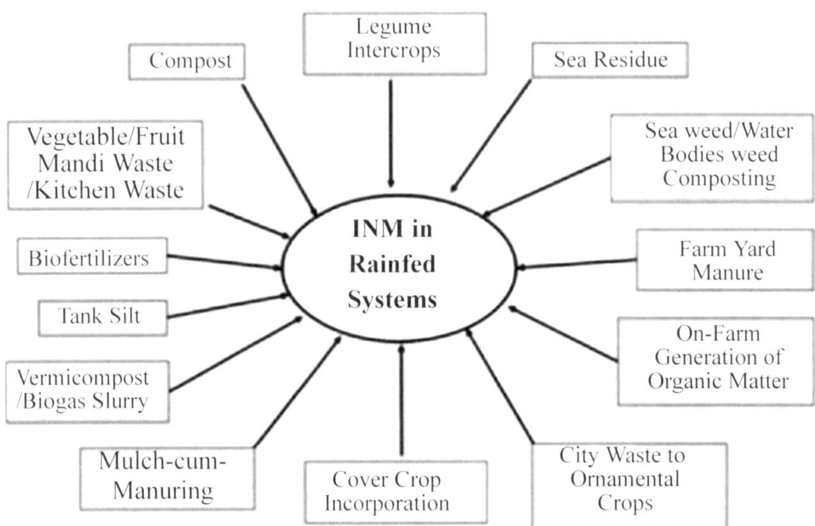

Fig. 2: Potential sources of organics along with fertilizers in rainfed INM strategies in India

Organic matter influences favourably physical and biological properties and productivity of the soil.

Organic manures consist of bulky farm yard manure (FYM), compost, crop residues concentrated oil cakes, poultry manure, and house waste. FYM is used as organic manure for high value crops. Crop residues can be recycled by compost and its nutrient content can be enhanced through addition of certain organic/inorganic amendments such as rock phosphate, microbial cultures, vermi-composting, mulching and direct incorporation. Based on their nutrient content it was observed that organic manures are less efficient than mineral fertilizers in the above context (Guertal, and Green, 2012)

A combination of crop residues and green leaf/green mauring can be used to maintain soil fertility in rainfed areas. In farms as well as homes large quantities of organic wastes are generated regularly. The nutrients in these residues can be effectively used for higher productivity using earthworms. The process of preparing valuable manure from all kinds of organic residues with the help of earthworms is called vermicomposting and this manure is called vermi-compost. It can be prepared by different methods in shaded areas viz., (i) On the floor in heaps, (ii) in pits (upto1m depth) and (iii) in an enclosure with wall (1m height) constructed with soil and rocks or brick material or cement.

***In-situ* generation of organic matter:** Green leaf manuring is one of the farming practices for increasing the organic matter in the soil. Green leaf manure plants such as sesbania, sunhemp, Gliricidia, Cassia, Leucaena can play an important role in rain fed farming systems for increasing soil fertility. The glyricidia/leaucaena plants can be grown on field boundaries. These plants work as nutrient banks besiding conserving soil through reduced soil erosion.

Gliricida is woody perennial and tender twigs can be used as mulch cum manure. It is legume, tolerant for pruning and its leaves contain 2.4% of N, 0.2% of P and 0.8% of K. It can be prorogated through seed, seedling, or stem. Gliricidia cuttings of one year old from mature branches 2-6cm diameter, and 30 to 100cm in length, which are brownish green bark colour are planted on field bunds in rainy season at 10cm, interval in zig zag manner. Alternatively, gliricidia seeds are soaked in water for 8-10hours and are sown in polythene bags filled with soil and watered regularly. Generally 3to4 month's seedlings can be planted on bunds in rainy season. After one year planting, harvesting green biomass can be started by looping the plants at 75cm above the ground. Pruning can be done at least thrice in a year i.e. June (before the sowing of rainy season crop and in November (before sowing of rabi crop)and one more in summer in march. Gliricidia plants planted 700m long bunds can provide 30kg N/ha.

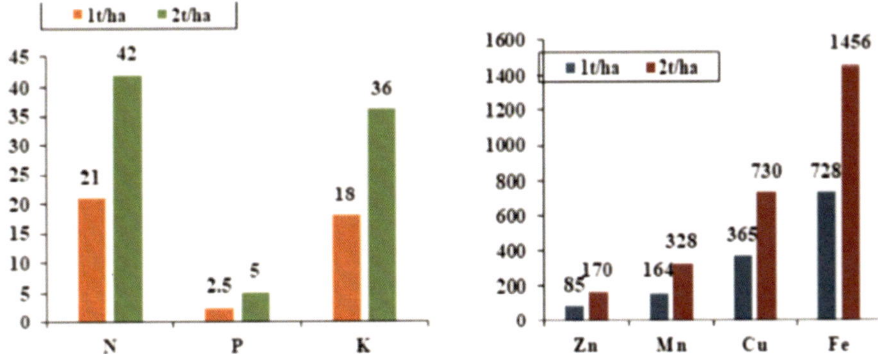

Fig 3. Macro and micronutrient addition through gliricidia green leaf manuring at different levels of manuring (*Source*: Srinivas Rao *et al.*, 2010)

The various on-farm components which would contribute to generation of organic matter are presented

Table 8: Availability organic sources in the country

Organic sources	Total availability yr1	Reference
Crop residues	500 - 550 Mt	NAAS (2012)
Municipal biosolid	48 Mt	Pappu *et al.* (2007)
Rice husk	20 Mt	Sengupta (2002)
Sugarcane bagasse	90 Mt	Sengupta (2002)
Groundnut shell	11 Mt	Sengupta (2002)
Sugarcane pressmud	9.0 Mt	Chanakva *et al.* (2006)
Poultry manure	6.25-8 Mt	THE HINDU (2009)
Coir pith	7.5 Mt	Vijaya *et al.* (2008)
Food/fruit processing industries	4.5 Mt	Chanakva *et al.* (2006)
Seri waste	5,000 t	Gunathilagaraj and Ravignanam (1996)
Willow dust	30,000 t	Chanakva *et al.* (2006)
Green manuring crop area	About 7 Mha	FAO (2005)

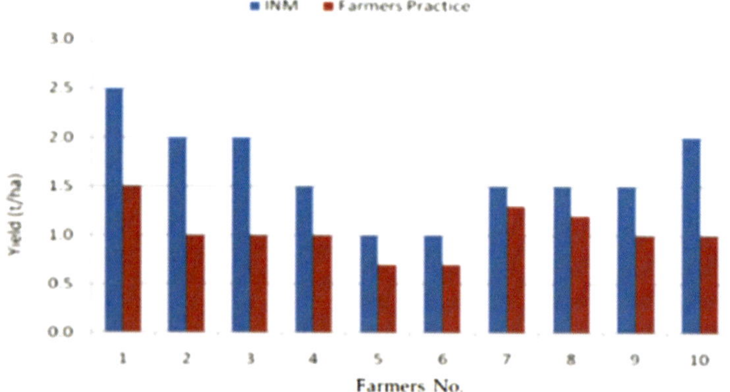

Fig 4. Influence of INM practices on yields in farmers fields

Biological inputs

The symbiotic N fixed between bacteria (*Rhizobium/Brady rhizobium*) and legumes contribute substantially (40kgN/ha/year). Non symbiotic and associate fixers (*Azotobacter and Azospirillum*) reduce N requirement of cereals or legumes by 20 kg/ha Phosphate solubilizing microorganisms and fungi solubilize inorganic phosphates and make them available to the plants in useable form and improve P use efficiency of plants. Vascular Arbuscular Mycorrhizae (VAM) help in increased uptake of nutrients such as P, S, Cu, Zn, etc.

Complementary use of organic manures with fertilizers helps in correcting micronutrient deficiencies in addition to other benefits. studies at research stations and on-farm sites have demonstrated the importance of farm yard manure (FYM), composted wastes and bio-fertilizers providing the nutrient requirement of crops and yield stability in dryland areas(Singh *et al.*, 2000). Fifty percent of fertilizer N can be replaced using FYM or compost. Application of FYM at 10t/ha along with recommended fertilizer doses stabilized the productivity of finger millet at about 3400kg/ha. The same treatment resulted in a crop yield index of 0.66 compared to 0.36 when only chemical fertilizer was used. Continuous application of chemical fertilizers resulted in decline in finger millet grain yield from an average of 2880kg/ha during initial 5years of the study to 1490kg/ha by the 19th year (Gajanan *et al.*,1999).

In vertisols, providing 50% of recommended nutrient dose through crop residues and remaining 50% through *Leucaena leucocephala* lopping enhanced the sorghum yield by 87,31 and 45% respectively compared to the application of 25kg N/ha of chemical fertilizers alone (AICRPDA,1999). In Alfisols and Vertisols, about 20kg N/ha could be supplemented the addition to green leaves of *Leucaena* or *Gliricidia*. (Subba Reddy 2002). Application of FYM in set rows resulted in an additional yield increase about 20,30,90 and 90% of sorghum ,sunflower, castor, and pigeon pea, respectively ..At the same time water holding capacity and organic C increased in the set rows by 8.5 and 5.7% respectively.

For maintaining soil fertility, green manuring is feasible in better rainfall areas when short duration legumes are grown after main rainy season crop (Subba Reddy, 2002). Apost season cover crop (horse gram [*Macrotyloma uniflorum* (Lam.Verdic] or cowpea [*Vigna unguiculata* (L.) can be raised using the post-season rainfall and plowed back into the soil before flowering (Katyal *et al.*,1994 in order to maintain soil fertility. A summary of INM recommendations for the principal crops is given in Table (2).

Biological Nitrogen Fixation (BNF) is economically sound practice to enhance soil fertility. Rhizobium inoculation is to be practiced to ensure adequate nodulation in leguminous crop plants as a process of Biological Nitrogen Fixation (BNF). There is need to use Rhizobium or Phopo bacterium for efficient use of nutrients. Recent studies from rain fed vertisols for 12 years showed that inclusion of grain legumes (pigeon pea, chickpea) or intercropped with either sorghum or maize enhanced the income and increased the productivity of succeeding crop

Mineral fertilizer

Mineral fertilizers need to be supplemented in INM system; they should be applied in optimum quantity just to meet plant demand without adversely affecting biological nitrogen fixation. These should be placed in furrow and cover with the soil. N dose should be split instead of one application. The rain fed farmers lands found deficient not only in nitrogen, phosphorus but also in sulfur, boron and zinc. Application of sulfur, boron and zinc along with nitrogen and phosphorus enhanced the yields (30 to120%) of field crops including sorghum, maize, castor, sunflower and groundnut. Complementary use of organic manures with fertilizer helps to correct micronutrient deficiencies.

Multi-nutrient deficiencies in dryland agriculture

Rainfed dryland soils are not only thirsty but also hungry. Soils of arid and semi-arid regions in India are diverse covering Vertisols, Alfisols, Oxisols, Inceptisols, Aridisols, b ;/'Entisols, etc., with rainfall variation between 400-1200mm. Length of growing season is in between 60-180 days. Besides, these soils are highly degraded, low in soil organic carbon and having multi-nutrient deficiencies (Table3). Some of the rainfed farming areas have extremely low soil organic carbon, as low as 0.15%. Most of the soils are deficient in N and P and deficiencies of K, Mg, S, Zn and B are emerging as major constraints in crop production (Table 7)

Table 9: Emerging nutrient deficiencies in dryland soils(0-15cm) under rainfed / dryland production system

AICRPDA center	Rainfall (mm)	Soil type	Production system	Limiting nutrients needs to supplied
Rajkot	615	Vertisol	Groundnut	N,P,S,Zn,Fe,B
Anantapur	590	Alfisols	Ground nut	N,K,Mg,Zn,B
Indore	950	Vertisols	Soybean	
Rewa	900	Vertic incept sols	soybean	Zn,N
Akola	825	Vertisols	Cotton	N,P,S,Zn,B
Kovilpatti	750	Vertic Inceptisols	Cotton	N,P
Bellari	500	Vertisols	Rabi Sorghum	N,P,Zn,Fe
Bijapur	680	Vertisols	Rabi Sorghum	N,Zn,Fe
Jhansi	1020	Inceptsol	Kharif sorghum	N
Solapur	720	Vertisol	Rabi Sorghum	N,P,Zn
Agra	665	Inceptisol	Pearl millet	N,K,Mg,Zn,B
Hissar	412	Inceptisol	Pearl millet	N,Mg,B
S.K.Nagar	550	Aridiaol	Pearl millet	N,K,S,ca,Mg,Zn,B
Bangalore	925	Alfisols	Finger millet	N,Ca,K,Mg,Zn,B
Arjia	650	Vertisol	Maize	N,Mg,Zn,B

(*Source*: Srinivasa Rao 2010)

Amongst plant nutrients essential for growth, nitrogen is of prime importance in crop production, though constitutes· only a fraction of one per cent of the total dry weight. Nitrogen is transported from roots to the leaves where the process of assimilation takes place, and it is transformed in to protein a substance, which constitutes an important constituent of protoplasm. Growth is usually rapid with abundance of carbohydrate and nitrogenous compounds. Nitrogen, that gets liberated from soil environment, makes it practically very difficult to be utilized by plants for their requirement. The reason for this phenomenon could be de-nitrification and volatilization of ammonia, leaching of nitrates, fixation of ammonia or sometimes biological immobilization by bacterial activities.

Nitrogen being one of the most important limiting nutrients for crop production, several workers studied the effect of nitrogen nutrition on the growth yield of crops and the quality parameters. There is general trend that due to increasing rate of fertilizer growth, yield increases up to some extent. Nutrient absorbed by plant during tillering phase does not exceed 25-30 per cent of its total nutrient absorbed up to the maturity stage. Initial higher concentration of nitrogen in crop is necessary because during this stage tiller development takes place.

Phosphorus

Phosphorus is one of the essential nutrients and also known as the 'key element' for plant life. It maintains the adverse effect of high application of nitrogen fertilizers. Several workers reported that application of phosphorus fertilizer has positive effect on crop in dryland situation. It is generally applied as basal dose and mostly the water soluble forms of phosphorus viz., Single Super Phosphate (SSP), Di Ammonium Phosphate (DAP) etc., are applied

Potash: This nutrient required in larger amounts than nitrogen because of its role in photosynthesis, translocation of sugar, protein synthesis etc... Therefore, its requirement is needed through overall growth period, and due to this reason initial basal application is recommended. Generally, potassium sulphate · (K2 SO 4), and muriate of potash (KCl) are used as sources of potassium

Sulphur: Sulphur is given through Ammonium Sulphate (23.7%), and Pyrites (53.5%) etc. · Generally the requirement is more in graminaceous plant as compare to leguminous because it play vital role in oxido- , reduction process of respiration, because of being a part of ferredoxin. Sulphur (S) is involved in amino acid and protein synthesis, enzymatic and metabolic activities in plants. Its deficiency is rapidly emerging in areas under ' oilseeds and pulses due to higher removal of S by crops. Fertilizers required for correcting the deficiency of all these ' nutrients especially sulphur for oilseeds and pulses, .

Micronutrients

Plant takes zinc in the form of Zn2+. Zinc is required in a large number of enzymes and plays an essential role in DNA transcription. is a need to ascertain and promote the use of various types of fertilizers required to correct the deficiency of all these nutrients especially zinc for cereal crops. Deficiency symptoms of Zn in cereal crops are more pronounced, hence some time reduction in yield· often correlated to low availability of zinc only. It is applied through soil but foliar feeding can also be done. Most important source is Zinc sulphate (35% Zn). It is applied in soil at the rate of 25 or 50 kg/ha.

Foliar Nutrition

Foliar fertilization refers to the supplementation of major, minor, beneficial plant hormones, stimulants and other beneficial substances to the plants by applying them through sprays. Foliar supplementation has several advantages such as meeting the nutrient demand of the crops through foliar sprays grown in moisture deficient soils in rainfed areas. During severe nutrient deficiency conditions, it facilitates rapid absorption of the nutrients and thereby minimizes the deficiency impact. It also helps in avoiding the nutrient fixation and immobilization due to various soil chemical interactions. Foliar fertilization has advantages in terms of low application rates, uniform distribution of fertilizer materials and quick response to nutrients. To counteract cyclic droughts, foliar fertilization with K in groundnut has been found to be beneficial . Foliar application of potassium on growth and yield of toria revealed that application of 100% N & P +75% K as basal + 2% KCl spray before flowering + 2% KCl spray at siliqua formation resulted in highest grain yield, straw yield, rain water use efficiency, net returns and B:C ratio (Pranjit *et al.*, 2015)

Table 10: Foliar Application of nutrients in dryland agriculture

District/Location	Crop	Treatment	Improvement in yield (%) over farmer's practice
Biswanath Chaiali	Papeseed	KCI2%	51
Jagdalpur	Paddy	Urea 2%	6
Arjia	Maize	$ZnSO_4$, 0.5%	22
Ballowal Saunkhri	Maize	KNO_3 1% and $ZnSO_4$, 0.5%	5-6
Bengaluru	Finger millet	KCI 2% and thiourea 250 g ha^{-1}	40
Agra	Pearl millet	Urea 2% + KNO_3 2%	71
S K Nagar	Pearl millet	Urea 1%and N@ 20kg ha^{-1}	35
Anantapur	Groundnut	KNO_3 2%	15
Rajkot	Cotton	KNO_3 2%	25
Parbhani	Soybean, cotton	KNO_3 2%	12

Impact of foliar sprays on yield of different crops during dry spells District/ Location.

The positive effect of foliar treatment on different crops resulted in significant improvement in yield (%) as compared to farmer's practice as shown in Table 10.

Nutrient Placements Positioning of fertilizers in soil at a specific place with or without reference to the position of seed is referred as placement. It is normally recommended when the quantity of fertilizers to be applied is small under resource constraints. The most common methods of placement are described below.

(i) *Plough Sole Method*: In this method, fertilizer is placed at the bottom of the plough furrow in a continuous band. During the process of ploughing, every band is covered as the next furrow is turned and is suitable for the areas where soil becomes quite dry to few centimeters below the surface and soils with a heavy clay pan just below the plough sole layer.

(ii) *Deep Placement*: It is the placement of nitrogenous fertilizers in the root zone soil to prevent loss of nutrients by runoff.

(iii) *Localized Placement:* It refers to the application of fertilizers into the soil close to the seed or plant in order to supply the nutrients in adequate amounts to the roots of growing plants. The common methods in this context are as follows:

(a) *Drilling*: In this method, the fertilizer is applied at the time of sowing by means of a seed-cum-fertilizer drill. This places fertilizer and the seed in the same row but at different depths in the moist zone. This method has been found to be suitable for phosphatic and potassic fertilizers. Use of seed-cum-fertilizer drill helps in saving time and labour.

(b) *Side dressing*: It refers to the spread of fertilizer in between the rows and around the plants. The common methods of side-dressing are 1) placement of nitrogenous fertilizers by hand in between the rows of crops like maize, sugarcane, cotton etc. to apply additional doses of nitrogen to the growing crops, and 2) placement of fertilizers around the trees like mango, apple, grapes, papaya etc.

(c) *Band placement*: It refers to the placement of fertilizer in bands. Band placement is of two types

(i) *Hill placement*: It is practiced for the application of fertilizers in orchards. In this method, fertilizers are placed close to the plant in bands on one or both sides of the plant. The length and depth of the band varies with the nature of the crop.

(ii) *Row placement*: When the crops like sugarcane, potato, maize, cereals etc., are sown close together in rows, the fertilizer is applied in continuous bands on one or both sides of the row, which is known as row placement. The major benefits of placement of fertilizers are (a) reduction in fixation of nutrients,(b) higher fertilizer use efficiency, (c) reduced leaching losses of nutrients, and (d) phosphates, being immobile, are better utilized with placement

Nano Fertilizers

The large-scale application of chemical fertilizers especially nitrogenous fertilizers contributes to global greenhouse gas (GHG) emissions as they are one of the major anthropogenic sources of nitrous oxide (N2O). Temperature anomalies prevailing in arid and semi-arid regions aggravate emissions from nitrogenous fertilizers. Nanotechnology has provided the feasibility of exploring nanoscale or nanostructured materials as fertilizer carrier or controlled release vectors for building the so-called smart fertilizers which would aid in enhancing nutrient use efficiency and minimize the cost incurred on fertilizers. The use of nano-fertilizers has benefits viz. site targeted delivery, reduction in toxicity and enhanced nutrient utilization of delivered fertilizers (which would help in lowering the quantity of fertilizer usage) and eventually contribute to the reduced GHG emissions.

Precision Nutrient Application and Variable Rate of Technology (VRT)

Precision Agriculture (PA), as the name implies, is useful technology for growing and fertilizing crops more precisely or efficiently, thereby retaining water and nutrients in the root zone. Three techniques with which PA can help achieve this objective are 1) the collection of spatial data from pre-existing conditions in the field (e.g., remote sensing, canopy size, or yield measurement), 2) the application of precise fertilizer amounts to the crop when and where needed, and 3) the recording of detailed logs of all fertilizer applications for spatial and temporal mapping. Thus, PA can help us determine exactly where to place nutrients and how much to apply, and then track the applied nutrients with accumulation logs and GIS maps. Variable-rate technology (VRT) allows fertilizer, chemicals, lime, gypsum, irrigation water and other farm inputs to be applied at different rates across a field, without manually changing rate settings on equipment or having to make multiple passes over an area. Sensor-based precision equipment like sensor dependent variable rate fertilizer applicator based on the variable rate of urea application system; integrated with spectral reflectance-based sensor (Green seeker) have been developed at CIAE Bhopal. An estimated 8-15% saving in urea is achieved with use of NDVI-based variable rate fertilizer applicator in wheat and rice crops (Singh *et al.*, 2017). Besides saving on input charges this technology also would reduce release of N2O, eventually lowering the GHG emissions.

7. Integrated pest management

Crop pest constitutes major constraint for increases yields in rainfed environment. Crop losses due to pest attack ranges from 10-30% depending on crop and environment. Excessive dependence on chemical pesticides led to the development of harmful side effects such as direct toxicity to the applicator, consumer, pest resistance to chemical pesticides, resurgence of pest species, outbreak of secondary pests, and destruction of non-target and beneficial organisms like predators and parasites. It also results in the accumulation of harmful residues in food products. To overcome such situations and minimize the damage to human and animal health, the concept of integrated pest management was initiated by several organizations

The integrated pest management can be defined as one or more management options adopted by farmers to maintain density of the potential pest populations below the thresh hold levels for enhanced productivity and profitability of the farming system as a whole, the health of farm family and its livestock, and quality of immediate and downstream environments.

Among various plant protection options, the watershed team has to choose to promote ecofriendly approaches for use by the farming communities. These include: diagnostic surveys and farmers interactions for determining economic importance of various pests, ,capacity building in management of pests, bio-safety measures, use of pest resistant measures/ practices; building knowledge on the role of cultural practices to avoid pest incidence, and natural enemies of pests and diseases, production and adoption of bio-pesticides at village level and need based application of chemical pesticides..

An important difference between Bio- intensive integrated pest management (BIPM) and conventional IPM is that BIPM is on provocative measures to redesign the agricultural system to the disadvantage of pest and its advantage of parasite and predator complex. BIPM include crop rotation to break the cycle of the pest carryover, deep ploughing to expose hibernating insects to hot summer temperature, soil solarization technique for control of soil borne diseases, growing crop varieties naturally resistant to pest and diseases, seed treatment with permitted preparations and bio-pesticides, choosing sowing times that prevent pest and disease outbreaks, improving soil health through cover crops, green manure, animal manures to get vigorous seedlings, encouraging pollinators and natural biological agents for control of pests, augmentation of release of parasitoids and predators such as *Trichogramma, chysopera* and *coccinelloids.* Application of permitted microbial pesticides, such as NPV(*nuclear poly-hydrosis virus*), granulosis virus (GV),*Bacillus thuringiensis* (Bt)and those of plant origin biopesticides (neem,vitex etc.,). This method also includes using physical barriers from insects and animals and also use of hormones such as pheromone attractants for trapping pests. The studies conducted on IPM and BIPM gave encouraging results in farmers fields. (Table 11)

Table 11: Economically important pests of major crops in India

Crop	Common name	Scientific name	ETLs	Existing control methods
Cereals				
Rirp	Stem borer	Scirpophaqa incertulus Walker	5% white ears/One eqq mass sqm'	IPM
	Brown plant hopper	Nilaparvata lugens stal.	10 hoppers per clump.	IPM
	Gall Midge	Orseolia oryzae wood-mason	5-10% silver shoots	Host plant resistance (HPR)
	Leaf folder	Cnaphalocrocis medinalis guen	10-15% webbed foliage	HPR
Wheat	Aphid	Schizaphis graminum (rondani)	5-10% of plants with infestation	HPR
Maize	Stem borer	Chilo partellus (swinhoe)	5-10% infestation	Chemical
	Shoot fly	Atherigona spp.	5-10% dead hearts	Chemical
	Earworm	Helicoverpa armigera hubner	25-30% damage to cobs	Chemical
Legumes				
Pigeonpea	Pod borer	Helicoverpa armigera (hubner)	5 eggs or 3 small larvae per plant	IPM
	Pod fly	Melanagromyza obtusa (malloach)	In all endemic locations	Chemical
	Leaf webber	Monica vitiulu (yeyer)	5 webs per plant	Chemical
	Pod sucking bugs	Clavigralla gibbosa spinola	One egg mass per plant	Chemical
Chickpea	Pod borer	Helicoverpa armigera (hubner)	3 eggs or 2 small larvae per plant	IPM
	Cutworm	Agrotis ipsilon (hufnagel)	5% plant mortality	Chemical
Soybean	Stem fly	Ophiomyia phasioli (tryon)	5% plant infestation	Chemical
	Girdle beetle	Obereopsis brevis (swed)	5% incidence	Chemical
	Hairy caterpillar	Spilosoma obliqua (walker)	5 larvae meter row	
Oil Seeds				
Groundnut	Leaf miner	Aproaerema midicella deventer	5 mines per plant at 30 days of crop age	IPM
	Tobacco caterpillar	Spodoptera litura (fab)	20-25% defoliation at 40days	IPM
	Thrips	Scirtothrips dorsalis hood	5 thrips/terminal at seedling stage	Chemical

Crop	Common name	Scientific name	ETLs	Existing control methods
	Aphids	Aphis craccivora kouch	5-10 aphids per terminal at seedling stage stage in dry spells onlyin rainy season	IPM
Sunflower	Gram pod borer	Helicoverpa armigera hubner	One larva per head	Chemical
Sesame	Leaf webber	Antigastra catalaunalis dub	2-5 webbs per plant	Chemical
Rapeseed	Aphids	Lipaphis erysimi (Kalt)	5-10 aphids per plant	Chemical
Cotton	American bollworm	Helicoverpa armigera hub.	5-10% boll infestation	IPM
	Pinkbollworm	Pectinophora gossipiella saund	5-10% boll infestation	IPM
	Whitefly	Bemisia tabaci genn.	8-10adults/leaf	IPM
	Spoted bollworm	Earias insulana boisd.	5-10% boll infestation	IPM

(*Source*: Ranga Rao *et al.*, 2009)

Table 12: IPM technologies for dryland crops

Location	Type of Crop and Pest	Treatment	Yield Kg ha^{-1}
Hydera	Castor (Swnltmprr)	Farmers' praciicc	350
		Hand picking	576
		Dusting turmeric powder	314
		Chemical	691
	Castor (Red hairy caterpillar)	Farmers' practice	640
		Hand picking	732
		Chemical	750
	Pigeonpea (Pod borer)	Farmers' practice	203
		Bio-inseclicide (extract from custard apple seed) at 5 mL L^{-1}	365
		Chemical	363
Warangal	Groundnut (Red hairy caterpillar)	Trap crop (Calntmplr minlean/ Jatmpha curcai)	1650
		Chemical	1420
Coonoor	Groundnut (Leaf webber)	Farmers' practice	650
		Trap crop (cowpea/castor)	800

8. Weed Management

- Weeds especially in early stages of crop growth will compete and share large quantity of nutrients and moisture and consequently leading to reduced crop growth and heavy yield losses. The studies(Anonymous 2020) revealed yield losses of 44% in pigeon pea, 36% in green gram, 50% in black gram and 42% in chickpea due to weed infestations over weed free plots. The weeds have to be managed within 30-40days in short duration crops like cowpea, green gram, black gram and chickpea. In long duration crops like pigeon pea, castor, management of weeds within 45-50 days after sowing of crops is optimum. d Grain sorghum yield increased following all pyrasulfotole plus bromoxynil as early post emergence treatments compared to the untreated check in 2009 (Fromme *et al*., 2012.) Atrazine or Propazine or Prometryne at 1.0 kg a.i./ha is recommended to control emerging weeds after the sowing of sorghum. There should be optimum soil moisture at the time of application of Atrazine. (Agropedia, 2010)

Among various practices, pre-emergence application of pendimethilin at 1.0-1,5kg/ha,and preplant incorporation of 1.0kg/ha fluchloralin were effective in controlling seasonal weeds in chickpea, lentil, pea, common bean, pigeon pea, black gram and green gram in Northern India.

9. Farm Mechanization

Proper tillage and precise placement of seed and fertilizer in the moist zone are the most critical steps for successful crop establishment in rain fed lands. Since sowing of crops must be completed in a shorter span of time, use of appropriate implements/mechanized tools is necessary to cover large area before the seed zone dries out. With current seeding practices, farmers are unable adhere to the sowing schedules since the power available in rain fed areas falls much shorter of the demand. Also, lack of time-saving devices suited to the draft power available with the farmers is another practical problem, effective weed control, hand tools like *khurpi* are still popular. The labour requirement in this context is 35-60 human labourers/ha. Many types of weeding tools are developed by different Institutions. The most popular type is CIAE wheel hoe having different sizes of sweeps and blades suited to crop row widths and available power. There is need to design suitable and efficient tractor drawn plant protection equipment to control pests in commercial crops like pigeon pea, cotton and horticultural crops. Harvesting of rainfed crops and horticultural crops consumes considerable human time and labour cost. Researches at different dryland research centers in India demonstrated the benefits of conservation tillage, mulch, residue incorporation, precision and timely seeding, weed control, plant protection, timely harvesting and post harvest operations(Table 12)

Table 12: Farm implements suitable for different production system

Production system	Centre/domain	Name of the operation	Improved implement	Remarks
Oilseed	Anantapur Groundnut	Sowing, fertilizer and placement	Eenati Gorru (Bullock drawn)	Suitable for light draft animals
		Harvesting of Groundnut	Asha Guntake Tractor drawn	4-5ha/dat.cutting width .8m,source of power.35HP tractor.Cost of operation Rs400.ha
			Groundnut digger	Field coverage:3-4 ha/day. Cutting width 1-2m. power.35HP tractor. Cost of operation Rs500/ha
	Rajkot (Gujarat)	Deep ploughing	Sub-soiler	Approx.price Rs10000
		Fertilizer,seeding and covering	Seed cum fertilizer drill	Bullock drawn and tractor drawn available
		Spraying Tractor operated sprayer	Tractor operated sprayer	Approximate cost Rs50000 cover.1.8ha/hout
		Threshing	Groundnut thresher	Approximate Price Rs80000
	Indore (Soybean)	Land preparation	Tractor drawn MB plough and rotovator	
		Sowing	Inclined plate planter	Tractor drawn
		Sowing and fertilize application	Seed cum fertilizer drill	Tractor drawn, approx. price: Rs. 27000 per un
		Interculture	Rotor combination weeders	Petrol/diesel engine operated
		Weeding	NRCS rotary weeder	Tractor drawn
	Rewa (Soybean)	Tillage and sowing	Dufan	Also suitable for sowing of paddy, soybean, wheat, chickpea etc.

Nutritious cereal	Bijapur (RABI SORGHUM)	Sowing and fertilizer application	Seed cum fertilizer drill	Bullock drawn
		Opening furrows	Multi opener	Tractor drawn
		Sowing and fertilizer application	Seed cum fertilizer drill	Tractor drawn Cost Rs 26000/u
		Incorporation of residues and green manure	Rotovator	Tractor drawn
	Solapur (rabi sorghum)	Land preparation,harrowing, interculture in orchards	Tractor drawn V blade harrow	
		Sowing of crops	Bullock drawn seed cum fertilizer drill	
			Tractor drawn multicrop planter	Rabi sorghum Rs25000/unit
			Bullockdrawn Jyothi planter-pune	
		Interculture in crop rows	Cycle hoe(Mannual)	
			Tractor drawn combination	
		Land preparation. Sowing of sorghum,pigeon pea and pearl-millet	Shivaje multi farming machine	Bullock drawn Rs10000/unit
	Hissar {Pearlmillet)	Sowing	Ridger seeder	Lister bottom pushes top dry soil to the sides for seeding in most soil in the furrow Rs20000/unit
		Interculture	Tractor drawn ridger cultivator	
	Bangalore (Finger millet)	Sowing and fertilizer application	Bullock drawn seed cum fertilizer drill	
		Seeding and ferilizer	Tractor drawn inclined and cup feed planter	

(*Source*: Mission-mode Project on Farm Mechanization 2003 and AICRPDA Report 2000-20007, Central Research Institute for Dryland Agriculture, Hyderabad)

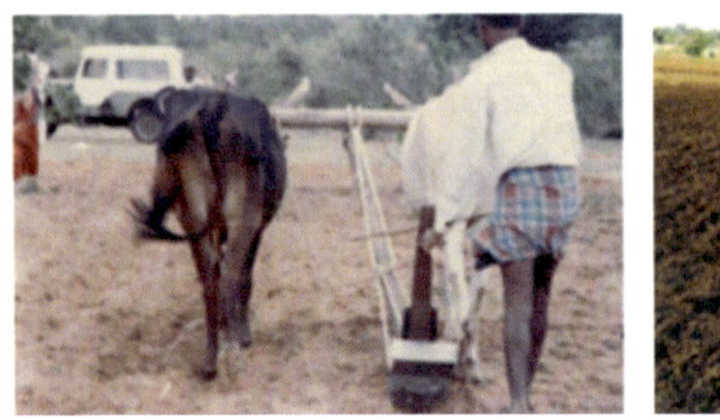

Drill

Plough planter

Two row planter

Three row planter

Six row planter

Nine row planter

Fig. 5: Different Machines developed in CRIDA suitable for dryland agriculture

Three row ridger planter cum BBF planter

Close up view of Manual weeder

Orchard Sprayers

Ground nut stripper

Castor pod sheller

Mini Dal Mill

Source: Farm Mechanization in Rainfed Regions: Farm Implements Developed and Commercialized. Central Research Institute for Dryland Agriculture, Hyderabad, Telangana State. 2014

Fig. 6. Diverse efficient farm equipments for dryland production

10. Alternate High Value Crops

Cultivation of crops for dyes, pharmaceuticals, and aromatics on drylands may enhance the economy of poverty stricken dryland farmer. Most of such plant species possess robust root system and possess the attributes to enhance the quality of of degraded lands. These crops are both perennials and annuals with the plants mostly of bushy stature. The advantages of over large perennials is that the former offers less competition to associated crops. The promising plants for cultivation in drylands are dyes like Indigo(*Indigofera tinctoria L), Henna(Lawsonia innermis L),* and Bixa*(Bixa Orellana L),*Medicinal Plants like Ashwagandha(*Withania somnifera*),Senna(*Cassia angustifolia* Vahl.),Muccina [*Muccina Pruriens*(L)DC] and aromatics like curry leaf [*Murrya koeingii* (L)Spereng].lemongrass[*Cymbophogan* winki (DC)Stapf], palmorosa (*Cymbopogan martini* (Roxb) Wats], and sweet basil (*Ocimum basilium* L.] hold potential.

Initial experiments at CRIDA, Hyderabad showed that Bixa could be successfully cultivated in areas receiving 700mm and above rainfall areas in medium deep red soils. About 500 plants can be grown in one ha. (4.5mx4.5m) and each plant yields 1-1.5kg of seed from third year onwards. Henna is another dye yielding plant that can be profitably raised in rainfed lands. It is grown mainly in Rajasthan and is slowly picking up in other states. Mesta is another promising crop that can be grown successfully as rainfed crop. All these crops can be grown as intercrops with annual crops like sesame, sunflower and pigeonpea.

11. Alternate land use systems

Any farm enterprise other than crop production is usually referred to as an alternate land use system. Examples are growing trees in arable crops in strips, tree farming, wood lots, pastures, grass lands, leyfarming,and agroforestry systems. Among these, growing arable crops with compatible tree species has been a traditional practice by the dryland farmers who view it as a way to meet the demands of food, fodder, fuel etc., Such a system offers sustainable productivity, covering land with vegetation and conservation of natural resources. General recommendations for alternate land -use systems based on annual rainfall and soil types are outlined as below.

Agri-silviculture is recommended for land capability IV with annual rainfall of 750mm. A large number of tree-crop combinations particularly of N fixing trees with sorghum, groundnut, castor and pulses were evaluated in Alfisols and Vertisols. Short duration dryland crops such as pearl -millet, blackgram and green gram combined with widely spaced tree rows of *Fatherbia albida* and *Hardwikia binate* have been found compatible in semi-arid tropical areas (Korwar,1992).

Silvi pastural system is recommended for the above kind of land capability class. It integrates trees with a perennial system or grass as pasture to improve the productivity as well as increase fodder availability. Native pastures with babul and *subabul* are common in dryland areas. Silvipastural systems involves growing trees with palatable grasses and legumes (stylos) with trees such as subabul, anjan and sisso were found to be more productive and profitable in drylands (Singh, 2002).

Agri-horticulture can be adopted in land capability classes II and III receiving >750 mm of rainfall. Agri-horticulture consists of fruit trees grown with arable crops are recommended. Promising fruit trees which can be grown successfully in drylands, are Jujubi, gooseberry, custard apple, guava,tamarind and mango in combination with arable crops such as cluster bean, cow-pea, groundnut and horse gram.

Alley cropping or hedge row farming is an agro-forestry practice in which arable crops are grown in between perennial hedge rows spaced at regular intervals. This system is recommended for land capability II to IV receiving 500 to 750 mm annual rainfall. Hedge rows are pruned during the cropping season to reduce their competition with crops. Short duration rainy season crops such as pearl millet and sorghum were found to be compatible alley crops while long duration species such as castor and pigeon pea are not. Wider alleys and shorter hedge row height trees used in wetter tropical areas were found to be better in semi-arid areas.

Trees on field boundary: Growing multi trees on field boundaries is common agro-forestry practice in drylands. Promising erect tree species are teak, subabul, palmyrah palm and babul. In this system the area along the field boundary is gainfully used.

12. Crop—Livestock integration in a farming system mode

Farming systems in rainfed areas by and large are complex and characterized by several environmental and socio-economic variables. It is the concept, which takes into account the component of soil, water, crops, livestock and other resources with the farm family at the centre, managing agricultural and related activities and even non-farm activities. The characteristics of the farming system are whole farm approach, farming based, holistic in nature, problem solving, gender sensitive, and farmers' participatory and decision-making, extensive on farm approach and interactive and iterative.

Live stock plays vital role in sustaining the livelihoods in drylands as they absorbs shock due to droughts. India has about 15% of world rruminants (cattle,buffalo sheep and goat)population with only 2.4 % of worlds geographical area.

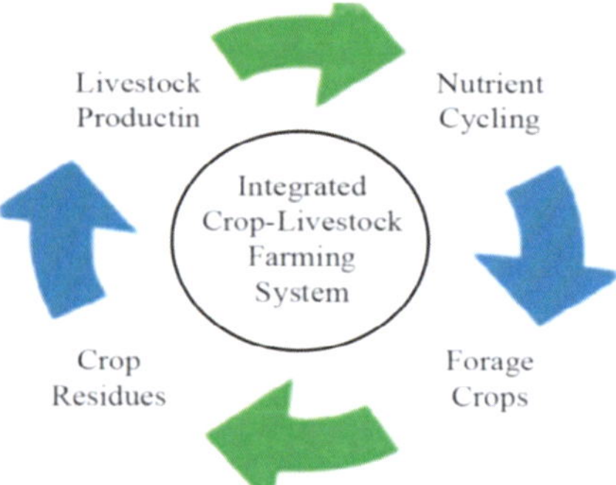

Fig. 7: Integrated Crop-Livestock farming system: Key aspects

An integrated crop-livestock farming system represents a key solution for enhancing livestock production and safeguarding the environment through prudent and efficient resource use. The increasing pressure on land and the growing demand for livestock products makes it more and more important to ensure the effective use of feed resources, including crop residues. An integrated farming system consists of a range of resource-saving practices that aim to achieve acceptable profits and high and sustained production levels, while minimizing the negative effects of intensive farming and preserving the environment. Based on the principle of enhancing natural biological processes above and below the ground, the integrated system is the combination that (a) reduces erosion; (b) increases crop yields, soil biological activity and nutrient recycling; (c) intensifies land use, improving profits; and (d) can therefore help reduce poverty and malnutrition and strengthen environmental sustainability. The waste products of one component serve as a resource for the other. For example, manure is used to enhance crop production; crop residues and by-products feed the animals, supplementing often inadequate feed supplies, thus contributing to improved animal nutrition and productivity. Integrating crops and livestock serves primarily to minimize risk and not to recycle resources. In an integrated system, crops and livestock interact to create a synergy, with recycling allowing the maximum use of available resources. Crop residues can be used for animal feed, while livestock and livestock by-product production and processing can enhance agricultural productivity by intensifying nutrients that improve soil fertility, reducing the use of chemical fertilizers

A high integration of crops and livestock is often considered as a step forward, but small farmers need to have sufficient access to knowledge, assets and inputs to manage this system in a way that is economically and environmentally sustainable over the long term. (Vinod Gupta *et al.* 2012)

The overall benefits of crop livestock integration can be summarized as follows:

- Agronomic, through the retrieval and maintenance of the soil productive capacity;
- Economic, through product diversification and higher yields and quality at less cost;
- Ecological, through the reduction of crop pests (less pesticide use and better soil erosion control); and
- Social, through the reduction of rural urban migration and the creation of new job opportunities in rural areas. This system has other specific advantages
- It helps improve and conserve the productive capacities of soils, with physical, chemical and biological soil recuperation. Animals play an important role in harvesting and relocating nutrients, significantly improving soil fertility and crop yields.
- It is quick, efficient and economically viable because grain crops can be produced in four to six months, and pasture formation after cropping is rapid and inexpensive.
- It helps increase profits by reducing production costs. Poor farmers can use fertilizer from livestock operations, especially when rising petroleum prices make chemical fertilizers unaffordable.
- It results in greater soil water storage capacity, mainly because of biological aeration and the increase in the level of organic matter.
- It provides diversified income sources, guaranteeing a buffer against trade, price and climate fluctuation

The possible options of farming systems that can be considered for different environments in drylands are

A. For the areas with annual rainfall < 500mm

- Linking arable farming with animal husbandry.

- Adopting arable farming limited to millets and pulses and adoption of agro forestry, silvipastoral and hortipastoral systems.
- Growing drought-tolerant perennial tree species that provide fuel, fodder and food.
- Adopting arid-horticulture to augment farm income.
- Emphasizing efficient management of rangelands and common village grazing lands, adopting improved strains of grasses, reseeding techniques, and developing fodder banks.

B. For the areas with rainfall 500-750 mm

- Increasing emphasis on oilseed and legume based intercropping systems in not so favourable tracts.
- Adopting high value (fruits, medicinal, aromatic bushes, dyes and pesticides) high tech (drip irrigation, processing, extraction and value added products) agriculture.
- Encouraging watershed approach in a farming systems perspective.
- Efficiently utilizing marginal and shallow lands through alternate land use systems with agriculture-forests-pasture system with a range of options.
- Increasing afforestation in highly degraded undulating lands.

C. For regions receiving rainfall between 750-1150 mm

- Developing Aquaculture in high-rainfall, double-cropped regions with rationalization of area under rice.
- Adopting intercropping systems and improved crop varieties of maize, soybean, groundnut and double cropping in deeper soil zones.
- Rainfed horticulture.
- Tree farming.

Potential important technologies, which can make significant enhancement in productivity of both the crop and livestock with in the systems can be adopted through:

- Increased fodder production as an intercrop with cereals, alley and relay cropping ,forage production on bunds,food-feed cropping systems etc.,

- Better utilization of available fodder resources bt timely harvesting and preserving it either as hay or silage
- Improving the feeding value of straw/stover by chopping, soaking with water, urea treatment, Strategic supplementation of concentrate, mineral mixture or tree foliage etc., for enhanced utilization
- Development of degraded, marginal lands through agro-forestry systems.
- Managing common property areas through user groups
- Establishing fodder Banks in areas where surplus fodder is available and transporting it to the fodder deficit areas
- Supplying improved seeds or saplings for fodder seed multiplication and reseeding of grazing lands
- Removing low grade animals through castration
- Regular culling of un-economical animals
- Protecting animals against harsh environment and adaption of preventive measures through health camps

Thus shaping of integration process depends on large extent on farm management. In order to achieve sustainable development in dryland areas,all components of farming system must be taken into account as well as the context within which resources poor farmer operate.

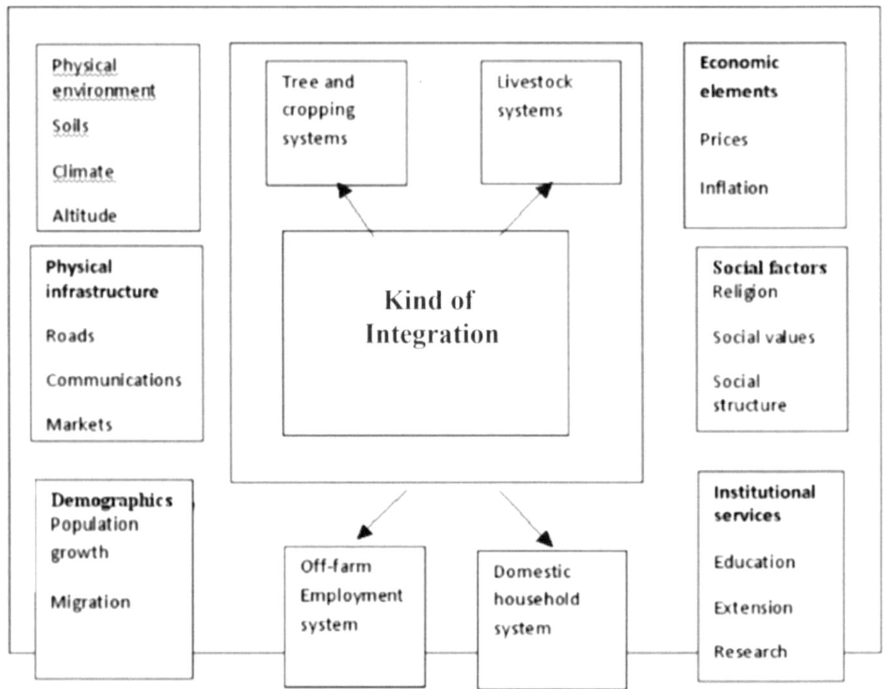

Fig. 8: Integration of livestock with arable crops in farm management

Integrating Dryland Crop Production Technologies through Watershed Management

The concept of watershed management is important for efficient utilization of resources. Since rainwater conservation and utilization is the corner stone of successful dryland farming, waters shed with distinct hydrological boundary is considered an ideal level for area development. Past experiences of watershed projects implemented in the drylands have led to improved water availability in terms of additional surface storage and enhanced recharge of ground water. Increased water availability in wells and storage facilities have led to an increased crop intensity by 50% over the period of 5 years.(CRIDA,1997). In turn, these efforts resulted in adoption of improved dryland technologies and marked economic gains. Farmers participation in technology development, innovations in transfer, and using Indigenous Technical Knowledge with active participation of gross root extension personnel, policy support by the government agencies results wider and higher adoption of dryland technologies.

6

Intercropping in Dryland Agriculture

Historical perspective: Planting of single crops or mono-cultural agriculture—is a recent invention of the industrial agricultural complex (Hirst, 2019). While unequivocal archaeological evidence is difficult to come by, it's believed that most agricultural field systems in the past involved some form of mixed cropping. That's because of even botanical evidence of plant residues such as starches or phytoliths of multiple crops are discovered in an ancient fields (Hirst, 2019). Mixed cropping, also known as polyculture, or co-cultivation, is a type of agriculture that involves planting two or more crop plants simultaneously in the same field so that they grow together. The primary reason for prehistoric multi or mixed cropping probably had more to do with the needs of the farmer's family and perhaps to avoid crop failure. It's possible that certain plants adapted to multi-cropping over time disappeared as a result of the domestication process (Hirst 2016).

Intercropping practice is a refinement of age old mixed cropping that was practised by the Indian farmers particularly on drylands, as an insurance against total crop failure. Intercropping is a method of growing two or more crops in a particular row proportions, with systematic agronomy including nutrient management. Intercropping practice may be practised in dryland areas with rainfall of less than 800 mm. Intercropping is considered as one of the strategies adopted under rain-fed and dryland conditions to minimize risk of yield reduction or total crop failure due to inadequate or uncertain moisture availability during crop (Sudhakar Reddy, 2019).

Types of intercropping system are divided into the following two kinds, based on the percent of plant population maintained for each crop in intercropping system.

1. **Additive series**: In this system, the main crop is sown with 100 percent of its recommended population in pure stand which is known as the base crop. Another crop known as intercrop (component crop) is introduced

into the base crop by adjusting or changing crop geometry. The population of intercrop shall be less than the recommended population for the pure stand. For example, growing of soybean/groundnut between two rows of maize crop in1:1 row ratio.

2. **Replacement series**: In this system of intercropping, both the crops are called component crops. By sacrificing certain proportion of population of one component, another component is introduced. For example, growing of maize + soybean in 2:2 row ratio . Various types of crop geometry for intercropping in maize has been presented as schematic diagrams. Paired row planting is an arrangement where two rows are paired by reducing the interspace so that the space between another two rows is increased. The intercrops are accommodated in that increased row area In the North-Eastern-Hill (NEH) Region of India, replacement series performs better compared to additive series.

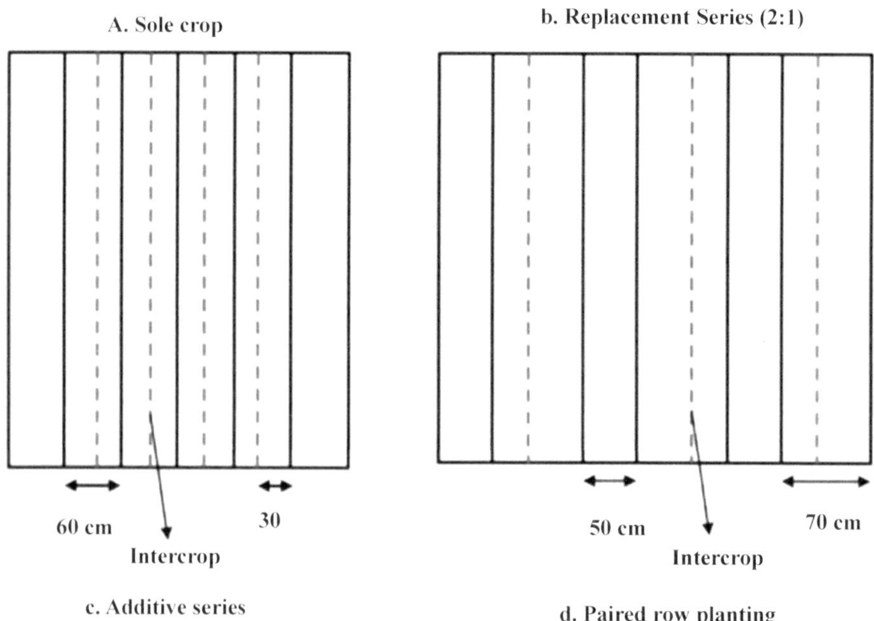

Fig.1: Additive Series (c) with normal spacing of the main crop and Replacement Series (b) with Paired-row Planting (d).

Mechanism of yield advantage in intercropping

The most important index of biological advantage is the relative yield total (RYT) introduced by Land van Den Bergh *et al* (1965) or land equivalent ratio by Willey (1979). 1. The mixture yield of a component crop expressed as a portion of its yields as a sole crop from the same replacement series is the relative yields of the crop and sum of relative yields of component crop is called Relative yield total (RYT).

The total land area required under sole cropping to obtain the same yield levels obtained in the intercropping is called Land equivalent ratio (LER). Both the expressions (RYT and LER) are similar

Land Equivalent Katio(LER)

This is the most frequently used efficient indicator. LER can be defined as the relative land area under sole crop that would be required to produce the equivalent yield under a mixed or an intercropping system at the same level of management.

$$LER = L_a + L_b = \frac{Y_{ab}}{Y_{aa}} + \frac{Y_{ba}}{Y_{bb}}$$

Where, La and Lb are LF.R of crop a and crop b, respectively; Y,h = yield of crop an in intercropping, Yh, = yield of crop b in intercropping, Y,, = yield of crop an in pure stand and YM, = yield of crop b in pure stand.

LER of more than I indicates yield advantage, equal to 1 indicates no grain or no gain or no loss and less than 1 indicates yield loss. It can be used both for replacement and additives series of intercropping.

LER is the summation of ratios of yields of intercropping to the yield of sole crops.

LER gives a better picture of the competitive abilities of the component crops. It also gives actual yield advantage of intercropping. In other words LER is the measure of production efficiency of different system by convening the production in terms of land acreage. LER gives an accurate assessment of the biological efficiency of intercropping

Example: Let the yields of groundnut and red gram grown, as pure crops are 1.200 and 1.000 kg/ha, respectively. Let yields of these cops when grown, as intercrop be 1.000 and 600 kg/ha, respectively. The land equivalent ratio of groundnut * red gram intercropping system is

$$\text{LER of groundnut} = \frac{\text{Yield of intercrop}}{\text{Yield of sole crop}} = \frac{1000}{1200}$$

$$\text{LER of red gram} = \frac{\text{Yield of intercrop}}{\text{Yield of sole crop}} = \frac{600}{1000}$$

$$LER\ of\ system = \frac{1000}{1200} = \frac{600}{1000} = 1.43$$

LE.R of 143 indicates that a 43 percent yields advantage is obtained when grown as intercrop compared to growing as sole crops. In other words the sole crops have to be grown in 1.43 ha to get the same yields level that is obtained from 1.00 ha of intercropping.

Effective Land Equivalent Ratio (ELER): Willey *et al* (2014) showed that any required ratio could be achieved by growing the intercrop of the land area and one of sole crops the remainder

Area Time Equivalent Ratio (ATER)

The LER (Land Equivalent Ratio) emphasizes the only land area without considering the time factor, for which the crop occupies the field. As time factor is not a part in the LER, researchers needed another expression considering the field occupancy by the crops in an intercropping to correct this constraint of the LER.

The developed the concept of Area Time Equivalent Ratio (ATER) has the duration of crops (starting from seeding to harvest). The ATER is calculated by the following formula.

$$ATER = \frac{(RYc \times tc) + (RY_p \times t_P)}{T}$$

where, RY = Relative yields of crop species "c" and "p" = Yield of intercrop ha 1 / Yield of sole crop ha t = duration (in days) for species "c" and "p" and T = duration (in days) for the intercropping system. However, the LER generally overemphasizes and the ATER. Researchers revealed the advantageous ATER values in different intercropping system

Table: 1. Area time equivalent ratio (ATER) in maize-legume intercropping systems

Intercropping System	Proportion	ATFR	Country
Cotton * Cow pea	-	1.13	Pakistan
Lupine + Wheat	75% + 100%	131	Ethiopia
Mai/e + Soybean	2:6	132	India
Maize + Black cowpea	2:2	131	India
Pearlmillet + Green gram	2:1	1.25	India
Wheat + (aha bean	-	1.28	Pakistan
Potato + Dolicho>	1:2.4	1.13	Kenya

(*Source:* Maitra, S. *et al.*, 2021)

Competitive Ratio (CR)

In an intercropping system, competitive ratio (CR) denotes the competitive ability of the component species. The CR expresses the number of times by which one component crop is more competitive than other (Willey *et al* 2014) actually represents the proportion of individual LERs of the crops considered in intercropping and also takes into account the ratio of the crops sown in a mixed stand.

The CR can be calculated by the following formulae.

CRa = (LERa/LERb) × Zba/Zab) (6)

CRb = (LERb/LERa) × (Zab/Zba)

where, CRa and CRb are indicative of the competitive ratios of the crop species "a" and "b" and LERa and LERb are the LERs of the crop species "a" and "b" respectively. Zab is the sown ratio of species "a" in mixture with "b" and Zba is the sown proportion of the species "b" in mixture with "a". If the value of CR is 1), there is a negative impact.

In this condition, the competition between intercrops in mixture is too high, and they are not recommended to grow as intercrops.

Table 2: Competitive ratio (CR) of finger millet + legume intercropping (4:1) systems

Intercropping Systems	Competitive Ratio (CR)	
	Finger Millet	Legumes
I'inger millet + Red gram (4:1)	0.28	3.59
Finger millet + Green gram (4:1)	0.71	1.41
Finger millet + Groundnut (4:1)	0.58	1.73
Finger millet + Soybean (4:1)	0.68	1.48

The values of CR of legumes appeared as >1, representing that legumes were more competitive than finger millet. Among the legumes, green-gram was found to be the least aggressive on affecting the growth of finger millet and thus it provided a balanced competition with finger millet.

The interaction between different components crops

The interaction may be 1. Competitive 2. Non-competitive and 3. Complementary

1. **Competitive Interaction**: - One species may have greater ability to yield higher by limiting the yield of the companion crop and hence, will gain at the expense of the other, which is called as competitive interaction or interference. Alternatively, when one or more growth factors of one

crop unit are limiting, the other species that is better equipped to use the limiting factors(s) will gain at the expense of the other and this is called as competitive interaction. In mixed crop communities, if the associated species are to share their growth from a limited pool of recourses such as light, water or nutrients, then it is non-competitive interaction or interference.

2. **Non-competitive:** - If the crops are grown in association and the growth of either of the concerned species is not affected, such type of interaction is called non-competitive interaction or interference. Or if these resources (growth factors) are present in adequate quantities, as a result of which, the growth of either of the concerned species is not affected, then it is non-competitive interaction or interference.
3. **Complementary**: - If one species is able to help the other, it is known as complementary interaction. Or if the component species are able to exploit to supply of growth factors in different ways (temporal or spatial) or if one species is able to help the other in supply of factor (like legumes supplying part of N fixed by symbiosis to non-legumes), it is complementary interaction or positive interference. This also referred to as Annidiation

Resource use in intercropping system

a) Solar Radiation

The taller crop in the intercropping systems intercepts most of the solar radiation while shorter component suffers. In some intercropping systems, both crops utilize solar radiation efficiently. In Groundnut + pigeon pea intercropping system, light interception is prolonged as red gram starts growing after harvesting of groundnut. If the component crops have different growth durations the peak demand for light occurs at different times. In maize + green grams intercropping system, green gram flowers in 35 days after sowing and is harvested 65 days after sowing, peak light demand for maize occurs at 60 days after sowing when green gram is ready for harvest. In such intercrops, there is less competition among component crops and higher solar radiation is intercepted in intercropping system than in pure crops. Proper choice of crops and varieties, adjustment of planting density and pattern are the techniques to reduce competition and increase the light use efficiency.

1. When one component is taller than the other in an intercropping system, the taller component intercepts greater share of light. As a result, the

growth rates of the two components will be proportional to the quantity of PAR they intercept provided other growth factors are no limiting and the crops are in their vegetative.

2. The inclination of leaves greatly influences the amount of light intercepted by the taller component and the amount that is available to shorter components. For example, one unit of LAI of prostrate-leaved perennial rye grass (*Trfolium ripens*) absorbed 50% of the incoming light whereas some LAI of erect leaved perennial rye grass (*Lolium pereme*) absorbed only 26% (Breugham, 1958).

3. Light interception in a sorghum (*Sorghum bicolor*) based intercropping system was studied by Selvaraj (1978). Light was measured influx units from 45th-90th day after sowing sorghum at 15-day interval. Light interception was expressed as percentages of light on the top of the canopy of each crop. The red gram/sorghum intercropping system makes better use of growth resources, particularly light, was brought out by Willey *et al.* (1980). Ideally the taller component should have more erect leaves and the shorter component more horizontal leaves. If these are to be planted in alternate rows, there would be less competition for light. Under intercropping situations, the component crops are grown in such a way that competition for light is minimized (Okigbo, 1981); this can be achieved by proper choice of crops and genotypes, the shorter components being harvested sufficiently early so that the later harvested component is not greatly affected.

b) Moisture and nutrients

Competition for moisture and nutrients may result in two types of effects on the less successful components. First, the roots of this component may grow less on the sides towards plants of aggressive component. Secondly, plants affected by competition for soil factors may have increased root/shoot ratio. The aggressive component generally absorbs greater quantity of nutrients and soil moisture. In legume and non-legume combination, the latter takes up large amount of P, K and S. As a result, the legume may show deficiency of these nutrients. Such effects, however, may be mitigated by appropriate fertilizer application. Among intercrops, sorghum and pearl millet are more competitive in extracting nutrients. Generally, the intercrop stands remove greater amount of nutrients than sole crop stands.

c) Other Interactions

Allelopathy is any direct or indirect harmful effect that one plant has on another through the release of chemical substances or toxins into the root environment. Some crops may be unsuitable to be grown as inter- crops because they may produce and excrete toxins into the soil, which are harmful to other components. Allelochemical produced from the leaves of *Eucalyptus globules* drastically reduces the germination of mustard sown underneath. On the contrary, negative allelopathy stimulates the growth of this associated crop by release of hormone like substances, is also possible (Ercisly *et al.*, 2005). But release of N from root nodules of legume is not considered to be a form an allelopathy. The chemical released by one species may inhibit species of plants other than the one releasing it (allow inhibition) or may inhibit more strongly plants of the producer species itself (auto inhibition). Toxic substances may be converted into active substances by some microorganisms (functional allelopathic). The type and quantity of allelochemical produced will vary depending on the environment and genetic makeup of the plant. Some allelochemical may be produced by the aerial portion of the plant and may reach the ground through raindrops, falling leaves or insects, inhibiting the growth of the species growing underneath. Allelochemical produced from the leaves of *Eucalyptus globules* drastically reduced the germination of mustard (*Brassica* spp.) seed sown underneath (Trenbath, 1976).

Many plants exude organic substances from their roots and some of these roots exudates act as allelochemicals inhibiting the growth of the neighboring species. Living roots of walnut (*Juggles nigra*), cucumber (*Curcumis sativa*) and peach (*Prunes persia*) are known to exude toxic substances which inhibit the growth of the plants growing near them.

Annidiations: Annidiations is complementary use of resources by exploiting the environmental supplies in different ways by the component of a community. Or it refers to complementary interaction which occurs both in space and time. a) Annidiations in Space: The leaf canopies of component crops may occupy different vertical layers with taller component tolerant to strong light and high evaporative demand and shorter component favoring shade and high relative humidity. Thus, one component crop helps the other. Multistoried cropping in coconut and planting of shade trees in coffee, tea and cocoa plantation use this principle Nimbolkar *et al.* (2016). Similarly, root system of component crops exploits nutrients from different layers of soil and thus utilizing the resources efficiently. Generally, one component with shallow root system and another with deep root system are selected for intercropping as in *Setaria* (shallow) + red gram (deep) intercropping system.

Annidiations in Time

When two crops of widely varying duration are planted, their peak demands for light and nutrients are likely to occur at different periods, thus reducing competition. When the early maturing crops are harvested, condition becomes factorable for the late maturing crops to put forth its full vigor. This has been observed in sorghum + red gram, groundnut + red gram and maize + green gram intercropping system.

Other Complementary Effect in Intercropping Systems

In an intercropping system, involving as legume and non-legume, part of the nitrogen fixed in the root nodule of the legume may become available to the non-legume component. The numerous reports of such beneficial effect of legumes on non-legumes are available in literature, Palaniappan *et al* 1976, (Sundarajan and Palaniappan 1979). The presence of rhizosphere micro flora and mycorrhiza, one species may lead to mobilization and greater availability of nutrients not only to the species concentrated, but also to the associated species. Another example is the provision of physical support by one species to the one intercropped is climbing species may improve the yield of the climber. e.g. coconut + pepper and maize + beans. The taller component acts as wind barrier protecting the short crop as in maize + groundnut, onion + castor, turmeric + castor.

Overall Effects of Competition

Three broad categories of competition can be recognized: 1. Actual yield of each species is less than expected. This is called mutual inhibition. This is rare. 2. The yield of each species is greater than expected. This is called mutual cooperation. This cannot be unusual. 3. One species yield is less and the other is more than expected. This can be termed compensation. The species which yields more than expected is believed to have greater competitive ability and called the dominant species. The other species is called the dominated species.

Legume effect in Intercropping Systems

Legumes are widely used for food, fodder, shade, fuel, timber, green manure and cover crops. Legumes either increase the soil N status through fixation, excretion or in absence an effective N fixing system. Thus they have potential for self-sufficiency for N, the nutrient most limiting to productivity. Nutrient self-sufficiently is desirable characteristics of agronomically sustainable cropping system. Inclusion of legumes in intensive cropping systems has many ramifications. They are less demanding on soil resources, many of them can tolerant some amount of shading, fix atmospheric N in root nodule, contributing part of N to associated crop and improving soil fertility, capable of extracting less soluble forms of soil P and K, thus, making it available to other crops and also complete better for higher valence cations like Ca and Mg due to more CEC of roots of legumes. But in low K soils they are likely to be deprived of their due share of K, especially in mixture with cereals. Hence with the overall view of maintaining soil fertility and economizing on fertilizer, it is beneficial to include legumes as components of intensive cropping systems

The quantity of N fixed by the legume component in cereal legume intercropping depends on the species, morphology, and density of legumes in mixture, type of management and competitive abilities of the component crops. Legumes of indeterminate growth are more efficient in terms of N fixation than the determinate types shading by the cereals reduces both the seed yield and N fixation potential of the companion legume.

In sequential process the preceding crop has considerable influence on the succeeding crop mainly due to i) changes in soil condition, ii) complementary effect such as release of N from the residues of the previous crop, particularly legume, iii) presence of allelopathic chemicals, iv) shift in weeds, v) temporary immobilization of N due to wide C: N ratio of these residues and vi) carry over effect of fertilizer, pest and diseases.

In terms of land use, the practice of inter-cropping is more productive than growing them separately, It is traditionally practiced in rain-fed regions of India since component crops give higher yields in combination, ensures land use efficiency per unit time, improves the soil fertility and uses resources more efficiently. The productivity of inter cropping systems is mainly dependent on complementarity of component crops. The network of ACRIPDA centres have developed efficient inter cropping system by optimizing the row ratios, and component crops to minimize the competition between the component crops for higher biological productivity (Table 5).

Table 3: Efficient inter-cropping systems and row ratios for rain-fed agriculture

Production System/ Centre/	AESR	Annual Rainfall (mm)	Soil Type	Predominant and profitable intercropping systems
I. Rice Based				
Phulbani	12.1	1174	Oxisols	Rice + Pigeonpea (4:1), Rice + Radish (4:2), Rice + Okra (4:2), Rice + Blackgram (1:2), Castor + Tuber crops (1:1), Maize + pigeonpea (2:1), Pigeonpea + radish (2:1)
Ranchi	12.3	1370	Oxisols	Rice + Pigeonpea (1:3), Rice + Sesame (1:1), Rice + maize (1:2), Pigeon pea + Okra (1:1)
Varanasi	4.3/9.2	1080	Entisols	Pigeonpea + Okra (1:2), Lentil + tomato (1:1), Lentil + mustard (4:1), Barley + mustard (6:1), Maize + blackgram (1:1), pigeonpea + blackgram (1:1)
Faizabad	9.2	1070	Entisols	Mustard + chickpea (1:4), pigeonpea + groundnut (1:5)
II. Maize Based Production Systems:				
Rakh Dhiansar	14.2	1100	Inceptisols	Maize + okra/Roundgourd (1:1), Gobi sarson + Oat (1:1), Wheat + mustard (4:1), Barley + chickpea (2:2) Chickpea + mustard (4:1)
Arjia	4.2	862	Vertisols	Maize + blackgram (2:2), maize + pigeonpea (1:1), Groundnut + sesame (6:2), castor + greengram (1:2), Chickpea + mustard (4:1), chickpea + safflower (2:1)
III. Oilseed Based Production Systems:				
Indore	5.2	964	Vertisols	Maize + soybean (2:2), Soybean + Pigeonpea (2:2)
Rewa	10.3	1048	Vertisols	Sorghum + pigeonpea (2:1), Soybean + pigeonpea (4:2)
Rajkot	2.4	592	Vertisols	Groundnut + castor (3:1), Groundnut + pigeonpea (2/4:1), Pearlmillet + castor (2/4:1)
Anantapur	3.0	643	Alfisols	Groundnut + pigeonpea (5:1), Groundnut +pigeonpea (mixed with other pulses like cowpea, greengram, horsegram and field bean) (7:1)
IV. Cotton Based Cropping Systems				
Akola	6.3	825	Vertisols	Sorghum + greengram (2:1), Sorghum + pigeonpea (2:1), Cotton + greengram (1:1), Cotton + pigeonpea (2:1), Pigeonpea + greengram (1:1)
Kovilpatti	8.1	780	Vertisols	Sorghum + blackgram (2:1), Sorghum + cowpea (2:1) Cotton + blackgram (2:2)

Production System/ Centre/	AESR	Annual Rainfall (mm)	Soil Type	Predominant and profitable intercropping systems
IV. Nutritive Cereal Based Cropping Systems				
a) Sorghum based				
Jhansi	12.4	936	Alfisols	Sorghum + pigeonpea (2:1), Rabi sorghum + chickpea + safflower
Solapur	6.1	561	Vertisols	Pearl millet + kidney bean/ horse gram (2:1), Sunflower + pigeonpea (2:1) chickpea + safflower (3:1), pearl millet + pigeonpea (2:1), pearl millet + moth bean (for shallow soils)(2/3: 1), Sorghum + pigeonpea (1:1)
Bijapur	3.0	622	Vertisols	Pigeonpea + pearl millet (1:3), chickpea + safflower (3:1), Chickpea + sorghum (1:2), pigeonpea + ground nut (1:3)
Agra	4.1	598	Oxisols	Pigeonpea + green gram (1:1), Chickpea + mustard (4:1), barley + chickpea (3:2)
Dantiwada	2.3	844	Loamy dessert	Pearl millet + greengram (3:1), pearl millet + cluster bean (2:1),
Hissar	2.3	769	Eridisols	Pearl millet + mung bean (2:1), pearl millet + cluster bean (2:1)
Bangalore	8.2	696	Alfisols	Finger millet+ pigeonpea (10:1), Finger millet + soybean (1:1), ground nut + pigeonpea (8:2)
Fodder based production system				
Rahuri	6.1	600	Clay loam soils	Sorghum + cowpea (2:2)
Jhansi	4.4	902	Alfisols	Guinea + siratro

(Subba Reddy and Maruthi1996)

Types of Inter-Cropping

While planting different species together, farmers should consider their density, architecture, required period for maturing, irrigation, sunlight, and nutrient needs. Intercropping involves a certain arrangement of plants that gives the basis for its classification. Thus, intercropping types fall into the row, strip, relay, temporal, mixed, guard, alley, trap, repellent, and push-pull cropping among others.(EOS, 2020).

Row intercropping: As the name suggests, plants are arranged in rows in this case. A common and beneficial combination is cereals with legumes like corn and beans. Row ratios may vary, including either single or multiple rows. A study recommends four maize rows with six soybean ones. The benefits of row cropping here include additional nitrogen fixation by legumes in symbiosis with the bacteria of the Rhizobium genus. It is nothing but growing two or more crops simultaneously where one or more crops are planted in rows.

Mixed intercropping: The intercropping practice involves sowing different species (two or more) in one terrain with no distinct arrangement in rows or in the same rows. In this case, the time to sow and harvest coincides. Mixed cropping gives additional protection to the primary culture from winds, frosts, droughts, and other severe weather conditions

Strip intercropping: Growing two or more crops simultaneously in different strips wide enough to permit independent cultivation but narrow enough for the crops to interact ergonomically. The practice is similar to row cropping, but the sections of land are wide enough to facilitate mechanical operations and to enable independent cultivations.

A popular pattern of strip cropping in the U.S. is growing wheat, corn, and soybeans alternatively, six rows each.

Fig. 2: Strip intercropping of wheat and soybean

Relay intercropping: Growing two or more crops simultaneously during part of the life cycle of each. A second crop is planted after the first crop has reached its reproductive stage but before it is ready for harvest. Inter-cropping may be divided into the following four groups (Singh 1990) Farmers practicing this technique **plant one or two more species at the same piece of land but at different times**, after one crop has flowered. Relay intercropping reduces temporal overlap in harvesting different species, but the second crop must be tolerant of the shade of the first one. Examples of the relay cropping system are cotton and corn or chickpea and upland rice.

i) **Parallel Cropping**: Under this cropping two crops are selected which have different growth habits and have a zero competition between each other and both of them express their full yield potential. e.g. 1) Green gram or black gram with maize 2) Green gram or soybean with cotton.

ii) **Companion Cropping**: In companion cropping the yield of one crop is not affected by other. In other words, the yield of both the crops is equal to their pure crops, since the standard plant populations of both crops are maintained. e.g.1) Mustard, wheat, potato, etc. with sugarcane 2) Wheat, radish, cabbage, sugar-beet etc., with potato.

iii) **Multistoried Cropping or Multi-tire cropping**: Growing plants of different height in the same field at the same time is termed as multistoried cropping. It is mostly practiced in orchards and plantation crops for maximum use of solar energy even under high planting density. e.g. 1) Eucalyptus + Papaya + Berseem, 2) Sometimes it is practiced under field crops such as Sugarcane + Potato + Onion 3) Sugarcane + Mustard + Potato 4) Coconut + Pineapple + Turmeric/Ginger

Alley cropping/Hedge row intercropping:

The alley cropping system suggests growing crops in-between trees, bushes, or hedges forming alleys. The aspects of alley cropping are as follows. Higher plants protect the lower ones from winds and shelter from extra sunlight as well as prevent soil erosion with their vigorous root systems.

It is a type of intercropping where arable crops are grown in between the alleys (interspace) formed by the two hedge rows or leguminous shrub rows. It is also called as **hedge row intercropping**. Depending upon the slopes, plant species involved, the alley width may vary from 2 to 5 m. In north-east, leguminous shrubs like *Crotolaria, Tephrossia, Casia, Acacia,* Cajanus cajan, Flemingia Indigofera etc are suitable as alley crops or hedge row crops. Ginger, turmeric, maize etc are grown in between the alley. The alley height is generally maintained at about 1 m by pruning periodically and the pruned biomass is either used as mulch or incorporated in to the soil as a source of nutrients. Intercropping in interspace of hedgerow is a proven and sustainable technology for the NEH Region. The hedge-row grown in the contours helps in developing natural terrace in few years' time. This system of cultivation reduces erosion and conserves natural resources.

Fig. 3: Alley Cropping or Agroforestry.

Fig. 4: Maize grown in between *Tephrosia* hedge rows

Trap Cropping

As the name hints, the intercropping technique **aids in trapping pests to protect the main culture**. Popular examples of trapping plants are mustard, marigold, among others. The basic idea is to attract insects or fungi to the sacrificial secondary crops, thus protecting the cash one. Blue Hubbard squash is reported to be an efficient remedy from squash bugs, squash vine borers, spotted and striped cucumber beetles.

Trap intercropping allows saving on pesticides with no chemical application at all or partial treatment of trapping areas. However, possible disadvantages of intercropping, in this case, include developing resistance to insecticides or uncontrolled pest nurseries. (EOS, 2020).

Intercropping has significant advantages over monoculture farming, which are aimed at boosting yields and more efficient usage of land and resources.

The most fundamental intercropping benefits include the following:

- **Increased profit**. Secondary crops provide more returns and ensure profit even when the primary crop fails.
- **Ergonomic usage of land**. Planting species in-between rows allows utilizing the soil in a more efficient way, unlike mono cropping when spaces between rows are unused.
- **Crop protection from biotic stresses**. Intercropping performs several functions, either repelling or trapping pests and attracting beneficial ones as well as protecting from winds or giving shelter from extra sunlight. Pest management in this case is characterized by the reduction of the use of chemical applications and thus saves costs.
- **Prevention of soil erosion**. Plants between rows and in alley intercropping, in particular, mitigate erosion with their roots.
- **Added nutrients for the main crop**. The leguminous family is known for nitrogen fixation and thus provides nitrogen for the neighboring species.
- **Reduction of fertilizer applications**. When intercropping cultures contribute to soil fertility, they spare the necessity to apply synthesized fertilizers.
- **More efficient use of natural resources** like water and solar energy as they are distributed to secondary crops as well.
- **Improved** weed management. Beneficial plants, not weeds, occupy vacant spaces between rows in intercropping.
- **Enhanced biodiversity and ecological stability**. The more agricultural species grow, the better it is for the environment.

Sustainability of intercropping systems

In rain-fed agriculture sustainability of yield becomes more important, than " sample mean " as the magnitude of yield is rainfall dependent. While considering rainfall yield index, it is observed, rice + radish (4:2) and rice + black gram (4:2) in Oxisols of Phulbani, Pigeon pea + okra (1:1), Maize + Okra in dry sub humid inceptisols of Ranchi, Sorghum + Cowpea (F) under delayed sowing conditions of Arjia, Pearl millet + pigeonpea in Vertisols of Solapur, Sunflower + pigeon pea in Vertisols at Indore and green gram + castor in semi-arid entisols of Danthiwada recorded high sustainability over other systems. (Table 4)

Table 4: Stability of inter cropping systems in rain-fed areas

Centre	Intercropping	Yield (Kg/ha)	SI
Rice based			
Varanasi	Rice + Radish (4:1)	9068	0.80
	Rice + Radish (2:1)	8217	0.72
	Rice + Okra (4:2)	3471	0.27
Phulbani	Sole Okra	7260	0.81
	Rice + Okra	6701	0.75
	Sole Rice	2803	0.28
Ranchi	Pigeon-pea + Okra (1:2)	3110	0.63
	Sole Okra	4025	0.84
Nutritive Cereal based			
Arjia	Sorghum + Cowpea (F)	2267	0.64
	Castor	1943	0.52
	Maize	935	0.16
Groundnut based			
Solapur	Pearl millet + Pigeonpea	875	0.62
Indore	Sunflower + Pigeonpea (4:2)	3303	0.68
	Maize + Pigeonpea	2895	0.39
	Sole Soybean	890	0.73

Generally, these advantages are more pronounced in stress environments. In addition, it helps to spread labour peaks, maintain soil fertility (with inclusion of legume) and stability in production. Furthermore, greater efficiency of resource utilization is expected from intercropping in a wide range of environments. In general, intercropping with additive series was found better than replacement series under most of drought situations (AICRPDA, 2003). The mean LER of additive series was 23% higher than replacement series in 54 out of 59 experiments taking sorghum, maize, pearl millet, pigeon pea, safflower and wheat as base crop. However, intercropping systems were more favorable in kharif than rabi season in Indian rain-fed regions probably due to replenishment of soil moisture during Kharif season (AICRPDA, 2002 - 2006).

Management of intercropping systems

1. Seedbed Preparation

The seedbed preparation is generally done as per the needs of base crop. Deep rooted crops respond to deep ploughing while for most of the cereals shallow tillage is sufficient. The crops with small seed require fine seedbed. Certain crops like cotton and maize are planted on ridges, while most of the other crops are grown on flat seedbed. The seedbed for sugarcane, as usual, is made into ridges and furrows. Sugarcane is planted in furrows and intercrops are sown on ridges. In groundnut + red gram intercropping system, flat seedbed is preferred for sowing the crops.

ICRISAT recommends broad bed and furrows for black soils of semi-arid regions for pure crops as well as intercrops grown under rain-fed conditions. Where the crop requirements are quite different as in rice + maize under rain-fed conditions and also in agroforestry, forestry, seedbed preparation is done separately for component crops. In rice + maize intercropping system, ridges and trenches are formed. Maize is planted on ridges and rice in trenches. In agro-forestry, pits are dug for tree species and a rough seedbed is prepared in interspace for the introduction of forage crops.

The varieties of component crops should be less competing with the base crop and the peak nutrient demand period should be different from the base crop. The difference in duration between between the components in intercropping should be a minimum period of 30 days (Maize + soybean; Sorghum + red gram; Toria + Gobhi sarson). Selection of compatible genotypes of component crops increases the complementarity of intercropping system.

2. Crop Varieties

The varieties of component crops should be less competitive with the base crop and the peak nutrient demand period should be different from the base crop. The difference in duration between the components in intercropping should be a minimum period of 30 days (Maize + soybean; Sorghum + red gram; Toria + Gobhi sarson). Selection of compatible genotypes of component crops increases the complementarity of intercropping system. The varieties selected for intercrop should have thin leaves, tolerant to shading and less branching since these crops are generally shaded by the base crop. If the base crop is shorter than intercrop, the intercrop should be compact with erect branching, branching, and its early growth should be slow.

3. Sowing

Sowing of base crop is done either as paired row (20/50 or 20/40 or 30/60), paired wider row or skip-row planting. The sowing of base crop and intercrop is also done in fixed ratios. The intercropping system of groundnut + pigeon pea is either in 5: 1 or 7: 1 ratio and sorghum + pigeon pea in 2: 1 ratio. In traditional cropping systems, systems, the component crops are grown with sub-optimum population. Base crop population is maintained at its sole crop population and intercrop population is kept at 80% of its sole crop population. When the difference in duration of component crops is less than 30 days, staggered planting is done to increase the difference in duration. The aggressive or dominant crop is sown 10 to 15 days after sowing the dominated crop.

4. Nutrient Management

Nitrogen is an essential constituent of chlorophyll and imparts a green color to the leaves. The higher availability of nutrients especially N, in the initial stage, helps in acquiring a definite advantage over other treatments in respect to growth. Nitrogen application also helps the plant in making quick growth and vigorous development regarding root and shoot dry weight. Supply of N in adequate amount and available form has a high degree of positive correlation with crop productivity.

When a cereal crop is grown in intercropping with a legume, the cereal may be benefitted by direct N transfer from the legume through BNF (Giller and Wilson 1991). As legumes have the capacity to fit in different cropping patterns and fix N from the atmosphere in the soil, they may offer opportunities to increase the productivity of the intercropping system (Jeyabalan and Kuppuswamy 2001). This may be due to soil fertility enhancement either through the supply of biologically fixed N or root excretion from the associated legume crop. When legumes are associated with cereal crop in intercropping system, a portion of nitrogen requirement of cereal is supplemented by the legume. The amount may be as small as a few kilograms to 20 kg/ha. Cereal + legume intercropping, is therefore; mainly advantageous under low fertilizer application. Considering all the factors, it is suggested that the nitrogen dose recommended for base crop as pure crop is sufficient for intercropping system with cereals + legume or legume + legume.

With regard to phosphorus and potassium, one-eighth to one-fourth of the recommended dose of intercrop is also added in addition to recommended dose of base crop to meet the extra demand. Basal dose of nitrogen is applied to rows of both components in cereals + legume intercrop system. Top dressing of nitrogen is done only to cereal rows. Phosphorus and potassium are applied as basal dose to both crops.

The amount of fertilizer to be applied in a cereal intercropping system is varied with the site and types of component crops. Application of 75% recommended dose of fertilizer (RDF) to maize and 50% of soybean reported to significantly increase the productivity of the intercrops, maize equivalent yield, and profit from the system over supply of 50% RDF to maize and soybean (Manasa *et al*., 2018)

In maize+cowpea intercropping system, cowpea was reported as the best cover crop which decreased soil disintegration than a maize-bean sequence (Kariaga, 2004). Intercropping of sorghum+cowpea reduced soil loss by 50% against growing them separately crops are well known for enriching the soil by supplying N through the process of biological nitrogen fixation (BNF), especially when N fertilizer is restricted (Fujita and Ofosu-Budu 1996). However, the nitrogen fixation in legume intercropping system depends on type of legumes grown, the crop morphology, plant density, cultivation practices followed, nitrogen fixating capacity and aggressiveness of component crops. The legume crop modifies the carbon: nitrogen (C:N) ratio and enhances the activity of soil enzyme, as a result conversion of unavailable to available form of nutrients is also increased.

Pulses also play an important role for improving the microbial environment in the soils (Meena *et al*., 2018). Some legume crops like soybean, common bean, cowpea, lablab, groundnuts etc. act as an important host for these microorganisms to perform biological nitrogen fixation. They are also reported to release a part of unused nitrate fixed through symbiotic nitrogen fixation to the soil (Herridge *et al*., 1995).

Interestingly, it was reported that about 50–60% of soybean N demand was met by biological N2 fixation (Salvagiotti *et al*., 2008). The results after twenty years of experimentation at Akola, Maharashtra revealed that significantly highest yield of cotton and green gram with build up of soil fertility and maximum economic returns were obtained with the application of 25 kg N ha-1 and 25 kg P2O5 ha-1 through Urea and single super phosphate, respectively and 25 kg N/ha through FYM over the years in cotton and greengram intercropping system [Gabhane *et al*. 2013)

Satyanarayana, (1997) recorded increase in the available phosphorus content of soil at harvest due to incorporation of subabul, FYM and RDF when compared to control in black soil of Bijapur (Karnataka) It was noticed that significantly increased uptake of N, P and in soybean with application of RDF + FYM at 5 t ha-1 or RDF + vermicompost at 2.5 t ha-1 as compared to RDF alone

5. Water Requirement

The technique of water management is the same for sole cropping and intercropping or sequential cropping. However, the presence of an additional crop may have an important effect on evapotranspiration. With proper water management, it is possible to grow two crops where normally only one crop is raised under rain fed condition. Intercropping system is generally recommended for rain fed situations to get the stable yields.

The total water requirement of intercrop does not increase much compared to sole cropping. At ICRISAT, the water requirement of sole sorghum and intercropping with pigeon pea was almost similar (584 and 585 mm, respectively).

6. Weed Management

Generally, it is believed that intensive cropping reduces weed problems. Weed infestation depends on the crop, plant density and cultural operation done. Weed problems are less in intercropping system compared to the sole crops. This is due to complete crop cover because of high plant density in intercropping which causes severe competition with weeds and reduce weed growth. The weed suppressing ability of intercrop is dependent upon the component crops selected, genotype used, plant density adopted, proportion of component crops, their spatial arrangement and fertility moisture status of the soil. Experiment carried out at ICRISAT, Hyderabad, indicated that there was 50 - 75 % reduction in weed infestation by intercropping of Pigeon pea + sorghum , which is extensively practiced in Karnataka, M. P .and A.P is known to reduce weed intensity.

In pearl millet + groundnut intercropping system type of weeds changes with proportion of component crops. As more rows of groundnut are introduced in place of pearl millet of rows, there is a striking increase in both numbers and biomass of the tall and competitive *Celosia*, especially in groundnut rows.

In certain situations, intercrops are used as biological agents to control weeds. Black gram, green gram, cowpea in sorghum and cowpea in banana reduce weed population and one hand weeding can be avoided by this method. In some intercropping systems like maize + groundnut, rice + tapioca, maize + tapioca, weed problem is like their sole crops. The growth habit of genotype used in intercropping has a great influence on weed growth. Weeds present in sole crops are different than those present in intercropping systems. Though weed problem is less, weed control measures are necessary in intercropping system. But the labour required for weeding is less. Second weeding is not necessary because of crop coverage. Chemical weed control is difficult in intercropping system because the herbicide may be selective to one crop, but non-selective to another.

Table 5. Herbicides suitable for different Intercropping systems

Herbicide	Time of application	Intercropping system
Pendimethalin or alachlor	Pre-emergence	Maize + greengram; Maize + cowpea
Trifluralin	Pre-plant incorporation	Maize + groundnut
Fluchloralin	- do- -	Sorghum + pulse
Alachlar	Pre-emergence	Maize + soybean ; Maize + cowpea ; Sorghum + pulse
Dinitramine /Ametryne	Pre-emergence	Sorghum + lablab
Prometryne /Terbutryne	Pre-emergence	Sorghum + redgram
Nitrofen	Pre-emergence	Sugarcane + groundnut;
		Sorghum + pulse

7. Management of Biotic Stresses

Pest and diseases are believed to be less in intercropping system due to crop diversity than sole crops. Some plant combination may enhance soil fungicide effects and as well as antibiotic effects through indirect effects on soil organic matter content. The spread of the diseases is altered by the presence of different crops. Little leaf of Brinjal is less when Brinjal is sheltered by maize or sorghum, as the insect- carrying virus first attacks maize or sorghum; virus infestation is less on Brinjal. Non – host plant in mixtures may emit chemicals or odor that affects the pests, thereby protecting host plants. The concept of crop diversification for the management of nematode population has been applied mainly in the form of decoy and trap crops. Decoy crops are non-host crops, which are planted to make nematode waste their infection potential. This is affected by activating larva of nematode in the absence of hosts by the decoy crops.

Table 6. Use of Decoy Crops to control nematode pests in intercropping

Crop	Nematode	Decoy crops
Brinjal	Meloidogyne incognita	Sesamum orientate
Tomato	Meloidogyne pratylenchus alleni	Caster, groundnut
Soybean	Pratylenchus sp	Crotalarias spcctabills

Trap crops are host crops sown to attract nematode but destined to be harvested or destroyed before the nematode manage to hatch. This is advocated for cyst nematode. The technique involves is sowing in pineapple plantations; tomatoes are planted and ploughed in to reduce root knot nematodes. There is also evidence that, some plants adversely affect nematode population through toxic action. Marigold reduces the population of *Pratylenchus* species

8. Soil erosion control

Plant cover in intercropping plays an important role in stopping energy from rainfall and prevent runoff which could cause soil erosion. It is known that, cereals have the capacity to stop erosion and legumes can fertilize soil by fixing biological N and together they play complementary role (Thyamini, 2010). Kariaga (2004) showed that in maize-cowpea cropping system, cowpea acts as a good cover and decreases run off than maize-bean system. Rana and Rana (2011) found that taller crops act as wind barrier for short crops, in intercrops of taller cereals with short legume crops. However, sorghum cowpea cropping system decreases erosion by 20-30% than sorghum mono crop by 45-55% compared to cowpea monocrop. However, Kinama *et al.*, (2007), Kinama *et al.*, (2011) found that, intercropping maize Senna and senna-cowpea reduced soil erosion compared to mono-cropped plots.

9. Agronomic relevance of Intercropping systems

On the fringes of modern intensive agriculture, intercropping is important in many subsistence or low-input/resource-limited agricultural systems. By allowing genuine yield gains without increased inputs, or greater stability of yield with decreased inputs, intercropping could be one route to delivering 'sustainable intensification' growth (Brooker *et al.*, 2015).

Proven intercropping systems of a dryland tracts offer advantages such as insurance against crop failure in the case of aberrant weather, sustainability of dryland crop production, reduction in the use of chemical fertilizers, reaping the benefits of symbiotic interaction and complementarity of the crop components of the system, conservation agriculture on drylands, effective weed, crop pest and disease management, availability of more than a crop for marketing and domestic consumption and efficient utilization of resources viz., light, water and crop nutrients (Brooker *et al.*, 2015)

Further, the practice of inter-cropping has been proven to provide a rich, biodiverse environment, fostering habitat and species richness for animals and beneficial insect species including butterflies and bees. There is evidence to suggest that poly-cultural fields produce higher yields as compared to mono-cultural fields in some situations, and almost always increase biomass richness over time. Polyculture in forests, heathlands, grasslands, and marshes has been particularly important for the regrowth of biodiversity in Europe (Hirst 2019).

Over-yielding occurs when the yield produced by an intercrop is larger than the yield produced by the component crops grown in monoculture on the same total land area. Over-yielding is calculated using the Land Equivalency Ratio

(LER), which is a measure of how much land would be required to achieve intercrop yields with crops grown as pure stands.

When the LER is greater than 1, over-yielding occurs and as such, the intercrop is more productive than the component crops grown as sole crops. When the LER is less than 1, no over-yielding is observed, when the sole crops are more productive than the intercrop.

Higher yield occurs for a variety of reasons, including the following:

- **Weed suppression and least susceptibility to biotic stresses** are due to the very nature of the intercropping system containing the complementary crop components
- **Complementary resource use.** A mix of diversity involved in plants will use resources more efficiently than plants that are all of the same type. Plants of varying types may also provide benefits to each other, such as fixed nitrogen from legumes

(https://www.umanitoba.ca/outreach/naturalagriculture/articles/intercrop.html)

Competition, allelopathy, and biotic stress management are some of the major issues involved in intercropping. Plant interaction demands particular control, which is a time-consuming process and a difficult task for a human eye. In this regard, satellite-based (Eos,2020) Crop Monitoring software provides precious information on the overall crop health in the field.

Vegetation indices aid in monitoring growth stages and detecting anomalies since each particular crop has specific values at a certain phenological stage. This correlation allows farmers to make weighted agricultural decisions. Thus, the sunflower is supposed to give a high NDVI index while flowering. Correspondingly, if it is low, which means that things might go wrong, and the area should be checked.

Recent advances in agronomy and plant physiology extend better understanding of the mechanisms of interactions between crop genotypes and species – for example, enhanced resource availability through niche complementarity. Ecological advances include better understanding of the context-dependency of interactions, the mechanisms behind disease and pest avoidance, the links between above- and below-ground systems, and the role of micro-topographic variation in coexistence (Brooker *et al*., 2015). This improved understanding according to Brooker *et al* (2015) can guide approaches for improving intercropping systems, including breeding crops for intercropping.

Stability and Sustainability of dryland crop production through Intercropping: Sesame is considered as one of the fragile dryland crops in terms of its adaptability. Nevertheless, when sesame intercropped with legumes, may prove more remunerative under dryland conditions. Diverse intercropping practices are followed in different regions based on the rainfall pattern, soil type and moisture availability during the cropping regime. In this context Ali *et al.*, (2007) found that the seed yields of mung-bean and sesame when mixed cropped under variable seeding rates were less than their sole crop yields, but combined yields or equivalent yields of mung bean and sesame from mixed cropping were more than the sole crop yield of either mung bean or sesame. Significantly, the highest equivalent yield of mung bean (1476.00 kg/ha) was obtained by 70% mung bean with 30% sesame seeding ratio which was statistically identical to 90% mung bean with 10% sesame and 80% mung bean with 20% sesame but the lowest seed yield (949.00 kg/ha) was found in sole sesame. Land equivalent ratios (LER) from mixed cropping were increased by 12 to 27 over sole crop yields. The highest LER (1.27) was obtained by 70% mung bean with 30% sesame. The highest net returns and BCR (2.57) were found in 70% mung bean with 30% sesame and the lowest net return and BCR (2.01) were found in sole sesame.

Intercropping studies with sesame in the arid zone of north-western India (Ram, 2020) with average rainfall of around 365 mm involving other dryland leguminous crops viz., black gram, green gram and moth bean under diverse nutritional regimes such as low nitrogen, medium phosphorus and high potassium content revealed green gram or mung bean as an intercrop not only resulted in higher monetary returns; but the system of sesame + mung bean or green gram was observed to be more sustainable. Under low and variable rainfall conditions of desert soils of western Rajasthan, sesame production and productivity have been associated with risk. The intercropping of sesame with mung bean was found to be sustainable, thereby minimising the risk of crop failure (Ram, 2020).

The economic performance of sesame, hybrid cotton and castor in sole and intercropping system was studied in two years from 2001 to 2003 (Bhat *et al.*, 2010) in medium to high rainfall zone of Gujarat. All intercropping systems of sesame + cotton had higher total productivity in terms of sesame equivalent yield, net return, return per rupee invested. The income equivalent ratio (IER) of sesame + cotton system was found to be higher than that of all the sole crops and also sesame + castor intercropping systems. It was reported that maximum advantage of the intercropping system was observed in a proportion of 3 rows of sesame + 1 row of cotton.

Willey *et al* (2014) conducted growth studies with the intercrops of sorghum/pigeon pea and millet/groundnut which showed how intercropping systems can achieve much larger yields than sole crops by using environmental resources more fully over time or more efficiently in space. Data on moisture stress illustrated that the advantages of intercropping can be even greater under stress conditions. Possible nitrogen benefits from legumes in intercropping systems were evident with particular reference to a study on maize/groundnut. Weed, pest and disease control were more perceptible with regard to some favourable effects of a sorghum intercrop on the incidence of pod borer and wilt disease in pigeon-pea. Evidence comes from this system for enhanced yield stability in intercropping systems. It is emphasized that intercropping is especially beneficial to the small farmer in the low-input/high-risk environment of the developing areas of the world but some brief comments are made on its applicability in more developed conditions. The beneficial interaction that is perhaps most widely applicable in intercropping systems is the better use of environmental resources. This is illustrated with reference to two intercropping combinations, sorghum/pigeon-pea and millet/groundnut, both of which have been studied in considerable detail at ICRISAT. Sorghum/pigeon-pea is one of the commonest combinations of crops in India and it is typical of many combinations throughout the world where a rapid-growing, early-maturing crop is grown with a slower-growing, later maturing one. The sorghum grows for three-four months, maturing about the end of the rainy season; the pigeon pea usually flowers just after the sorghum harvest and it grows for a further two-four months, surviving mainly on the residual soil moisture. The farmer's main objective with this combination is to grow a reasonably good yield of the staple cereal (i.e. as near as possible to a sole sorghum yield). The pigeon pea is introduced to provide some 'bonus' pulse yield, but only to the extent that it does not seriously jeopardize sorghum yield; Traditionally, the farmer has achieved these objectives by sowing predominantly sorghum, with only the occasional rows, or plants, of pigeon pea. However, while this method maintains sorghum yield it produces very little pigeon pea on the farmers' fields.

Sorghum growth and yield were observed to be very good and intercrop yield was only a little less (5%) than the sole crop. Thus, despite the much higher proportion 86% of pigeonpea than in traditional systems, the farmer's primary objective of maintaining virtually a full cereal yield was fulfilled; this has been ascribed to the high sorghum population that was maintained in the intercropping system. Slow initial growth of the pigeon pea crop was even further reduced by the sorghum intercrop. At the time of sorghum harvest, the dry matter yield of the intercrop pigeon pea averaged only 16% of the sole pigeon pea. From then on, however, the effect of the high population of the pigeon pea became apparent and the crop was able to make a relatively rapid recovery to produce a dry matter yield equivalent to 53 % of the sole crop, a much higher proportion than in traditional systems. A further feature of this intercrop pigeon pea was that, because the competition from the sorghum reduced only its vegetative growth, it achieved a harvest index (30%) appreciably higher than the sole crop (22%). This improved efficiency of dry matter partitioning helped the pigeon pea to produce a substantial seed yield, equivalent to 72% of the sole crop. Looking at the system in total, therefore, we can see that for a sacrifice of only 5% in sorghum yield for a 72% yield of pigeon pea.

Millet/groundnut is a combination used on lighter soils and it is found in both India and West Africa. Unlike the sorghum/pigeon pea combination, it is typical of crop combinations, where there is little difference between the growing periods of the two crops but some difference in canopy height; more specifically, of course, it is typical of the cereal/low canopy legume combinations that are so prevalent in many parts of the world. The yield objectives of farmers seem to vary a good deal but the important groundnut cash crop is usually the major component, with the millet reduced to a minor role. For most of the growing period, accumulation of dry matter in the intercrop groundnut was less than the 75 % sole crop 'expected' yield, indicating that its growth was being depressed by the millet. But it was able to recover from this effect towards the end of the season, especially after the millet harvest, and at final harvest actual yield was similar to 'expected' yield. In contrast, dry matter accumulation in intercrop millet was more than twice its 25 % sole crop 'expected' level and at final harvest yield was 62% of the sole crop. Combining these dry matter yields into a relative yield total gave an overall advantage for intercropping of 36%; for seed yields the advantage was a little less (25%) because of small decreases in the harvest indices of both millet and groundnut. The manner in which resources were utilized more efficiently in these two intercrop combinations is indicated by the light interception pattern and the efficiency with which intercepted light was converted into dry matter. In the sorghum/pigeonpea combination, the intercrop was clearly able to combine much of the capacity of the sorghum to intercept light early in the season with at least some of the capacity of the pigeonpea to intercept it later. For each crop, however, the efficiency with which intercepted light was converted into dry matter was the same for intercropping as for sole cropping (Natarajan & Willey 1985). Thus in this combination higher yields were achieved in intercropping because of greater light interception and not because of greater efficiency of conversion; in fact, the combination displayed the classic 'temporal' complementarity of resource use that has traditionally been associated with combinations of early- and late-maturing crops. Willey *et al* (2014) found in the millet/groundnut combination, although some temporal difference was observable between the crops, by the end of the season the total amount of light intercepted by the intercrop was virtually identical with that 'expected' from the interception patterns of the sole crops. Thus, in contrast to the sorghum/pigeonpea combination, the greater yield from intercropping was brought about not by greater interception but by greater efficiency of conversion. This effect has been ascribed to a better dispersion of light over a larger area of leaf in the intercrop, and perhaps to some complementary interaction between the C4 millet and the C3 groundnut canopies (Reddy and Willey 1981). Whatever the actual mechanism, this combination provides an excellent example of the kind of 'spatial' complementarity of resource use that can occur in intercropping. As to the use of other resources in these two combinations, there-is some increase in the extraction of water from the soil profile compared with the sole crops and an improvement in total water-use efficiency because a greater proportion of the evapotranspiration passes through the crop as transpiration instead of being lost as evaporation from the soil surface (Natarajan & Willey 1985, Reddy & Willey 1981). In recent experiments with millet/groundnut there is also some evidence of a greater production of dry matter per unit of water transpired. For nutrient use, the pattern has been identical for both combinations in that any increase in yield over sole cropping is associated with an equal increase in nutrient uptake (Natarajan & Willey 1985, Reddy & Willey 1981); for this resource, therefore, it seems likely that in some situations where higher yields from intercropping will have to be at least partly paid for by greater fertilizer inputs. A further aspect of resource use that is of considerable interest is how the advantages of intercropping are affected by the availability of resources. The studies carried out by the ICRISAT have concentrated on the effects of nutrient and/or moisture stress. For the nutrient effects, it was consistently observed that greater intercropping advantages in the cases where fertility is lower (IRRI 1975).

In Botswana of Southern Africa, Lightfoot and Tayler (2008) carried out the field experiments of intercropping involving sorghum and cowpea in different environments over three years. Though cowpea yields in intercropping were lower as compared with the sole crop performance, land equivalent ratios were greater than one and resulted in higher gross income and cash returns.

Protecting wheat during the transition to organic production and managing soil nitrogen during the organic production phase represent major impediments to increasing organic acreage. Bruke *et al.*, (2015) have developed a system that incorporates weed control and the use of a pea intercrop as a source of nitrogen to address these impediments by intercropping winter wheat with pea crop. Intermediate and long-term impacts include advances in soil improvement and nitrogen delivery to organic systems through pea with winter wheat, and the system was found to be viable transition and production tactic. Further, these studies have also elucidated that the practices of intercropping with forage triticale can reduce weed pressure.

To date most intercropping research has focused in the west, on grain-legume intercrops in organic systems. However, recent work in Canada and Australia has demonstrated massive potential for spring canola + pea intercrops to intensify current dryland cropping systems through winter pea and winter canola intercrops (Madsen and Pan,2020)

In the semi-arid regions of Morocco and elsewhere in the dryland regions, climate change has resulted in more irregular rainfall and more frequent extreme weather events. Lentil is a major food-legume crop in Morocco that features prominently in local and North African cuisines. Moroccan farmers cultivate lentils by sowing two rows of the legume close to each other and leaving a wider strip of land fallow on each side. An experiment was conducted by with intercropping and relay cropping lentils with dryland spring crops (Fig.5) such as chickpea, sesame, onion, and quinoa. Quinoa, considered a climate smart crop, does not need supplementary irrigation when relay-cropped with lentils, except in years of extreme drought. Quinoa was found to be the most successful intercrop with lentils. Since harvest times are spaced out, this system ensured more regular income for farmers, while providing long term soil cover thereby preventing soil erosion. More importantly, this experiment showed that integrating quinoa does not affect the lentils' yield, since quinoa does not significantly compete for water and nutrients. In addition, introducing a secondary crop of quinoa with lentils clearly boosts the land's productivity, profitability, and resilience. (CGIAR 2022).

Fig. 5: Lentil + Quinoa and Quinoa alone after harvesting Lentil crop

Sida hermaphrodita is known as Virginia fanpetals and *Virginia mallow*, is a perennial forb native to the eastern United States, which produces white flowers in summer. The branching stem of *Sida hermaphrodita* is 1 to 4 meters tall, and up to 3 cm in diameter. The cultivation of perennial biomass plants on marginal soils can serve as a sustainable alternative to conventional biomass production via annual cultures on fertile soils. *Sida hermaphrodita* is a promising species to be cultivated in an extensive cropping system on marginal soils in combination with organic fertilization using biogas digestives. In order to enrich this cropping system with nitrogen (N) and to increase overall soil fertility of the production system, its potential was tested for intercropping with leguminous species (Nabel *et al.*, 2018). In a 3-year outdoor mesocosm study, the *S. hermaphrodita* plants were intercropped with the perennial legume species *Trifolium pratense, T. repens, Melilotus albus*, and *Medicago sativa*. Their effects on plant biomass yields, soil N, and above ground biomass N were studied. As a control for intercropping, a commercial grass mixture was used without N_2-fixing species as well as a no-intercropping treatment.

The results indicated that the total biomass yield got increased in intercropping treatments. However, grass species competed with *S. hermaphrodita* for N, more strongly than legumes. Legumes enriched the cropping system with fixed atmospheric nitrogen (N_2) and legume facilitation effects varied between the legume species. *T. pratense* increased the biomass yield of *S. hermaphrodita* and increased the total biomass yield per mesocosm by 300%. Further, the total above ground biomass of *S. hermaphrodita* and *T. pratense* contained seven times more N compared to the mono-cropped *S. hermaphrodita. T. repens* also contributed highly to N facilitation. It was concluded that intercropping of legumes, especially *T. pratense* and *T. repens* enhanced the yield of *S. hermaphrodita* on marginal soils for sustainable plant biomass production. (Nabel *et al.*, 2018)

Promotion of favourable soil microbial activity in dryland crops: Intercropping is an important practice in promoting plant diversity and productivity. Compared to the accumulated understanding of the legume/non-legume crop intercrops, very little is known about the effect of this practice when applied to native species on soil microbial communities in the desert ecosystem.

Plant species identities caused a significant effect on microbial community composition in monocultures but not in intercropping systems. Monoculture weakened the rhizosphere effect on fungal richness. The composition of bacterial and fungal communities (β-diversity) was significantly modified by intercropping, while bacterial richness (Chao1) was comparable between the two planting patterns (Zhang *et al*., 2021). Zhang *et al* (2021) reported that network analysis revealed that Actinobacteria, α- and γ-proteobacteria dominated bulk soil and rhizosphere microbial co-occurrence networks in each planting pattern. Intercropping systems induced a more complex rhizosphere microbial community and a more modular and stable bulk soil microbial network. Keystone taxa prevailed in intercropping systems and were Actinobacteria-dominated. Overall, planting patterns and soil compartments, not plant identities, differentiated root-associated microbiomes. Intercropping can modify the co-occurrence patterns of bulk soil and rhizosphere microorganisms in desert ecosystems.

Intercropping, a classic strategy for maximizing plant diversity, is usually used in agroecosystems to suppress replant disease and improve crop productivity (Kumar *et al*., 2013). This practice generates a mosaic of niches for microorganisms and soil resources, thereby increasing underground biodiversity and improving the ability to restore its original function after interference (Kumar *et al*., 2013). Gong *et al*. (2019) found that intercropping improved rhizosphere soil fertility and enzymatic activity by modifying rhizosphere microbial communities, which significantly differed from those in monoculture systems. Appropriate intercropping of non-legume and legume crops is known to be beneficial and is practiced globally (Solanki *et al*., 2019). N-fixing bacteria can assimilate the atmospheric N, which, in turn, reduces and even avoids the need for N-fertilizer (Pelzer *et al*., 2012). More generally, the soil microbiome may contribute to the added value of intercropping systems, because different microbial species often interact with each other, forming a complex network (Agler *et al*., 2016; Hartman *et al*., 2017). The complexity of microbial networks promotes ecosystem multifunctionality related to nutrient cycling (Wagg *et al*., 2019). Therefore, it is necessary to comparatively investigate the diversity, composition, and co-occurrence patterns of soil microorganisms (bulk soil and rhizosphere) in monoculture or intercropping systems.

Živanov *et al* (2018) studied the effect of dual legume intercropping on grain yield of normal-leafed pea during 2015 and 2016 in order to reduce lodging, to improve grain yield stability and to analyse the competitiveness of field pea with annual legumes and wheat. The research involved six species grown as sole crops: normal-leaved pea (*Pisum sativum* L. (Partim)), semi-leafless pea (*Pisum sativum* L. (Partim), faba bean (*Vicia faba* L. (Partim)), white lupin (*Lupinus albus* L.), fenugreek *(Trigonella foenum-graecum* L.) and wheat as a control (*Triticum aestivum* L. emend. Fiori et Paol) and intercrop mixtures of normal-leaved pea with other five species. Grain yield (t ha-1), yield components, land equivalent ratio, relative crowding coefficient and aggressivity value for grain yield were monitored. The highest grain yield of normal-leaved pea (2.87 t ha-1) was obtained from the mixture with semi-leafless pea in 2015 and from the mixture of wheat + normal-leafed pea (5.26 t ha-1) in 2016. The lowest number of pods and seeds per plant was formed by normal-leaved pea as a sole crop in 2015 (5.2 and 19.2, respectively). The obtained results showed that a thousand seed weight differed between treatments; however, the differences were not significant. The highest land equivalent ratio (1.40), relative crowding coefficient (4.44) and the positive value of aggressivity (0.19) were observed in the mixture of fenugreek + normal-leaved pea. The results demonstrated that semi-leafless + normal-leaved pea and fenugreek + normal-leafed pea are the most beneficial mixtures for grain production, while faba bean was a less suitable component for intercropping with normal-leafed pea.

Agroforestry – a Robust Intercropping System for Drylands:

Agroforestry which is the interaction of agriculture and trees, including the agricultural use of trees. It comes in many forms where trees, the primary crop, are managed in combination with other crops and/or animals to create a more diverse agriculture community with the potential to improve environmental quality and yield for all components of the system. Agroforestry systems create options to include some of the plant species that are not available in traditional tree plantations or other crop and animal production settings.

Intercropping with tree species usually employed in forestry also known as alley cropping represents mixed plant species' cultivation systems that can potentially reduce pressure on land and water resources by generating higher crop yields and by increasing resource use efficiencies through exploitation of complementarities between species.

In intercropping systems, it is one component, that is the perennial tree/shrub component which is more stable in terms of adaptability and sustainability than the other companion component such as an annual or biennial plant species despite seasonal variations (Singh and Reddy, 1988). The plant combinations are highly beneficial, especially for the more delicate crop, with the trees acting as wind breaks, habitat for beneficial insects, and trellises. Benefits also accrue in the soil, since "deep-rooted plants can draw up subsoil minerals and nutrients for shallow-rooted ones."

Agroforestry-intercropping systems have been developed as an alternative to conventional mono-cropping systems to address environmental, social and economic issues in a wide array of agricultural contexts. The crucial support systems like soil health, air and water quality, groundwater recharge, natural control of pests, etc., are diminishing. Therefore, need has been realized to conserve the natural resources and protect the deteriorating environment so that the much needed growth in agriculture is maintained sustainably. Tree based mixed systems are reported more productive than monoculture. The silvo-arable agroforestry for Europe has shown that one hectare planted with alternate strips of poplars (*Populus deltoids*) and wheat, produced the same output as 0.9 ha of wheat and 0.4 ha of poplar (Brelivet, 2006). Tree based systems have been observed to be more productive, since the resources otherwise unavailable to crops would be accessible.

In China 3% of the arable land was intercropped in in 2014. Agroforestry was practiced in that country on 1% of the arable land in 2014. Both intercropping and agroforestry did not decline during the period 2009–2014. Both practices show associations with labour availability (+) and machinery power (−) (YuHong *et al.*, 2017)

In agroforestry systems multi-storied intercropping is considered the best practice, in which crops with different heights were selected. Mostly this kind of intercropping is adopted in plantation crops. Another system observed is alley intercropping, which is a kind of agroforestry system where intercrops are grown in the alleys of trees and hedges etc.

It is suggested that in the intercropping systems soil moisture is the primary factor affecting the crop yields followed by light (Gao *et al.*, 2013). Deficiency of the soil nutrients also has a significant impact on crop yields particularly when linearly fast growing crop types are included in the system. Low canopy crop types seem to have an advantage in agroforestry based inter cropping systems. For example, compared with soybean, peanut was more suitable for intercropping with apple trees to obtain economic benefits in the region (Gao *et al* 2013). Gao *et al* (2013) suggested that agronomic measures such as regular canopy pruning, root barriers, additional irrigation and fertilization also should be applied in the intercropping systems. Intercropping on 571 million acres globally by 2050 would sequester 17.2 gigatons of carbon dioxide and, after a total investment of US$147 billion, save $22.1 billion over the 30-year span(https://www.theenergymix.com/2019/12/09/tree-intercropping-would-save-17-2-gigatons-of-carbon-by-2050-2/).

In the present context, growing annual crops or economically yielding plant species with perennial trees is a holistic concept that involves various organisms sharing habitat and its abiotic and biotic components having ecological and economic interactions in their spatial and temporal dimensions. Conservation of natural resources and optimization of productivity are vital in the functioning of the system. With small holdings it is a livelihood strategy; but in respect of larger holdings the dimension gets enlarged into a commercial venture. Rural farmers and people at large have been practising tree planting in their farms and home-steads, to meet household requirements for fuel, fodder, poles, timber, fruits, and non-timber forest produce. The objective of agroforestry is to take advantage of the complimentary relationships between trees, crops, and livestock in such a way that the productivity, stability, and sustainability of the total system exceed most of the single cases. Currently, the area under agroforestry (including farm forestry) in rain-fed areas is over 6 million hectares and nearly 10million hectares are covered with rubber, cashew, coconut, mango, and other species. As is the case in other areas, agroforestry in rain-fed areas also promotes sustainable agriculture, enriching the soil through enhancing positive microbial activity leading to fixation of atmospheric Nitrogen improvement of water holfing capacity of soil and drainage, and bringing about efficient nutrient cycling, thereby providing for vertical expansion to optimize land productivity and diversity in output to meet domestic needs and improving economy of farmers.

Most of the trees are drought resistant, providing fuel wood, food, fibre, fodder, and other products despite crop failure. Woody components integrated in dry land ecosystem can take advantage of the invaluable services of climate-moderation provided by adjacent natural and restored ecosystems. Most often, this perennial integration is specific to topographic and edaphic factors. Traditional agroforestry practices benefit biodiversity through in situ conservation of tree species on farms, reduction of pressure on remaining forests, and the provision of suitable holding for plant and animal species on farmland. Agroforestry systems/practices of dryland areas may be classified as traditional and advanced agroforestry systems

The leguminous trees meet the nitrogen requirements of crops grown in association, besides providing rich organic matter and improving the soil structure preventing land degradation. (Korwar *et al*., 2014) Most of the nitrogen fixing trees commonly planted include species of *Acacia, Albizia, Hardwickia, Gliricidia, Leucaena, Pongamia* and *Sesbania* (Solanki and Ramnewaj 1999). Several tree species have been tried in different parts for their potential for integration into arable systems. Some experiments were conducted on leguminous trees (e.g. *Faidherbia* and *Acacia*) at the Central Research Institute for Dryland Agriculture (CRIDA), Hyderabad; at Indian Grassland and Fodder Research Institute (IGFRI), Jhansi; National Research Centre for Agroforestry(NRCAF) Jhansi and on *Gmelina arborea* at Raipur. Results of the experiments conducted under rain-fed conditions at CRIDA on yield of intercrops with two densities of trees (156 trees/ ha) of *Faidherbia albida* from 5to12 years of tree age. The yield data revealed that there was increase of +57 and +32 % sorghum and groundnut productivity levels respectively at 156 trees per ha (Korwar *et al*., 2014).

Growing leguminous trees on maize farms is reported to have boosted and stabilised maize yields, as per the 12-year study in Malawi and Zambia (Appiah, 2012). The researchers behind the study, from the Kenya-based World Agroforestry Centre and the University of Pretoria, South Africa reported that it was the first analysis of long-term crop yield trends in cereal-legume agroforestry systems in Southern Africa. Maize is a common staple food crop in the region and is grown in more than half of cultivated land in countries across the region, according to the study, published in the *Agronomy Journal* in September. But factors such as rapid population growth, continuous cultivation, soil degradation, and climate change threaten maize yields. Given that intercropping (the practice of growing two or more crops in close proximity) generally increases maize yields, the researchers decided to explore whether the gains made could be sustained over the long-term in rain-fed maize-legume agroforestry systems.

The researchers planted the legume *Gliricidia* in three non-irrigated sites — one in southern Malawi and two in eastern Zambia. The yields from maize–*Gliricidia* intercropped farms were then compared with those from both fertilised and unfertilised monoculture maize farms in the same three sites. Itwas reported that maize farms with legume trees had, on average, a 50 per cent increase in yields and that the yields were stable, compared with those grown with or without fertilisers (Source: http://www.scidev.net/en/sub-suharan-africa/news/intercropping-boosts-maize-yields-by-50-per-cent-.html)

It has been reported that altering the tree geometry by increased *Eucalyptus* tree row spacing (or alley width), effectively improved the intercrop yield during the 4-years' period, especially with the triple row arrangement, which produced 73 and 66 % of the sole cowpea yields in 2003 and2004, respectively (Prasad *et al.* 2010a). Cowpea (*Vigna unguiculata*) yields in other wider-row spacing varied from 50 to 62 % of the sole crop yields in the third year and 39 to 59 % of the sole crop yields in the fourth year of tree plantation.

Working with *Leucania leucocephala* trees, it was observed that increasing the tree row spacing beyond 3 m was found to increase the intercrop yield till second year but not in the third year due to the increased growth of the *Leucania* tree canopy resulting in complete shade to the intercrop (Prasad *et al.*,2010b).

Among the trees that exhibit promise as feed stocks for biofuels, *Pongamia pinnata* L., (karanj, Indian beech) has been employed for agroforestry investigations (Korwar *et al.*, 2014). Pongamia (*Pongamia pinnata* L.) tree which forms an important component of afforestation, stands uniquely in agroforestry programmes as compared to several other tree components of the system. The uniqueness of Pongamia in terms of its inclusive attribute along with soil enhancing character, renders it suitable for intercropping with in the tree plantations.

Pongamia is an inclusive plant species, which allows companion cropping of diverse kinds of annual crop types. Pongamia offers an added advantage for intercropping systems, since the companion crops / plant species that grow in association with Pongamia have shown higher biomass yield including higher grain yields due to the availability of soil Nitrogen fixed by Pongamia in addition to the low incidence of biotic stresses due to its furano-flavanols viz., karanjin and pongomol. Of several tree components studied, Pongamia appears to be by far the best in terms of its superior drought resistance and inclusive habit. Nevertheless, in the current situation of absence of organized commercial plantations of Pongamia in India, the practice of inter-cropping in the stands of Pongamia is practically non-existent at the farmers' level, despite the impressive potential of the system.

The ongoing investigations, however, have revealed (Prasad, 2021 A) that inter-cropping of drought resistant annual crops viz., castor, pigeon pea, cowpeas, green gram, sesame, cluster beans, horse gram, niger, sunflower, fodder grasses etc., was recommended depending upon the local conditions in inter-row spaces of the plantation in the first three years. Inter-cropping of grain crops may not be possible after three years of establishment of the Pongamia plantation due to its canopy expansion. However, foliage yielding crops like forage grasses and essential oil yielding grassy types like Lemon grass, *Cymbopogon citratus*, Citronella etc., could be grown even after pongamia trees develop extended canopy after 4 years.

Banerjee *et al* (2013) conducted intercropping studies in Pongamia and Neem plantations with pigeon pea the intercrop. It was reported that most of the growth and yield attributing characters as well as yield of pigeon pea were higher under Pongamia plantation as compared to that of neem plantation. Moreover, Pongamia was also significantly better than neem in improving soil physicochemical properties resulting into higher intercrop yield.

Table 7: Yield attributes and yield of pigeon pea under sole and intercropping systems (Mean of two years' data) (Banerjee *et al.*, 2013)

System	No. of Pods/ plant	No. of seeds/ pod	100 seed weight (g)	Yield (t/ha)
Sole Pigeon pea	495.0	5.0	9.25	1.65
Neem + Pigeon pea	450.0	5.0	8.75	1.40
Karanj (Pongamia) + Pigeon pea	461.0	5.0	9.0	1.51
CD (P=0.05)	19.24	NS	0.20	0.13

Table 8: Residual Soil fertility under sole and intercropping systems (Banerjee *et al.*, 2013)

System	Organic C (%)	pH	Available N (Kg/ha)	Available P Kg/ha)	Available K (Kg/ha)
Sole Neem	0.43	5.0	198.0	29.1	156.7
Sole Karanj (pongamia)	0.45	5.1	199.8	30.8	160.7
Neem+Pigeon pea	0.51	5.2	235.4	33.8	170.5
Pongamia + Pigeon pea	0.52	5.3	240.5	34.6	175.0

Inamati *et al* (2015) reported that in the case of soybean grown as intercrop, the pod dry weight increased significantly when intercropped with Pongamia of different spacings, ranging from 1.5m and 3.0 m.

The studies conducted by the ICAR-Central Agroforestry Research Institute (Dhyani *et al.*, 2015) under their Al India Coordinated Research Project at the Agroforestry Centres at Hyderabad and Coimbatore have shown that castor can be successfully intercropped with Pongamia planted at 5m x 5m spacing resulting in remunerative yield levels of castor bean. It is further stated that other crops such as cowpea, groundnut and black gram could be successfully intercropped in the stands of Pongamia spaced at 6m x 6m.

The work carried out at ICRISAT has shown (Wani *et al.*, 2009) that intercrops like sorghum, pearl millet, pigeon pea, soybean, mung bean, chickpea, sunflower, safflower could be grown with *Jatropha* and pearl millet and pigeon pea could be successfully grown along with Pongamia. An additional income of INR 5000.0 to INR 16000.00 per ha can be obtained on low-quality soil.

Kaushik *et al* (2016) reported that the yield (grain and fodder) of different crops was not affected significantly by the Pongamia trees during initial four years of plantation. The mean grain yields of crops viz. cowpea (9.47q/ha), cluster bean (9.13q/ha), dhaincha (8.57 q/ha) and mung bean (9.50q/ha) were slightly less in agri-silvicultue system as compared to sole cropping. Similar trend was also observed for fodder yield. Pongamia growth (height and diameter) was more in agri-silviculture as compared to sole plantation. Maximum height of 300.00 cm and diameter of 89.20 mm was recorded when Pongamia was intercropped with cowpea, whereas it was 281.20 cm in height and 80.90 mm in diameter in sole plantation. Agri-silviculture system also improved the organic carbon and available N, P, K as compared to sole cropping. The lower net returns from agri-silviculture system of Pongamia + cowpea (INR 7178/ha), Pongamia +cluster bean (INR. 7725/ha), Pongamia +dhaincha (INR 7254/ha) and Pongamia + mung bean (INR 7100/ha) were mainly due to the fact that during initial years, installation of Pongamia plantation results in expenditure, without any economic return. It is evident from the results that the cost of establishment of plantation can be met through intercropping during the gestation period of Pongamia plantation.

Rao *et al* (2015) evaluated pigeon pea, horse gram, black gram and castor grown as intercrops along with Pongamia as the principal component. The study revealed that cultivation of pigeon pea as the intercrop with Pongamia was the best practice with a spacing of 8m x 6 m followed by 6m x 6m to achieve the best monetary returns.

Table. 9: Intercrop yield with *Pongamia* during two years. (Rao *et al.*, 2015)

Intercrop	Pongamia equivalent yield(Q/ha)	
	1st year (2007)	2nd Year (2008)
Red gram (pigeon pea)	19.06	19.28
Horse gram	9.92	9.95
Black gram	8.63	9.40
Castor	8.38	6.9
Sole Pongamia (Control)	1.45	
GM	9.49	9.64
LSD	1.19	1.20

Prasad (2018) raised genetically elite clonal plantation of Pongamia spaced at 5m x 5m on a typical dry land of 11 acres' area of low fertility in the zone of annual rainfall of the range of 367 to 450 mm at a distance of 50 km from Hyderabad City at Tunki Bollarum Village. During the years from 2013-2014 to 2014-2015, castor and pigeon pea crops were grown as the intercrops respectively, along with the pure solid stands of castor and pigeon pea in the adjoining area over an area of 0.5 acres each. No chemical plant protection measures were needed in the intercropped plots. At the time of planting Pongamia grafts in the rainy season of the year 2012, around 10 kg of farm yard manure, 1 kg of neem cake and 3kg of single super phosphate were added in pits per plant.

No chemical fertilizers were added in the intercropped plant populations of castor and pigeon pea, while appropriate production technologies were adopted in the solid pure stands of pigeon pea and castor. The yield of castor as intercrop was of the order of 1.5 t/ha as against the solid crop which yielded 1.6 t/ha. The yield of pigeon pea was of the order of 1.6 t/ha as intercrop as against 1.4 t/ha obtained from pure stand of pigeon pea crop. The seed yield of Pongamia trees of the age of 3years and seven months was of the order of 1.40 t/ha under intercropping. The Pongamia populations without intercropping yielded 1.45 t/ha.

It is observed that pigeon pea was more stable over seasons and environmental factors than several other grain crop components grown along with Pongamia in intercropping systems. Investigations carried out by Singh and Reddy (1988) under dryland agricultural conditions also revealed remarkable stability of pigeon pea. In other words, in addition to the high level of drought resistance and adaptability of Pongamia, which is the principal tree component of the intercropping system, inclusion of the annual crop type viz., pigeon pea which also exhibited very good stability offered added advantage in terms of stability of the Pongamia intercropping system.

Intercropping on 571 million acres globally by 2050 would sequester 17.2 giga-tons of carbon dioxide and, after a total investment of US$147 billion, save $22.1 billion over the 30-year span(Paul Hawken, 2019) on Project Drawdown A New Context for Global Warming

Fig. 6: (https://www.theenergymix.com/2019/12/09/tree-intercropping-would-save-17-2-gigatons-of-carbon-by-2050-2/)
Source: Arun4202/Wikimedia Commons

7

Farming Systems for Dryland Agriculture

In India since ages, the dryland farmers adopted a kind of mixed farming involving traditional annual crops and livestock consisting of one or two cows / bullocks and few poultry birds. This traditional system though of small scale provided some kind of risk minimisation particularly in the years of scanty rainfall causing agricultural drought. (https://coabnau.in/uploads/1587050523_Agriculturalheritage.pdf)

Discouraged by their marginal impact on farm economy, dryland farmers switched over to the lure of 'productive' mono-cropping with certain commercial crops like cotton etc. The continuous mono-cropping resulted in impoverishment of soils and low crop yields which took farmers into the labyrinth of poverty.

The current scenario of dryland agriculture in India is beset with defects of instability of production caused by deficiency of water availability, poor soil fertility, paucity of right kind of farm inputs and over and above, adherence to some kind of mono-cropping at the cost of other farm components viz., animal production including poultry and fisheries. (Dillon and Hardaker,1993) Currently, 63 percent holdings are below 1 ha accounting for 19 percent of the operated area while over 86 percent of holdings are less than 2 ha account for nearly 40 percent of the area (APCAS, 2010).

The small land holders are better contributors to the total production (78%) but weak in terms of generating adequate income and sustaining their own livelihoods. Small holdings (below 0.8 ha) generate inadequate income to keep a farm family out of poverty despite high productivity (Chand *et al.*, 2011). Thus there is a need to upgrade the income levels of small dryland farmers, through a system approach.

The terminology 'Farming system' refers to the strengthening of farming activity with a 'system' approach. This approach envisages incorporation of ingredients of biodiversity in not only in crop production but also into other components of farming, which have to be integrated with traditional crop production leading to synergy in farming, ensuring livelihood enhancements in rural areas. One of the key components of farming system approach is to generate the needed inputs for agriculture at the farm itself, thereby cutting down significantly cost of farm production. (FAO, 1995)

The farming systems' approach is considered more relevant and remedial for small and marginal dryland farmers in view of the resilience of the location-specific integrated technological inputs involved. The integrated technologies in this context can help the poor dryland farmer to overcome diverse problems such as declining resource use efficiency and degradation of farm productivity (Venkateswarlu and Gopinath, 2017).

The Integrated Farming System (IFS) shall provide the key ingredients of sustainable farming viz., (i) reduction on farmers' dependency on external inputs by recycling of organic residues including the agricultural and animal wastes generated at the farm, (ii) reduction of the cost of production through enhanced input use efficiency, (iii) risk minimization associated with mono-cropping, (iv) conservation of natural resources, (iv) provision of food security and (v) employment generation (Reddy, 2009).

In other words, the effort is towards achieving desirable changes in farm production by tilting the existing system towards meeting the needs of food, feed, fibre, fuel and farm inputs in a farmer centric manner, while enhancing the agro-ecosystem.

Two approaches of farming system viz., holistic and innovative are considered to be the powerful tools to enhance the income levels and employment opportunities of the farm family. The holistic approach deals with improving the productivity of the existing components in totality, while the innovative approach aims at improving the profitability of the farming system with user-perception based introduction of new components (Singh and Ravishankar, 2017) In this context, a well-coordinated research in farming system is essential to investigate the options available to overcome the constraints at each site specific situation and develop farmer centric models.

The methodologies involved in farming systems approach envisage an effective integration of several farm related components such as annual crops, perennial tree and forage components including horticultural ingredients and animal production including poultry and fisheries, in addition to small scale off-farm enterprises in order to optimise resource use and income levels. The effectiveness and efficiency of the farming system depends on the strength of the natural resources viz., soil and water, farm size, availability of recyclable farm residues and labour (Venkateswarlu and Gopinath, 2017). Additionally, other diverse factors such as market prices, level of post-harvest management and infrastructure also have bearing on the success of the farming system.

Current farming systems in place in India:

In India, crop and livestock farming systems are practised to certain extent; but the much needed recycling of residues within the farm needs strengthening (Singh and Ravishankar, 2017). The results obtained under the aegis of the ICAR-Indian Institute of Farming Systems Research (Singh and Ravishankar, 2017) revealed the existence of 35 kinds of farming system with involvement of as many as five components. In the sub group of crop and livestock system, crop and dairy is predominantly practised by 48% of marginal farmers. The subsequent order of farming system practised was that of crop-dairy-goat system by 11% of the cases studied.

Diversification of Farming Systems: Bio-intensive complementary cropping systems consisting of use of land configurations to accommodate two or more crops of synergistic nature at a time offered promise. The bio-intensive system of raising maize for cobs and cowpea for vegetable in 1:1 ratio on broad-bed-furrow system and *Sesbania* in furrows during the rainy season was found promising. In the post rainy season mustard was raised in furrows along with three rows of lentil or 3 rows of green gram on broad beds leading to remunerative income levels (Singh and Ravishankar, 2017). The above system resulted in the impressive yield levels of 24 t/ha of rice equivalent with productivity of 50.2 kg of grins/ha/day and profitability of Rs.500/ha/day (Gnagwar and Ravishankar,2013a).

The complimentary effects could be exploited in the broad bed – furrow (BBF) system as furrows served as not only the seat of moisture conservation; but also as mechanisms of drainage during heavy rains in the monsoon season.

Diversified farming system models could be synthesised using the primary data collected from farm households on basic characteristics and secondary data available from the research results obtained from on-station and on-farm experiments. The cropping systems could be modified to include grain legumes, oilseed crops, vegetables and fruits to meet the needs of the farmers' families (Singh and Ravishankar, 2017). To enhance the income levels one could plan mixed plantation of fruits like mango and guava which facilitate intercropping with vegetable crops and fodder crops about which income stabilization could be achieved (Singh and Ravishankar, 2017).

Family Farming Model for small farmer

Integrated Farming Systems' (IFS) approach should start in a watershed area with analysis of farmers' knowledge, problems and priorities in a given hydrological unit. The approach should relate to the land use that results in an efficient, optimum and sustainable use of natural resources. In this approach the focus should be on conservation, development and optimization of the use of natural resources like soil, water and vegetation to cut-down the soil erosion by management of flood-situations, alleviate drought, improve water availability to increase agricultural production including food, fodder, fuel and fibre on sustainable basis. (Venkateswarlu and Gopinath, 2017). Such efforts help in not only easing the problems of farmers; but also help in generating employment and in developing economic resources and income resources of the farming community of the village (Gill, 2010)

Since most of the small dryland farmers possess approximately around one hectare of farm land, a farming model consisting of an annual or biennial crop like castor/ cereals like sorgum/pearl millet /pigeon pea/ horse gram etc. (0.78 ha), fruit yielding trees such as custard apple or Ber (*ziziphus muaritiana*) / hardy leguminous plants like guar (cluster bean, *Cyamopsis tetragonoloba*)/ okra/ brinjal (egg-plant) etc., in an area of 0.14 ha, dairy (one or 2 cows), goat (around 10), and ducks of 25 numbers with a boundary plantation of subabul *(Leucaena leucocephala)* could be remunerative with stable income levels. The farmer of the above category may need some loan to buy cattle / goat etc., which should be available from the public sector banks as per the prevailing governmental policy.

The advantage associated with the above model is that it provides the farmer's family with milk (approximately 300 litres), eggs (800 nos.) and fish (90-100kg) thereby taking care of the nutritional security of the family, apart from promising stable agricultural income from the above crops/ trees which are suited to cultivation on drylands. The animal wastes and the plant residues may be recycled and processed to be used as organic manures on the farm, thereby obviating the need for chemical fertilizers. Since the issue involved is dryland agriculture, it goes without saying that moisture conservation including rainwater harvesting are absolutely necessary.

Ziziphus mauritiana (Ber fruit tree)

In the zones with rainfall of the order of 500-700 mm, the farming systems should be based on livestock with promotion of grasses and trees which can thrive on lower quantum of water to meet the fodder, wood and fuel needs of the farmers. In the regions with 700 to 1100 mm rainfall, crop production, horticulture and livestock based systems can be adopted depending upon the soil type and marketability (Venkateswarlu and Gopinath, 2017). Runoff-water harvesting in the watershed based farming system should be the priority in the above zones.

In the areas receiving more than 1100 mm rainfall (for example low lands of eastern states growing unirrigated rice crop), the IFS should integrate rice production with fisheries.

A model farming system for small holders with 1.12 ha of area in Alfisol has been developed by the Central Research Institute for Dryland Agriculture (CRIDA) covering arable crops with oilseeds and pulses, vegetables, green manuring bushes on bunds, fruit trees on the lower side and grasses on the upper topo-sequence in a micro-watershed. A cumulative volume of 955 cubic meters of water was harvested in 500 cubic meters' farm pond. Castor, pigeon pea, horse gram, pearl millet, vegetables and baby corn were raised in 0.2 ha by using harvested rain water, with production cost of Rs. 4,500 and a net profit of Rs. 12,850. Twenty ram lambs were raised utilising crop residues and fodder crops. Economic analysis of the model after six years (2005-2011) revealed that the cost of production of crop component was Rs. 15,164/ ha and net profit from the crops was Rs. 31,556/ha. Inclusion of ram lambs contributed a net-profit of Rs. 20,800. Over-all net profit from the farming system module was Rs. 52, 356/ha, the highest in comparison regularly prevailing cropping systems of the zone viz., sorghum + pigeon pea system of inter-cropping with net returns of Rs. 12,340/ha and sole castor crop with net returns of Rs. 3,550/ha. The individual components of the system viz., agroforestry, vegetables, forage grasses and bushes contributed 38%, 10%, 27%, 7% and 17% respectively to the overall income of the system. (CRIDA, 2011)

Table 1: Farming System Model for a small farmer (1.12 ha) in Southern Telangana with mean rainfall of 500 mm

Farming System	Net Income (IRs/ha/year)
Improved Farming System	
Annual crops viz., Pigeon pea, pearl-millet, castor, horse gram	31556
Vegetables viz., okra, brinjal, cluster beans	
Fruit trees (Custard apple)	
Fodder crops (*Stylosanthes, Cenchrus*, Sorghum)	
Ram / Lamb : 20 numbers	20800
Total	**52356**
Existing farming /Cropping System	
Sorghum and Pigeon Pea	12340
Castor	3550

(Venkateswarlu and Gopinath, 2017)

The above data are indicative of the efficacy of the farming system approach in enhancing the income levels of small farmers.

An integrated farming system module comprising 35.4% of area under cereals, 25.7% area under pulses, 21% area under oilseeds, 17.3% area under commercial crops and 1.2 % area under fodders together with back-yard poultry of six birds per house hold was found to be ideal for small and marginal farmers of Dharwad Region of Karnataka (TAR-IVLP, 2005).

The poultry component played a major role in stabilizing farmers' income during drought years. In Arjia region of southern Rajasthan, IFS module of maize, pulses, oilseeds and forage grasses together with *in-situ* rain water management and bio-fencing gave 22.4% higher profitability (AICRIPDA-Arjia 2006)

The success of the farming system depends particularly when the focus is on the small farmer, his family and their livelihood security. The farmer in question should be convinced about enhancing the biodiversity of his farming activity conducive to enhancement of his income levels as against the mono-cropping systems that have been in vogue. This calls for a detailed analysis of each houschold for their resources, investment capacity and labour availability.

Crop based farming system: As is common with several small and marginal farmers, crops are the main source of his livelihood. In such cases, the farmer could be incentivised to include animals which could be used for agricultural operations. The animals can be fed from agri-wastes and the dung can be used a manure as well as fuel.

In the case of high rainfall zones, fish production can be encouraged. The fish production could be integrated in the ponds, which are manured by crop residues and animal products.

The IFS model on dryland vertisols (Kovilpatti) in Tamil Nadu indicated that crop, four goats, twenty poultry birds, six sheep and a dairy animal recorded the highest net returns of the order of IRs. 17,598/ha/year (Venkateswarlu and Gopinath, 2017) as shown in Table 2. The conventional system of annual crop cultivation alone resulted in net income of only meagre sum of Rs.2057/ha/year and the sustainability index of the system was not encouraging. In the other systems, the animal wastes from diverse sources indicated were collected and applied to the field to improve soil fertility and crop yields. The employment increased from185 man days in Crop alone system to 389 man days in the "crop, goat, sheep poultry & dairy" integrated system (Table 1). The data also revealed that the above system showed the highest sustainability index and reveals the importance of sheep rearing integrated with crop production in offering gainful employment in rain-fed areas.

Crop production together with sheep farming involving cotton resulted in annual net returns of IRs. 27,500 per hectare as compared to growing cotton alone which resulted in annual net returns of Rs. 8,700 per hectare in Warangal district of Telangana state (TAR-IVLP 2003). On the other hand, the net income from dryland grown groundnut system was of the order of Rs. 5,059/ ha, while the enhancement of the income level to Rs. 22,000/ha was achieved with a system of groundnut with one dairy animal and 8 to 10 sheep in the case of a marginal farmers of the eastern-central dry zone of Karnataka (Shankar *et al* 2007). The system involving crop and dairy recorded higher net income in the case of both small and large farmers. However, the system of crop and sheep rearing was more economical for medium level farmers as shown in Table 3.

Table 2. Performance of different farming system models at Kovilpatti, Tamil Nadu.

Farming System	Net-income (IRs./ha/yr)	Benefit-cost Ratio	Employment (mand ays/ha/yr)	Sustainability Index
Crop alone	2057	1.28	185	-23.0
Crop, goat and poultry	8274	1.92	297	12.3
Crop,goat,poultry &dairy	14208	1.72	343	46.1
Crop, goat, poultry & sheep	7265	1.47	343	6.6
Crop,goat,sheep poultry & dairy	17598	1.75	389	65.3

(Venkateswarlu and Gopinath, 2017)

Table 3: Net income (IRs. /ha) from the existing farming systems among different categories of farmers in Karnataka

Farming System	Small Farmers	Medium level Farmers	Large Farmers
Crop and Sheep	6,250	9000	-
Crop & dairy	10,400	17125	22938
Crop +dryland horticulture	19,000	-	43000
Crop, dairy & sheep	12,500	20667	26000
Crop, dairy & piggery	20,000	-	-
Crop, dairy & dryland horticulture	27,100	42,800	100,900
Crop, dairy, dryland horticulture & sheep	31,000	49,250	105,000

(Venkateswarlu and Gopinath, 2017)

Enhancing Cropping intensity of IFS through short-duration pulses and oilseed crops

Increased cropping intensity should be the key strategy for gains from the system. In this context, short duration pulses, oilseeds and other high value crops will certainly find their niche in any IFS programme as sequential or inter crops with high yield stability (Suresh *et al* 2017). Under the technology assessment and refinement programme of the National Agricultural Technology Project (NATP), an integrated farming system module comprising cereals (30%), pulses (26%), oilseeds (21%), commercial crops (17%) and fodders (1% area) along with backyard poultry (6 birds per household) was found to be ideal for small and marginal farmers in Dharwad region of Karnataka (TAR-IVLP 2005)

In the typical scarce rainfall zone of Anantapur in Andhra Pradesh, higher net returns of the order of IRs. 43,360/ year in 2 hectares were recorded by integrating poultry with groundnut crop cultivation, while much higher income levels of the order of IRs. 40,606 were achieved with the integration of dairy (with three buffaloes) with groundnut production. The income level in the case of groundnut crop alone was of the order of IRs. 14,872/ha (Reddy 2005).In this system, groundnut crop offers the advantage of valuable fodder through its haulms.

According to Shankar *et al.*, (2008) integrating sericulture and fodder cultivation with the farming system of crop-dairy-sheep raring and poultry generated higher net income of IRs. 1,15,584/ha followed by the system involving crop, dairy, sheep, goat and fodder cultivation which resulted in income of IRs.1,04,078/ha, as compared to other systems under rainfed conditions of Chamaraj nagar district of Karnataka.

Around 120 farmers of the rainfed regions of Karnataka, small farmers earned higher net returns of IRs. 31,000/-from combination of crop, dairy, dryland horticulture and sheep rearing compared to other farming system modules (Nagaraja *et al.*,2004).

Rudraradhya (2002) working in the southern transitional zone of Karnataka, had developed a farming system model on a micro-watershed basis through the integration of crops, horticulture, agro-forestry, silvi-pasture, green manure, hedge plants, kitchen garden, apiary, cattle, fish and poultry components, besides rainwater harvesting from the nearby farm building. The advantage of this model lies in the vivid demonstration of synergies realised through the integration of food, fodder, fibre, fruits and fish components which resulted in provision of nutritional and economic security on a sustainable basis.

In the rice based rain-fed farming in Odisha, the farmers' income can be significantly augmented with the inclusion of fish production using the conserved excess water at the downstream of rice field (James *et al.*, 2005)

Barik *et al* (2010) working on rainfed farming systems in Odisha, reported that integrated farming systems (IFS) in combination with crops such as rice, off-season tomato and cauliflower along with poultry, paddy straw based mushroom production and vermi-composting was superior to farmers' practice of growing rice alone and / or rice followed by green gram. The above IFS resulted in higher net returns of Rs. 44,230/ha as against a sum of Rs.15, 820 obtained from Rice – Green-gram system, resulting in higher sustainability index of 54.58% as against 9.65% obtained from rice-green gram system.

At the Tripura Centre, efforts were dedicated in developing a farming system model combining agriculture with horticulture, forestry and livestock-rearing on the land of one hectare. The elements of the above system were cereal crops, pulses, oilseed crops and horticultural crops such as mango and pineapple, vegetable crops, livestock elements like duckery, piggery, fisheries in the structures developed to harvest water. The results indicated that the multi-enterprise system is five times more profitable than the traditional mono-cropping system of rainfed rice cultivation (Sharmah *et al.*, 2019).

Baishya *et al* (2007) also reported that multi component based systems such as (i) crop, dairy, fishery and poultry; (ii) crop, dairy, and goatery and (iii) crop, dairy, goatery and pigeon rearing were the more remunerative systems of farming for marginal, small and medium level farmers of Kamarup district of Assam respectively in that order.

Agroforestry based farming systems:

Studies across the world suggest agricultural intensification has been responsible for net gains in human well-being and economic development, but with an increasing cost of degradation of natural resources. The traditional agroforestry systems have been increasingly recognized for its contributions to the sustainable intensification of food production while providing several additional benefits to society.

Agroforestry which is the interaction of agriculture and trees, including the agricultural use of trees. It comes in many forms where trees, the primary crop, are managed in combination with other crops and/or animals to create a more diverse agriculture community with the potential to improve environmental quality and yield for all components of the system. Agroforestry systems create options to include some of the plant species that are not available in traditional tree plantations or other crop and animal production settings.

Agroforestry represents mixed plant species' cultivation systems that can potentially reduce pressure on land and water resources by generating higher crop yields and by increasing resource use efficiencies through exploitation of complementarities between species.

In intercropping systems, it is one component, that is the perennial tree/shrub component that is more stable in terms of adaptability and sustainability than the other companion component such as an annual or biennial plant species despite seasonal variations (Singh and Reddy, 1988a).

Agroforestry-intercropping systems have been developed as an alternative to conventional mono-cropping systems to address environmental, social and economic issues in a wide array of agricultural contexts. The crucial support systems like soil health, air and water quality, groundwater recharge, natural control of pests, etc., are diminishing. Therefore, need has been realized to conserve the natural resources and protect the deteriorating environment so that the much needed growth in agriculture is maintained sustainably. Tree based mixed systems are reported more productive than monoculture. The silvo-arable agroforestry for Europe has shown that one hectare planted with alternate strips of poplars (*Populus deltoids*) and wheat, produced the same output as 0.9 ha of wheat and 0.4 ha of poplar (Brelivet, 2006). Tree based systems have been observed to be more productive, since the resources otherwise unavailable to crops would be accessible.

The lack of standard protocol for precise estimation of biomass and carbon storage of traditional agroforestry systems might have undermined the actual potential of such systems in climate change adaptation and mitigation (Das *et al*., 2020). There is an urgent need to revive the long historical tradition of tree growing on farms to promote the ecological, economic and social wellbeing of farmers in particular and people in general (Pandey 2007). Many tree species like *Pongamia pinnata* are inclusive by nature and promote positive growth and development of vegetation around including the annual inter-crops without reduction of their grain or fodder yield (Prasad, 2021A). More-over, fuel wood is the main source of energy in rural areas, more particularly landless rural inhabitants, who depend upon trees to meet their energy needs. Additionally, the perennial vegetation including grasses and trees impart stability to farming by virtue of their characteristic ability of withstanding vagaries of weather and at the same time offering a kind of protection from abnormal weather to their companion annual crops (Prasad, 2021A).

According to the investigations carried out at the ICAR-Central Arid Zone Research Institute Jodhpur. the returns on per rupee invested in arid rainfed ecosystems are much higher as compared to that of the arable crops, when trees are associated with annual crops and grasses in silvi-pasture, agri-horticultural and agro-forestry systems (Venkateswarlu and Gopianth, 2017). The systems involving agri-silviculture systems are generally recommended for dryland zones with land capability class IV, receiving annual rainfall of 700 to 750 mm, where short duration dryland crops such as pearl-millet (bajra), black gram, green gram and moth bean are recommended to be grown in widely spaced rows of trees African Winter Thorn (*Faiderbia albida)* and Anjan (*Hardwickia binata) found* suitable to semi-arid tropical areas (Korwar 1992). Similarly, arable crops such as castor and pigeon pea could be grown with African Winter Thorn in marginal land under rainfed conditions with the advantage of enriched site due to the deciduous nature of the tree species (Bheemaiah *et al* 1992).

Though many tree species used in agro-forestry programmes are inclusive by nature, the competition among the trees and annual crops in an intercropping system can't be ruled out. Nevertheless, simple and cost effective operation of root pruning in the middle of rainy season (August) may be carried out using the available implements like even country plough, which will regulate the branching of the tree species and reduce the competition between the hedge-rows and crop leading to enhancement of the crop yield (Korwar and Radder, 1994). Such root pruning operation may be carried out once in a month. This kind of root pruning operations were carried out successfully in the arid zone of Gujarat in the case of agri-silvicultural system consisting of neem trees (*Azadirachta indica*) and Indian Tree of Heaven (*Ailanthus excels),* with 59.3% and 29.3% more income (Venkateswarlu and Gopinath, 2017)

In the context of silvi-pastural development, Siris (*Albizzia lebbek),* desert teak (*Tecomella undulata),* mopane *(Colophosperum mopane*), gum arabic tree (*Acacia Senegal), umbrella thorn (Acacia tortilis),* jujube (*Zizyphus numularia)* and wild jujube (*Z. rotundifolia*) are sum of the tree species found to be compatible with grass component. (Venkateswarlu and Gopinath, 2017)

According to Samra (2004) among pasture legumes, blue pea (*Clitorea ternatea)* and Indian bean (*Lablab purpureus*), exhibited good compatibility with Sewan grass (*Lasiurus sindicus*) and anjan grass *(Cenchrus ciliariis)*

The system of integration of sorghum and cowpea crops together with *Cenchrus glaucous* as intercrops in Indian gooseberry (*Emblica officinalis*) together with goat rearing resulted in highest productivity and economic returns (Radhamani, 2001)

Next important system is agri-horticultural pattern that would provide positive economic returns to the farmers. Considering the aridity index in typical dryland zones, fruit yielding ber (*Zizyphus mauritiana*) grown along with pearl millet and pigeon pea at Solapur, pigeon pea and black gram at Rewa, castor at Dantiwada in Gujarat and cluster beans at Hyderabad proved to be remunerative (Venkateswarlu and Gopinath, 2017). Osman *et al* (1989) reported ber fruit yield of 40 kg / tree along with 100 kg of horse gram and 450 kg of cowpea cultivated in the interspaces of ber.

Low canopy crop types seem to have an advantage in agroforestry based inter cropping systems. For example, compared with soybean, peanut was more suitable for intercropping with apple trees to obtain economic benefits in the region (Gao *et al* 2013). Gao *et al* (2013) suggested that agronomic measures such as regular canopy pruning, root barriers, additional irrigation and fertilization also should be applied in the intercropping systems.

Mango and cashew trees offer remunerative horticultural enterprise for drylands of Karnataka. The system of growing ragi (Finger millet) and groundnut or horse-gram along in the mango and cashew orchards has become popular in the district of Kolar in Karnataka.

In some cases, cashew and tamarind trees are grown interspersed in mango orchards there. (Shankar *et al*, 2007)

On the other hand, guava based system in Meghalaya and lemon based system in Assam combining domesticated fruit trees and forest trees resulted in three-fold higher net returns (Bhatt and Misra, 2003).

In addition to monetary returns, the tree based systems improve soil fertility through build-up of organic matter and nutrients in the soil (Singh *et al*., 1997) as given in the Table below. Studies carried out by Arunachalam *et al* (2002) revealed that multiple use species such as bamboo help in soil binding during restoration of abandoned agricultural lands after shifting cultivation (*Jhum* fallows) in north-eastern India.

Pandey (2007) reported that agro-forestry systems are promising management practices to increase carbon in above ground parts of vegetation as well as soil carbon stocks to mitigate the adverse effects of greenhouse gas emissions (Table 4)

The studies of Sathaye and Ravindranath (1998) found that the average carbon sequestration potential of Indian agro-forestry is around 25 t / ha over 96 million hectares.

Considerable evidence is accumulating (Turner and Ward, 2002) that the agro-forestry systems have the potential for improving the water use efficiency by reducing the unproductive components of the water balance such as run-off, soil evaporation and drainage.

Table 4: Changes in soil properties (0-30 cm) under different tree-crop combinations in five years

Land Use System	Organic Carbon)%)	Available N (kg/ha)
Crop based system	+0.07	+10
Eucalyptus based System	+0.12	+21
Acacia based System	+0.20	+31
Populous based System	+0.17	+25

(Pandey, 2007)

Farming System for Nutrition (FSN)

In spite of remarkable progress achieved in agricultural production, considerable part of Indian population suffers from under nutrition and malnutrition. Swaminathan (2012) and Nagarajan *et al* (2014) designed a farming system for nutrition, comprising specific steps, including cultivation of nutrient rich varieties of millets, sweet potato etc. There have been efforts to address the issue of rural employment, food security and poverty amelioration by central and state governments, social organizations and farmers' groups. A critical appraisal to mitigate the farm distress including under nutrition in Wardha district has been made by Rukmani and Manjula (2009). Farming system interventions in this regard had two objectives viz., (1) to increase the productivity and profitability of the crops that are of commercial value and to enable purchase of nutritive food, (2) promote pulses and demonstrate bio-fortified crops like iron fortified pearl-millet and sorghum to address protein and micronutrient deficiencies. In this context multisite ACI trials were conducted with cotton, soybean, pigeon pea, quality protein maize, sorghum, micronutrient dense pearl millet in strip cropping and multiple cropping etc., during monsoon season of 2013. These interventions not only increased the crop yield, but reduced the cost of cultivation and at the same time fetched higher income to the farmers.

Pongamia pinnata for Intercropping with diverse crops:

Pongamia (*Pongamia pinnata* L.) tree which forms an important component of afforestation, stands uniquely in agroforestry programmes as compared to several other tree components of the system. The uniqueness of Pongamia in terms of its inclusive attribute along with soil enhancing character, renders it suitable for intercropping with in the tree plantations.

Pongamia is an inclusive plant species, which allows companion cropping of diverse kinds of annual crop types. Pongamia offers an added advantage for intercropping systems, since the companion crops / plant species that grow in association with Pongamia have shown higher biomass yield including higher grain yields due to the availability of soil Nitrogen fixed by Pongamia, in addition to the low incidence of biotic stresses due to its furano-flavanols viz., karanjin and pongomol. Of several tree components studied, Pongamia appears to be by far the best in terms of its superior drought resistance and inclusive habit (Prasad, 2021A). Nevertheless, in the current situation of absence of organized commercial plantations of pongamia in India, the practice of inter-cropping in the stands of Pongamia is practically non-existent at the farmers' level, despite the impressive potential of the system.

The ongoing investigations, however, have revealed (Prasad, 2021) that inter-cropping of drought resistant annual crops viz., castor, pigeon pea, cowpeas, green gram, sesame, cluster beans, horse gram, niger, sunflower, fodder grasses etc., was recommended depending upon the local conditions in inter-row spaces of the plantation in the first three years. Inter-cropping of grain crops may not be possible after three years of establishment of the Pongamia plantation due to its canopy expansion. However, foliage yielding crops like forage grasses and essential oil yielding grassy types like Lemon grass, *Cymbopogon citratus*, Citronella etc., could be grown even after pongamia trees develop extended canopy after 4years.

Banerjee *et al* (2013) conducted intercropping studies in Pongamia and Neem plantations with pigeon pea the intercrop. It was reported that most of the growth and yield attributing characters as well as yield of pigeon pea were higher under Pongamia plantation as compared to that of neem plantation (Table 5). Moreover, Pongamia was also significantly better than neem as well as control situation in improving soil physicochemical properties resulting into higher intercrop yield. (Table 6).

Table 5: Yield attributes and yield of pigeon pea under sole and intercropping systems (Mean of two years' data)

System	No. of Pods/plant	No. of seeds/ pod	100 seed weight (g)	Yield (t/ha)
Sole Pigeon pea	495.0	5.0	9.25	1.65
Neem + Pigeon pea	450.0	5.0	8.75	1.40
Karanj (Pongamia) + Pigeon pea	461.0	5.0	9.0	1.51
CD (P=0.05)	19.24	NS	0.20	0.13

(Banerjee *et al.*, 2013)

Table 6: Residual Soil fertility under sole and intercropping systems

System	Organic C (%)	pH	Available N (Kg/ha)	Available P (Kg/ha)	Available K (Kg/ha)
Sole Neem	0.43	5.0	198.0	29.1	156.7
Sole Karanj (Pongamia)	0.45	5.1	199.8	30.8	160.7
Neem+Pigeon pea	0.51	5.2	235.4	33.8	170.5
Pongamia + Pigeon pea	0.52	5.3	240.5	34.6	175.0

(Banerjee *et al.*, 2013)

Inamati *et al* (2015) reported that in the case of soybean grown as intercrop, the pod dry weight increased significantly when intercropped with Pongamia of different spacings, ranging from 1.5m and 3.0 m.

It is observed that pigeon pea was more stable over seasons and environmental factors than several other grain crop components grown along with Pongamia in intercropping systems. Investigations carried out by Singh and Reddy (1988) under dryland agricultural conditions also revealed remarkable stability of pigeon pea. In other words, in addition to the high level of drought resistance and adaptability of Pongamia, which is the principal tree component of the intercropping system, inclusion of the annual crop type viz., pigeon pea which also exhibited very good stability offered added advantage in terms of stability of the Pongamia intercropping system.

The studies conducted by the ICAR-Central Agroforestry Research Institute (Dhyani *et al.*, 2015) under their All India Coordinated Research Project at the Agroforestry Centres at Hyderabad and Coimbatore have shown that castor can be successfully intercropped with pongamia planted at 5m x 5m spacing resulting in remunerative yield levels of castor bean. It is further stated that other crops such as cowpea, groundnut and black gram could be successfully intercropped in the stands of pongamia spaced at 6m x 6m.

The work carried out at ICRISAT has shown (Wani *et al.*, 2009) that intercrops like sorghum, pearl millet, pigeon pea, soybean, mung bean, chickpea, sunflower, safflower could be grown with *Jatropha* and pearl millet and pigeon pea could be successfully grown along with pongamia. An additional income of IRs. 5,000 to IRs.16, 000 per ha can be obtained on low-quality soil.

Kaushik *et al* (2016) reported that the yield (grain and fodder) of different crops was not affected significantly by the Pongamia trees during initial four years of plantation. The mean grain yields of crops viz. cowpea (9.47q/ha), cluster bean (9.13q/ha), dhaincha (8.57 q/ha) and mung bean (9.50q/ha) were slightly less in agri-silvicultue system as compared to sole cropping. Similar trend was also observed for fodder yield. Pongamia growth (height and diameter) was more in agri-silviculture as compared to sole plantation. Maximum height of 300.00 cm and diameter of 89.20 mm were recorded when Pongamia was intercropped with cowpea, whereas it were 281.20 cm in height and 80.90 mm in diameter in sole plantation. Agri-silviculture system also improved the organic carbon and available N, P, K as compared to sole cropping. The lower net returns from agri-silviculture system of pongamia + cowpea (IRs. 7178/ ha), pongamia +cluster bean (IRs. 7725/ha), Pongamia +dhaincha (IRs. 7254/ ha) and Pongamia + mung bean (IRs. 7100/ha) were mainly due to the fact that during initial years, installation of Pongamia plantation results in expenditure, without any economic return. It is evident from the results that the cost of establishment of plantation can be met through intercropping during the gestation period of Pongamia plantation.

Rao *et al* (2015) evaluated pigeon pea, horse gram, black gram and castor grown as intercrops along with pongamia as the principal component. The study revealed that cultivation of pigeon pea as the intercrop with Pongamia was the best practice with a spacing of 8m x 6 m followed by 6m x 6m to achieve the best monetary returns. (Table 7).

Table. 7: Intercrop yield with Pongamia during two years.

Intercrop	Pongamia equivalent yield(Q/ha)	
	1st year (2007)	2nd Year (2008)
Red gram (pigeon pea)	19.06	19.28
Horse gram	9.92	9.95
Black gram	8.63	9.40
Castor	8.38	6.9
Sole Pongamia (Control)	1.45	
GM	9.49	9.64
LSD	1.19	1.20

(Rao *et al*., 2015)

Agroforestry systems have been developed as an alternative to conventional mono-cropping systems to address environmental, social and economic issues in a wide array of agricultural contexts. As research on the biological properties of these systems tends to demonstrate their potential, fostering their integration in agricultural landscapes requires an in-depth understanding of local stakeholders' perceptions (Laroche *et al*., 2020).

It was observed that social factors seem to have more impact than biophysical factors on the decision to integrate agroforestry intercropping systems in intensive and extensive agricultural landscapes. The relative value given to the decision factors varies greatly across stakeholders' categories and areas. Agroforestry intercropping systems designed to meet crop production needs or landscape aesthetic purposes are perceived as more suitable in both agricultural contexts than the tree-oriented design (Laroche *et al*., 2020)

Livestock Based Farming Systems

The livestock based farming systems in rain-fed agriculture are complex and generally based on traditional socio-economic conditions, notwithstanding the fact that Indian agriculture does have the long tradition of interactive mixed livestock systems (Birthal and Rao, 2002). Over several centuries Indian farmers had been using cattle for milk production as well as draught purposes in agriculture. In rural area of India, where bulk of livestock are raised and maintained, poultry, sheep, goat, pig and poultry have been used for meat, while cattle and buffaloes were exclusively reared for milk production and farm drought purposes. Crop residues and by-products have been the major sources of feed for livestock. By recycling the major amounts of crop residues, the livestock make land available for growing crops. The livestock systems provide dung as manure and domestic fuel and thus protect the ambience from over use of chemicals including petro-fuels in farming.

The modern Livestock based farming systems demand a thorough understanding of production factors viz., livestock, capital, feed land and labour and processors (description, diagnosis, technology design, testing and extension) that affect animal production besides intricacies of rainfed agriculture (Venkateswarlu and Gopinath, 2017). Livestock production in rainfed farming can be improved by growing fodders as intercrops in cereals, relay and alley cropping. Additionally, the activities like forage production on farm bunds, improving the feed value of stover by chopping, soaking in water, urea treatment, need based supplementation of concentrate and urea molasses help in enhancing the productivity of livestock in the farming system (Venkateswarlu and Gopinath, 2017).

All the important operations of scientific maintenance of livestock such as artificial insemination, culling-out low grade animals through castration, adoption of vaccination and deworming through health camps together with establishment of fodder banks in the areas of their surplus production help in significantly improving the productivity of milch animals (Mishra 2002).

Studies carried out revealed that urea treated straw significantly enhanced the milk yield of cows and buffaloes in Ranga Reddy district of Telangana. Urea molasses mineral block and mineral supplementation too enhanced the quality and quantity of milk by 25 to 30% and 58% respectively in cows and buffaloes. The enhanced net-returns form the mineral supplementation of animal feed was of the order of 187% as compared to the farmers' practice of grazing alone which was Rs. 2,156 (Venkateswarlu and Gopinath, 2017).

According to the Food and Agriculture Organization of the United Nations, China is currently the world's largest producer of farm-raised fish. A recent study found that fish farming has been enjoying healthy growth rates in South America, Africa, Norway, Chile and Egypt. The United States is presently considering joining the party, as Hubbs-SeaWorld Research Institute and Pacific6 Enterprise have proposed building a large-scale open-water fish farm off the coast of San Diego.(https://www.reference.com/business-finance/advantages-disadvantages-fish-farming-ba93f18bb50c3c0f).

- Pond Systems: One of the oldest types of fish farming, pond systems involve raising fish in a natural or manmade pond, ditch or canal.
- Open Net Pens: Open net pens look like small cages that float on the surface of a large body of water. They're usually made of metal, wood or bamboo, and they support a mesh enclosure that floats beneath the surface of the water. The fish are raised in these enclosures.
- Submersible Net Pens: Submersible net pens are similar to open net pens except that they exist completely underwater. They consist of underwater mesh pens where the fish are raised.
- Recirculating Systems: Among the most environmentally friendly systems of aquaculture, recirculating systems allow fish farmers to raise fish in a system of large indoor pools. A piping system pumps clean water into the tanks and filters wastewater out, making the tanks mostly self-cleaning.

Many types of aquaculture also tend to be slightly more environmentally friendly than the raising of other types of livestock. The effect that fish farms have on the environment often depends on the particular farm in question and on the methods it uses. Open net and submersible net fish farms are arguably the least environmentally friendly types of aquaculture, as there's little in the way of equipment to separate the water inside the net from the surrounding water outside. (https://www.reference.com)

Fish based farming system offers the following additional advantages:

- A farmer can often integrate fish farming into the existing farm to create additional income and improve its water management.
- Fish growth in ponds can be controlled: the farmers themselves select the fish species they wish to raise.
- Effective land use: effective use of marginal land e.g. land that is too poor, or too costly to drain for agriculture can be profitably devoted to fish farming provided that it is suitably prepared.
- Potential benefits of land-based fish farming systems include minimized threats of cultured fish escaping and competing with wild populations,
- Improved control of diseases and parasites, true management of water quality (temperature, oxygen rate, nutrient and suspended solids content), and
- Better control of nutrient releases to the environment

The investigations carried out by the Central Institute of Freshwater Aquaculture (CIFA) Bhubaneswar indicated that farming systems viz., fish-duck and fish-pig each in one ha pond yielded benefit-cost ratios of 1.42, 1.62 and 1.44 respectively.

A pond based IFS in 1.25 ha at Bhubaneswar comprising field and horticultural crops, fishery, poultry and duckery, apiary, mushroom, diary and agroforestry generated employment of 573 man days with benefit: cost (B:C) ratio of 2.18 ensuring better standard of living for small and marginal farmers (Behera and Mahapatra, 1999)

Behera *et al* (2008) also reported an employment generation of 248 man-days from one acre of IFS comprising field and horticultural crops, fishery, dairy and agro-forestry in the northeaster coastal plain zone of Odisha. This IFS gave higher net returns of the order of 80% than the conventional practice of growing crop component only.

In Chattisgarh, Ramrao *et al* (2006) developed a mixed farming module of 1.5 acre for marginal farmers that doubled the employment generation (316 man-days) and resulted in five-fold annual income over crop cultivation (Table 5)

Table 5: Income and Expenditure of different Mixed Farming Models for Marginal land holder of 1.5 acre (Ramrao *et al*., 2006)

Treatment	Expenses (Rs)	Net Income (Rs.)	Cost: Return Ratio	Employment (days)
Crop (1.5 acre)	12396	7843	1.63	165
Crop, 2 bullocks, I Cow	18920	14184	1.75	273
Crop, 2 bullocks, 1 buffalo	19188	18260	1.95	273
Crop, 2 bullocks, 1 cow, 1 buffalo,	21341	21462	2.00	291
Crop, 2 bullocks, 1 cow, 1 buffalo,10 goats,	23294	29400	2.26	308
Crop, 2bullocks, 1cow,1 buffalo,10 goats, 10 poultry and 10 ducks	24899	33076	2.23	316

Ramrao *et al* (2008) also reported further that IFS with 4 bullocks, 2 cows, 2 buffaloes and 25 goats along with poultry and duckery was the most beneficial system which can augment income levels significantly (Table below)

Cropping with goat and milch cow recorded higher system productivity 0f 16978 kg/ha and total income of Rs. 1,16,098/ha /year (Jayanthi, 2017).

Table 6: Income and Expenditure of different Mixed Farming Models for Middle level to large farmer (3 ha) in Chhattisgarh

Treatment	Expenses (Rs)	Net Income (Rs.)	Cost: Return Ratio	Employment (days)
Crop	77940	48255	1.67	907
Crop, 4 bullocks, 4 Cows	95294	67340	1.70	1158
Crop, 4 bullocks, 4 buffaloes	96366	83645	1.88	1158
Crop, 4 bullocks, 2 cows, 2 buffaloes,	95830	75493	1.78	1158
Crop, 2 bullocks, 2 cows, 2buffaloes,25 goats,	100713	95338	1.94	1202
Crop, 4 bullocks, 2cows,2 buffaloes,25 goats, 30 poultry and 30 ducks	105527	106367	2.00	1226

(Ramrao *et al.*, 2008)

While the data and findings presented above offer concrete proof for the economic viability and profitability of the Mixed farming systems presented, the systems could always be modified and refined with new animal and crop systems depending upon the needs of the site specific situations. The important consideration, however should be the replicability and economic sustainability of the mixed farming systems involving crops and livestock.

8

Climate Change and Dryland Agriculture

Climate change refers to long-term shifts in temperatures and weather patterns. These shifts may be natural, such as through variations in the solar cycle. But since the 1800s, human activities have been the main driver of climate change, primarily due to burning fossil fuels like coal, oil and gas. (https://www.un.org/en/climatechange/what-is-climate-change)

While it is possible to achieve enhanced food production to meet the demands of growing populations, the prevailing negative environmental trends, if continued, might cause crises in many parts of the world (Molden, 2007). The effects of climate change on agriculture can take place through changes in average temperatures, rainfall, and climate extremes (e.g., heat waves); changes in pests and diseases; changes in atmospheric carbon dioxide and ground-level ozone concentrations; changes in the nutritional quality of some foods; and changes in sea level. (Anonymous, 2014A).

Climate change effects on agriculture are unevenly distributed across the world. Future climate changes will most likely affect crop production in low latitude countries negatively, while effects in northern latitudes may be positive or negative. A range of policies for climate change adaptation may reduce the risk of negative climate change impacts on agriculture.

Despite technological advances, such as improved varieties, genetically modified organisms, and irrigation systems, climate is still a key factor in agricultural productivity, as well as soil properties and natural communities. The effect of climate on agriculture is related to variabilities in local climates rather than in global climate patterns. Consequently, in making an assessment, agronomists must consider each local area.

There is an opinion among populations that climate change is a recent phenomenon. This, however, is without basis. During the period from 1953 to 2018 there had been unprecedented and startling deviations from normal weather patterns in India which caused terrible rains, causing irrecoverable financial loss of the order of IRs. 400,000 crores (Four lakh crores of rupees) as per the estimates of National Disaster Management Organization of India. Around 50% of the cited financial loss occurred in the last ten years alone of the above period (Maitreya, 2022). Due to the disastrous effects of climate change, several hundreds of crores of people faced untold adversities since the year 2001. Around 83,000 people lost their precious lives only due to the adverse effects of climate change. There has been untold damage to Indian agricultural sector resulting in staggering financial loss of IRs. 13 lakh crores, which forms almost 6% of the national GDP as per the estimates of the State Bank of India (Maitreya, 2022).

Climate change effects on agriculture are unevenly distributed across the world. Future climate changes will most likely affect crop production in low latitude countries negatively, while effects in northern latitudes may be positive or negative. A range of policies for climate change adaptation may reduce the risk of negative climate change impacts on agriculture. Climate change causes global warming and hence the two have been treated together.

The United Nations Organization (UNO) reports that burning fossil fuels generates greenhouse gas emissions that act like a blanket wrapped around the Earth, trapping the sun's heat and raising temperatures.

Examples of greenhouse gas emissions that are causing climate change include carbon dioxide and methane. These come from using gasoline for driving a car or coal for heating a building, for example. Clearing land and forests can also release carbon dioxide. Landfills for garbage are a major source of methane emissions. Energy, industry, transport, buildings, agriculture and land use are among the main emitters.

Role of Greenhouse gasses

The greenhouse effect increases the temperature of the Earth by trapping heat in our atmosphere. This keeps the temperature of the Earth higher than it would be if direct heating by the Sun was the only source of warming (Royal Society, 2010). When sunlight reaches the surface of the Earth, some of it is absorbed which warms the ground and some bounces back to space as heat. Greenhouse gases that are in the atmosphere absorb and then redirect some of this heat back towards the Earth (Treut *et al*, 2007). The greenhouse effect is a major factor in keeping the Earth warm because it keeps some of the planet's heat that would otherwise escape from the atmosphere out to space. In fact, without the greenhouse effect the Earth's average global temperature would be much colder and life on Earth as we know it would not be possible (UK Met Office 2011) The difference between the Earth's actual average temperature 14° C

(57.2° F) and the expected effective temperature just with the Sun›s radiation -19° C (-2.2° F) gives us the strength of the greenhouse effect, which is 33 degrees C. (Treut *et al.*, 2007)

The increasing levels of greenhouse gases (GHGs) in the atmosphere have been attributed as a major driving force for rapid climate change. The main GHGs contributing to this phenomenon are CO2, CH4 and N2O. GHG emissions from agriculture sector alone constitute about 17% of the net CO2 eq missions

In 2000, Indian agriculture contributed 386.1-million-ton (Tg) CO2 eq. The agriculture sector primarily emitted CH4 (14.7 Tg) and N2O (137.3-thousand-ton, Gg). The sources accounting for emissions in the agriculture sector are enteric fermentation in livestock, manure management, rice cultivation, agricultural soils and burning of agricultural crop residue. The bulk of the GHG emission from the agriculture sector was from enteric fermentation (65%) followed by rice cultivation (23%) and the rest were contributed by manure management, burning of crop residue and application of N fertilizer to soil.

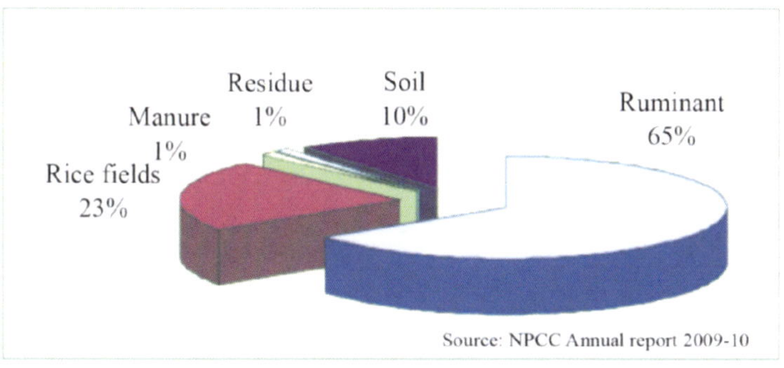

Fig. 1: Relative contribution of various sources to GHG emission in 2000

Greenhouse gas concentrations are at their highest levels in 2 million years

And emissions continue to rise. As a result, the Earth is now about 1.1°C warmer than it was in the late 1800s. The last decade (2011-2020) was the warmest on record.

Many people think climate change mainly means warmer temperatures. But temperature rise is only the beginning of the story. Because the Earth is a system, where everything is connected, changes in one area can influence changes in all others.

The consequences of climate change now include, among others, intense droughts, water scarcity, severe fires, rising sea levels, flooding, melting polar ice, catastrophic storms and declining biodiversity.

In a series of UN reports, thousands of scientists and government reviewers agreed that limiting global temperature rise to no more than 1.5°C would help us avoid the worst climate impacts and maintain a livable climate. Yet based on current national climate plans, global warming is projected to reach around 3.2°C by the end of the century. The emissions that cause climate change come from every part of the world and affect everyone, but some countries produce much more than others. The 100 least-emitting countries generate 3 per cent of total emissions.

The production of crops and animal products today releases roughly 13 percent of global greenhouse gas emissions, or about 6.5 gigatons (Gt) of carbon dioxide equivalent (CO2e) per year, without counting land use change. Even assuming some increases in the carbon efficiency of agriculture, emissions could plausibly grow to 9.5 Gt of CO2e by 2050. When combined with continuing emissions from land use change, global agriculture-related emissions could reach 15 Gt by 2050. By comparison, to hold global warming below 2° Celsius, world annual emissions from all sources would need to fall to roughly 21-22 Gt by 2050 according to typical estimates. To reach this target, agriculture must greatly reduce its greenhouse gas emissions— even while boosting production.

Relative contribution of various sources to GHG emission in 2000

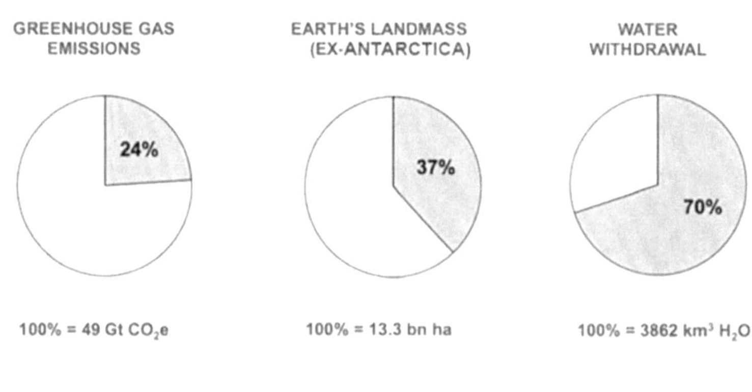

Fig. 2: Agricultural Share of Global Environmental Impact
(*Source*: World Resources Institute, 2010)

Methane Emission

Livestock rearing, an integral part of Indian agriculture is the major contributor of methane. Although the livestock includes cattle, buffaloes, sheep, goat, pigs, horses, mules, donkeys, camels and poultry, the bovines and small ruminants are the most dominant components, and the major source of methane emission. Methane emission due to enteric fermentation in 2000 was estimated to be 10.1 Tg. Buffalo and indigenous cattle, which are the main milk-producing animals in the country, contributed 44 and 42% total methane emission from livestock sector (Cross bred cattle emitted 8% and the small ruminants emitted about 7% of methane.

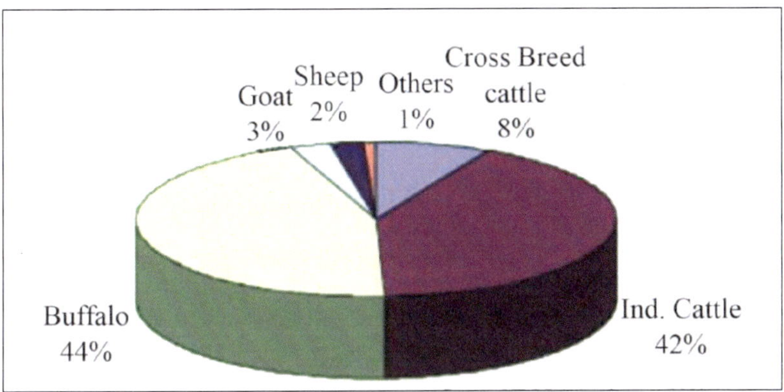

Fig 3: Relative contribution of methane emission by different categories of ruminant

In India, rice is cultivated under diverse management conditions, depending on availability of water. Methane emission due to rice cultivation was estimated to be 3.5 Tg. Continuously flooded rice emitted maximum methane (1111 Gg) followed by flood prone (827 Gg) and single aerated (598 Gg) rice cultivation.

N2O Emission

During 2000, Indian agriculture produced 137.3 Gg of N2O-N. Nitrogenous fertilizer application contributed 68% of that emission followed by manure and crop residue. Because of increased use of N fertilizer, N2O emission is also increasing over the years

When water warms up it expands. At the same time global warming causes polar ice sheets and glaciers to melt. The combination of these changes is causing sea levels to rise, resulting in flooding and erosion of coastal and low lying areas.

Heavy rain and other extreme weather events are becoming more frequent. This can lead to floods and decreasing water quality, but also decreasing availability of water resources in some regions. The European Union in their report (https://ec.europa.eu/clima/climate-change/climate-change-consequences_en) the following consequences of climate change are listed.

Climate change affects all regions around the world. (https://ec.europa.eu/clima/climate-change/climate-change-consequences_en) Polar ice shields are melting and the sea is rising. In some regions, extreme weather events and rainfall are becoming more common while others are experiencing more extreme heat waves and droughts. These impacts are expected to intensify in the coming decades.

Rainfall is the key variable influencing crop productivity in agricultural crops in general and rainfed crops in particular. Intermittent and prolonged droughts are a major cause of yield reduction in most crops. Long-term data for India indicates that rain fed areas witness 3–4 drought years in every 10-year period. Of these, 2–3 are in moderate and one may be of severe intensity. Analysis of number of rainy days based on the IMD grid data from 1957 to 2007 showed declining trends in Chhattisgarh, Madhya Pradesh, and Jammu Kashmir. In Chhattisgarh and Eastern Madhya Pradesh, both rainfall and number of rainy days are declining which is a cause of concern as this is a rain-fed rice production system supporting large tribal population who have poor coping capabilities.

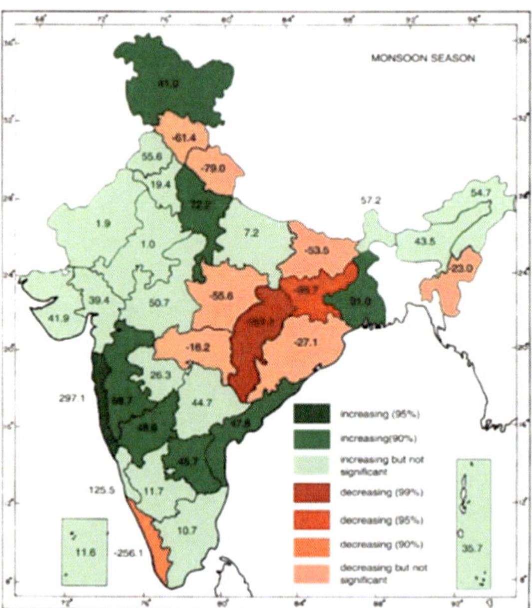

Fig. 4. Increase/decrease in summer monsoon rainfall (mm) in India during last century (Guhathakurta and Rajeevan 2007)

Guhathakurta and Rajeevan (2007) reported that all India summer monsoon (June to September) rainfall does not show any significant trend during the last century. However, three subdivisions viz., Jharkhand, Chhattisgarh, Kerala show significant decreasing trend and eight subdivisions viz., Gangetic West Bengal, West Uttar Pradesh, Jammu & Kashmir, Konkan & Goa, Madhya Maharashtra, Rayalaseema, Coastal Andhra Pradesh and North Interior Karnataka show significant increasing trends. However, the amount and distribution of rainfall is becoming more erratic which is causing greater incidences of droughts and floods. Temperature is another important variable influencing crop production particularly during rabi season. A general warming trend has been predicted for India but knowing temporal and spatial distribution of the trend is of equal importance. Lal (2001) reported that annual mean area-averaged surface warming over the Indian subcontinent is likely to range between 3.5 and 5.5 °C by 2080s

Table 1: Projected mean temperature changes over the Indian subcontinent (Guhathakurta and Rajeevan 2007)

Year	Temperature change(°C)		
2020	Season	Lowest	Highest
	Annual	1.00	1.41
	Kabi	1.08	1.54
	Kharif	0.87	1.17
2050	Annual	2.23	2.87
	Rabi	2.54	3.18
	Kharif	1.81	2.37
2080	Annual	3.53	5.55
	Rabi	4.14	6.31
	Kharif	2.91	4.62

(i) Consequences for Europe

- **Southern and central Europe** are seeing more frequent heat waves, forest fires and droughts.
- **The Mediterranean area** is becoming drier, making it even more vulnerable to drought and wildfires.
- **Northern Europe** is getting significantly wetter, and winter floods could become common.
- **Urban areas**, where 4 out of 5 Europeans now live, are exposed to heat waves, flooding or rising sea levels, but are often ill-equipped for adapting to climate change.

(ii) Consequences for developing countries

Many poor developing countries are among the most affected. People living there often depend heavily on their natural environment and they have the least resources to cope with the changing climate.

(iii) Risks for human health

Climate change is already having an impact on health:

- There has been an increase in the number of heat-related deaths in some regions and a decrease in cold-related deaths in others.
- We are already seeing changes in the distribution of some water-borne illnesses and disease vectors.

(iv) Costs for society and economy

Damage to property and infrastructure and to human health imposes heavy costs on society and the economy.

Between 1980 and 2011 floods affected more than 5.5 million people and caused direct economic losses of more than €90 billion.

Sectors that rely strongly on certain temperatures and precipitation levels such as agriculture, forestry, energy and tourism are particularly affected.

(v) Risks for wildlife

Climate change is happening so fast that many plants and animal species are struggling to cope. Many terrestrial, freshwater and marine species have already moved to new locations. Some plant and animal species will be at increased risk of extinction if global average temperatures continue to rise unchecked.

Global climate change has already had observable effects on the environment. Glaciers have shrunk, ice on rivers and lakes is breaking up earlier, plant and animal ranges have shifted and trees are flowering sooner.

Effects that scientists had predicted in the past would result from global climate change are now occurring: loss of sea ice, accelerated sea level rise and longer, more intense heat waves.

According to Cline (2010), two factors viz., (i) carbon fertilization and (ii) irrigation are very significant and important with regard to global warming which is the outcome of climate change.

(i) Carbon fertilization: The extent to which carbon fertilization could alleviate any adverse effect of global warming on agriculture is the central issue in the analysis of severity of these effects. Carbon dioxide is an input in photosynthesis, which uses solar energy to combine water and carbon-dioxide to produce carbohydrates, with oxygen as waste product. In addition, higher concentration of carbon-dioxide reduces plants' stomatal openings causing less loss of water to respiration. The C3 crops viz., rice, wheat, soybean, fine grains, legumes and large number of trees, benefit from additional atmospheric carbon-dioxide. Such benefits are very much limited in the case of C4 crops viz., maize, millet, sorghum and sugarcane. Research on experiments (FACE) indicate that with carbon dioxide to550 to 575 ppm the yield increase 11% for C3 crops and 7% for all five major food crops, which is one third to one quarter of the direct effect of CO2 modelled in the assessment for Europe and the USA (Darwin and Kennedy 2000). Later, Long *et al*., (2006) reported that FACE studies indicated that at 550 ppm carbon dioxide concentration, yield increases amount to13% of wheat as against 31% in laboratory studies, 14% instead of 32% of soybeans and 0% instead of 18% for C4 crops. Among the major crops, C3 species account for approximately 3/4th and C4 for1/4th of total value. This study investigates the emissions during the period 2070-2099, during which at the mid-point (2085), the central scenario used in the study places the atmospheric CO2 concentrations at 735 ppm (IPCC,2007). The carbon fertilization rises less than linearly with atmospheric concentrations (Long *et al*, 2006). Thus the central estimate of carbon fertilization effect by the 2080s is set at 15% increase in yield, which is smaller than the values arrived at in the past studies (Cline 2010). Using the atmospheric concentration of 735 ppm of CO2 by the 2080s, and applying the world future average temperature and precipitation for land weighted by farm land area – 16.2degrees C and 2.4mm per day, the output per hectare gives an estimate of US$333.2 without Carbon fertilization and US$ 376.6 with carbon fertilization, which amounts to an increase of 13%. (Cline, 2010).

(ii) Irrigation: Evapotranspiration (i.e., combined loss of moisture from soil through evaporation and plants through stomatal transpiration) increases with temperature. The need for irrigation rises as conditions become drier. It rises as function of the difference between evapotranspiration and precipitation. Since global warming will increase both temperature and precipitation, the implications of soil moisture and the need for irrigation depend on the outcome of race between rising temperature and rising precipitation (Cline 2010). As shown in a test case for USA, the incidence of irrigation is positively relate to temperature and negatively related to precipitation. In other words, higher incidence of irrigation for higher tempertures and lower incidence of irrigation with higher precipitation. The climate analysis for USA indicate that baseline global warming by the 2080s would cause the farm land-weighted averages for annual temperatures to rise by 5.4degrees C and the corresponding averages for precipitation to decline by 4.3 mm per year (Cline, 2010). In other words, the incidence of irrigation would need to rise by 20.3% as consequence of climate change. That means the race between temperature and precipitation would be won by temperature (Cline, 2010).

The climate models (Cline, 2010) place the change in the global mean temperature from1961-90 to 2070-99 at an average of 3.0 degrees C and a range of 1.3 degrees C to 4.5 degrees C. The higher degree of warming reported for land areas, 4.95 degrees C weighted by land area and 4.43 degrees C by farm area. This reflects that realized surface warming by given future date would be expected to be greater over land than for oceans (Cline, 2010). This is an important distinction for exercises examining agricultural impact in response to temperature change, since it means that the relevant temperature change for agriculture will be higher than the global mean temperature.

Considering the situation in respect of Canada, Reinsborough (2003) emphasized presence of an upper bound on benefits of climate change due to perfect adjustment to climate change without any costs other than change in land value. The subsequent work by Mendelsohn and Reinsborough (2007) confirms the above finding.

Considering that prime agricultural land today is in neither tropical or high – latitude areas, the prognosis implies lesser increase in precipitation for current agricultural areas than for global land based means (Cline, 2010).

According to the Mendelsohn, Dinar and Sanghi's model of 2001 (Cf. Cline 2010) model for India reported by Cline (2010), the reductions in revenue output potential ranging from about 35% in southern regions to about 60% in the northern regions. Irrespective of carbon fertilization, the losses would be severe for India. It is argued that the favorable impact of higher precipitation due to climate change is much smaller and at times negligible than the unfavorable impact of higher temperatures (Mendelsohn, Dinar and Sanghi, 2001).

At global level, even if the moderate reduction of 3% assuming the prevalence of carbon fertilization, the large disparity of results across countries would mean much more serious losses for many countries and regions. Generally, the developing countries would be affected badly than the industrial countries (Cline, 2010).

There are 21 countries, regions of large countries in which the production capacity would decline by more than 33% without carbon fertilization. The seven of the above in which the declining trend is glaringly evident even with carbon fertilization. If the thresh-hold is set at 20% or even greater, there shall be 53 countries and regions with severe losses without carbon fertilization and 29 would suffer decline even with carbon fertilization (Cline, 2010).

According to Mendelshon *et al.*, (2000) and Cline (2010), the results indicate that Africa (excluding Egypt and other North Africa) and Latin America are two developing regions most vulnerable to climate change and global warming. Though on average Asia is less vulnerable; this masks the divergence between the more favorable results for China in particular and more unfavourable results for India, reflecting the difference in their latitudes. The results show an inequitable distribution of the effects of climate change which includes global warming, due to the fat that the poor countries tend to be located in lower latitudes, where temperatures are already at or above optimal levels.

Cline (2010) warned that it would be a serious mistake to downplay the risks of future agricultural losses due to global warming on the grounds that technological innovations such as new and novel genetic crop varieties would offset any negative climatic effects. A close look at the pace of yield increases in the past two decades, combined with attention to prospective rise in global food demand and the conversion of substantial portion of agricultural land from food to energy crops, suggests that there is little margin for complacency about erosion in agricultural potential due to global warming. AT 2019)

Modeling studies by the Indian Council of Agricultural Research (ICAR) under the Network Project on Climate Change (NPCC) indicate that in the medium term (2010-2039), impacts due to climate change on food production are in the range of 4.5 to 9 per cent, while in the long term (2070-2099) the production is likely to decrease by as much as 25 per cent.

Since agriculture makes up roughly 15 per cent of India's GDP, a 4.5 to nine percent negative impact on production implies a cost of climate change to be roughly up to 1.5 per cent of GDP per year. For every degree increase in temperature throughout, in the growing season, the production of wheat in the country may reduce by 4-5 million tones. Rise in minimum temperature is happening during *kharif* season @ 0.19^0C 10 yr^{-1} in most part of the country which has a negative impact on paddy yields over 50% of cultivated area. (Bapujirao *et al.*, 2015)

Based on the recent growth rates in production and contribution to agricultural GDP, horticulture, dairy, meat and poultry are considered 'sunrise' sectors. These sectors are also equally vulnerable to global warming. Horticulture crops are particularly sensitive to temperature, unseasonal rainfall, hailstorms and pest and diseases caused by climate variability. The yields of apple in Himachal Pradesh are declining in the traditional areas because of rise in temperature and inadequate chilling requirement (NPCC, 2007). Farmers are shifting the crop to higher elevations on the hills in order to realize acceptable yields. Mango is another typical example of a premier crop in India suffering production losses due to climatic variability. In addition to the indiscriminate use of Paclobutrazol, unseasonal rains and temperature fluctuations are mainly attributed to the flowering, fruit drop and quality problems in alphonso mangoes. Grapes, oranges and pomegranate in Maharashtra and kinno in Panjab and Himachal Pradesh repeatedly suffer heavy losses due to hail storms. Climate change impacts pollination and yields, as high temperature affects pollinators.

Global warming also affects the livestock sector. Rise in THI in many parts of the country causes the dairy animals spend more energy for maintenance with negative implications on milk production. Modelling estimates by NDRI indicate that milk production might decline by 15 million tons by 2050. High producing crossbred cows and buffaloes will be affected more. The annual loss in milk production at all-India level due to heat stress was estimated at Rs.2662 crores. (Upadhyay *et al.*, 2009). While the commercial poultry is highly vulnerable to heat wave and we saw huge mortality of poultry birds during summer heat waves till a decade ago, the industry has responded quickly by improving the designs of the poultry farms and installation of foggers etc. They are also bringing new innovations in feed incorporating supplements that enhance the coping ability of birds to heat stress.

Besides direct effects on crops, climate change is likely to impact natural resources like soil and water. Increased rainfall intensity in some regions would cause more soil erosion leading to land degradation. Increased temperatures will also increase crop water requirement. A study by ICAR-CRIDA, Hyderabad on major crop growing districts in the country for four crops, viz., groundnut, mustard, wheat and maize indicated a 3% increase in crop water requirement by 2020 and 7% by 2050 across all the crops/locations. Irrigation requirement in arid and semi-arid regions is estimated to increase by 10% by every 1°C rise in temperature. The frequency of extreme weather events like hail storms is also increasing in recent period.

Hail storms usually cause localized damage and generally occur in Panjab and Himachal Pradesh regularly. However, during 2014 and 2015 rabi seasons

(February – March), widespread hail storms occurred in most parts of the country, i.e. Gujarat, Maharashtra, Karnataka and Telangana causing heavy damage to field crops and horticulture.

Climate change (CC) does not impact every one equally. For example, rain-fed agriculture is more vulnerable because of the dependence on monsoon. A vulnerability mapping of Indian agriculture to climate change by ICAR-CRIDA (Ramarao *et al*, 2013) indicated that many districts in the western Indian states like Rajasthan, Gujarat and Karnataka are highly vulnerable to droughts, while parts of eastern UP and Bihar are sensitive to flood damage. Small and marginal farmers are more vulnerable due to their poor coping abilities. Extreme events like hail storms, severely impact horticulture crops like grapes, oranges, mango, banana causing flower and fruit drop and sometimes farmers loose the entire crop for the season. Agricultural labourers loose wage income when crops fail but no systematic study has been made on the impact of CC on landless.

In view of the economic impact of climate change on agriculture sector and its implications on food security and farmers' welfare, adaptation and mitigation measures are urgently required at different levels. Individual farmers have to be provided with innovative technologies and products that enable him to reduce farm level production declines and minimize his risks, while States and Nations have to put in place the required policies that sustain aggregate level production. This is to check price volatility and avoid huge relief payments on compensation of losses. Adaptation, particularly has a prominent role in developing countries like India. This is particularly critical in mitigation. The recently concluded Paris Agreement has mandated partner countries to reduce GHG emissions by certain percentage so that the temperature rise can be contained below 2^0C. The establishment of Green Climate Fund is an important step in this direction.

Economic Losses of Climate Change in Agriculture

Economic losses from natural disasters are rising globally, and agriculture sector is highly vulnerable to these disasters. According to the United Nations Office for Disaster Risk Reduction (UNISDR 2018), disaster-hit countries experienced direct economic losses to the tune of US$ 2908 billion during 1998–2017. Of the total losses, 77 percent were due to climate related disasters. Climate change impacts are more pronounced on agriculture sector in the recent past. Government of India's economic survey (2018) estimated that the annual loss of US$ 9-10 billion was due to the adverse effects of climate change.

Right from the Himalayas to the coastal South Asian countries must always be prepared to combat the effects of global warming (Stern, 2006). As predicted, the South Asian zones may experience a warming effect of 2° to 6° C during the 21 century (Ravindranath, 2007). Specifically, Indian sub-continent and other continents are highly vulnerable to all kinds of existing climate change issues.

Impact of climate change in rain-fed agriculture

Floods: The main causes unprecedented floods in India are due to movement of cyclonic disturbances from Bay of Bengal and Arabian Sea on to the land masses during monsoon and post-monsoon seasons – In 2008, millions of people were forced to live in shelter-houses in Bihar due to floods. About 20 million people faced the similar problem in Mumbai in 2005. In Delhi and Haryana, millions over worth of properties were destroyed when Yamuna river was flooded. (NITI Ayog, 2021) (Fig.5).

The Indian summer monsoon rainfall varied from 604 to 1020 mm. These inter-seasonal variations cause floods and droughts, which are the major climate risk factors in Indian Agriculture.

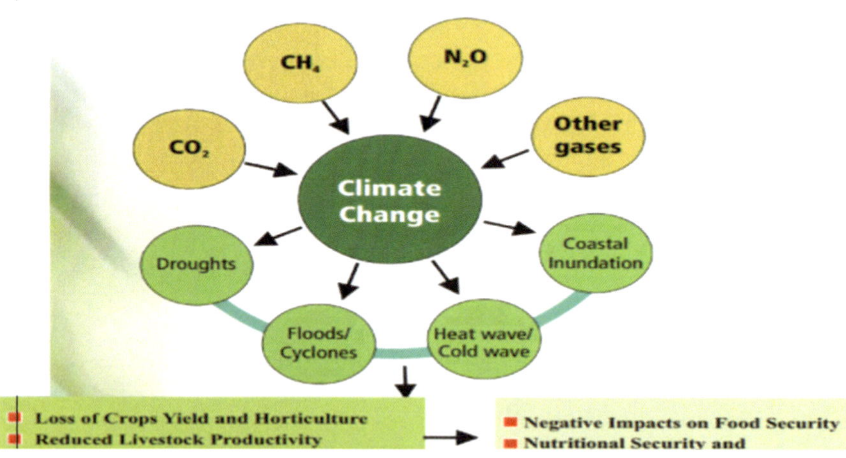

Fig. 5: Causes and impact of climate change on agriculture and allied sector

Droughts

Drought is a normal, recurrent feature of climate associated with deficiency of rainfall over extended period of time to different dryness levels describing its severity. Drought being subtle, its progress is insidious, and its affects can be devastating. During the period 1871-2009 there were 24 major drought years, defined as years with less than one standard deviation below the mean All

India Seasonal Rainfall (the deviation below –10%): 1871,1873, 1877, 1899, 1901, 1904, 1905, 1911, 1918, 1920, 1941, 1951, 1965, 1966, 1968, 1972, 1974, 1979, 1982, 1985, 1986, 1987, 2002 and 2009. The most recent major drought in 2009 expressed 50 country's continued vulnerability to droughts and the food-grain production in kharif is likely to fall by 1-015 million tons.

Arid regions in South Asia, including dry regions of Rajasthan in India and some regions of Pakistan, are facing severe drought. However, due to frequently occurring different kinds of droughts (late onset, mid-season and terminal), 2/3 of India's agriculture areas are under rain-fed condition. Western Rajasthan, parts of Haryana, Uttar Pradesh, Maharashtra, Southern Bihar, Madhya Pradesh, Southern Gujarat, Northern parts of Andhra Pradesh, and Karnataka are regularly facing dryness, and these regions are highly vulnerable to drought (Bhadwal *et al*., 2007). The arid and semi-arid zones are vulnerable to the losses of economic activities and livelihoods due to the changes in rate of precipitation.

Heat Waves

Heatwaves: Frequency and intensity of heatwaves are raising in India adversely affecting all allied sectors of agriculture including dairy, poultry, fishery, etc. Low water availability coupled with heatwaves have severe consequences on food security of the country. The heat wave conditions during 2003 May in Andhra Pradesh and 2006 in Orissa are examples that have affected the economy to a greater extent. Drinking water crisis for humans and livestock besides drying of long-standing horticulture orchards was seen. India and its neighbouring countries experienced a severe and longest heatwave from mid-May to mid-June in 2019. Churu in Rajasthan state, India documented a record of high temperature up to 50.8° C (123.4° F), which is almost missed by fraction of degree i.e., 51.0° C (123.8° F) highest set in 2016. As of 12 June 2019, the second largest heatwave period (32 days) ever was documented.

Extreme Events (Cyclones and storms)

Another troublesome indicator of global climate change such as storms, cyclones, landslides, etc., extremely affect South Asian countries. Around nine million people in the world were severely affected due to the destructive cyclones (UN assessment). The super cyclone in 1999 hit Orissa took toll of over a million lives besides properties loss in the coastal areas (Ahluwalia and Malhotra, 2006). Similar is the case with the coastal state of Andhra Pradesh during Hud-Hud cyclone in 2014.

Melting of glaciers

Future situations of South Asia may become adverse due to melting of glaciers and snow in the Himalayan regions. The Himalayan glaciers are rapidly melting down because of high temperature and if it continues in future, the fresh water stock for survival will be at great risk (Bajracharya, et.al. 2007). Increased air temperatures have resulted in the shrinkage of the Chhota Shigri Glacier by about 12 percent in the last 13 years. In addition, 12 percent shrinkage of the main stem of the Gangotri Glacier has occurred in the last 16 years. This implies a rapid rate of shrinkage of many of the glaciers in the Himalaya in recent years. From model results, it was determined that, for an increase of +3 °C air temperature, the ELA will move upward about 400 m, and the AAR of the Chhota Shigri and Gangotri Glaciers will decrease from 0.4 to about 0.10 and 0.15, respectively. The model thus suggests that, if there is an increase of +3 °C air temperature during the summer, 80 to 90 percent of the surface area of Chhota Shigri and Gangotri Glaciers will be in the ablation area and therefore in the process of melting. This will further increase the melt water discharges, rapid shrinkage, and rapid snout recession already seen at the end of the last century. Increases in the size and number of glacier lakes in the Himalaya clearly reflect the increasing influence of global climatic changes in the region (Vohra, 2005)

Cold waves

The Northern states of Punjab, Haryana, U.P., Bihar and Rajasthan experience cold wave and ground frost like conditions during winter months of December and January almost every year.(The frequency of National Symposium on Climate Change and Rainfed Agriculture, February, 18-20, 2010, CRIDA, Hyderabad,). India 39 such weather related events has significantly increased in the recent past due to reported climatic changes at local, regional and global scales.

Site-specific short-term fluctuations in lower temperatures and the associated phenomena of chilling, frost, fogginess, and impaired sunshine may sometimes play havoc in an otherwise fairly stable cropping/farming system of a region.

Sectorial impacts in India

'Agriculture' is the main occupation for 50 percent of population in India. Agriculture and allied sectors contribute 15.4 per cent of the Indian GDP (OECD, 2017). Farming activities are carried out by the selection of crop which is specific to suit climate, soil type, resource availability, etc. Therefore, farming production and productivity is completely dependent on climatic conditions

(Srinivasarao *et al.*, 2016; Bal and Minhas, 2017). Weather disruptions, like changes in temperature, precipitation and solar radiation, affect the agriculture ecosystem including livestock, arable and hydrology sectors. As per the global report prediction, a loss of 10- 40 per cent in crop productivity is estimated for 2100

Field crops

An average of 30 per cent decrease in crop yields is expected by mid-21st century in South Asian countries. Bangladesh, India, Pakistan and Mozambique are highly susceptible to erratic changes in rainfall and temperature (World Bank, 2011). For example, in India, an increase in temperature by 1.5° C and decrease in the precipitation of 2 mm, reduces the rice yield by 3 to 15 percent (Ahluwalia and Malhotra, 2006). Climatic changes driven by increasing Green House Gases (GHGs) possibly affects the yield and productivity of agricultural crops from region to region. According to the Met Office (United Kingdom's National Weather Service), the normal crop yield is anticipated to decrease by 50 per cent in Pakistan. The production of maize in European countries is expected to increase by 25 percent in ideal hydrologic conditions (en.wikipedia.org). The drastic changes in climate alters the progressive stages of pathogens that eventually affect the growth and yields of crops severely, and also could lead to an increase in pest and insect population, ultimately devastating the overall productivity.

Impacts of Carbon Fertilization

Elevated carbon dioxide level in the lower atmosphere is another important manifestation of climate change. In some crops, higher carbon dioxide levels in the air have a fertilization effect and result in production of more bio-mass and grain yield when water is not limiting and temperatures remain within the tolerance thresholds. In major rainfed crops grown during kharif season, it is found that water limitation is a major yield limiting factor than temperature. Therefore, there is a possibility that these crops would benefit from elevated CO2 coupled with higher rainfall as predicted in most scenarios.

Influence on Growth

Studies at CRIDA, Hyderabad through Open Top Chambers (OTCs) revealed that the response of different rainfed crops varies significantly to elevated CO2 (CRIDA 2009-2010). In general, the response of C3 crops was found to be higher than C4 crops. The increment in total biomass was 46% in pulses (C3) and 29% in oilseeds at vegetative stage with 600 ppm CO2, whereas with C4 cereals it was only 15% (Vanaja *et al.* 2006). Similarly, the response of black gram (*Vigna mungo* L. Hepper) to increased levels of CO2 (600 ppm) was significantly higher when moisture stress was imposed. This was possible due to greater proportioning of assimilates to the roots than to shoots under stress.

Influence on Flowering

Flowering is a critical milestone in the life cycle of plants, and changes in the timing of flowering may alter processes at the species, community and ecosystem levels. Therefore, flowering response to elevated CO2 was studied in castor bean, a monoecious plant bearing both male and female flowers on the same spike. The flowering in castor is sensitive to temperature. Our study revealed that elevated CO2 levels significantly reduced the duration of "days to initiation of flowering" and increased the number of female flowers in the spike (Vanaja *et al.* 2008) and there by improved the seed yield. Past studies indicate that carbon metabolism exerts partial control on flowering time, and therefore may be involved in elevated CO2 induced changes in flowering time. More studies are needed on the impacts of climate change drivers on plant development processes.

Influence on Seed Yield

The grain yield of pigeon pea improved from 22.8 g/plant at ambient to 42.4 g/plant at 700 ppm thereby showing an increment of 85.9% with enhanced CO2. In black gram, the grain yield recorded an increment of 129.3% at elevated CO2. The pod number per plant in pigeon pea showed an increase of 97.9% over ambient control making it an important yield contributing component. The flower to pod conversion and retention of flowers as well as pods in pulses is moisture, temperature and nutritional stresses there by reduction in sink size which results in reduced grain yield. A significant increase in the harvest index was observed in pulse crops under elevated CO2 due to improved partitioning efficiency

Influence on Quality Parameters

The experiments conducted with Open Top Chambers also revealed that there was no significant change in the content and quality of castor bean oil at two elevated CO2 levels of 550 and 700 ppm. However, the total oil yield was significantly higher owing to higher seed yields (Vanaja *et al.* 2008). In groundnut an increase of 3.3% oil content was observed at 550 ppm CO2 level when compared with ambient control and the change in protein content was insignificant. In edible oils, unsaturated fatty acids viz., oleic,linoleic and linolenic acids play a vital role for the human nutrition view while in the non-edible oil like castor it is the ricinoleic acid which is important for its industrial utility

Pests and diseases

Of the several environmental factors, temperature extremes play critical role in influencing the population of insect pests. Our experience indicated that there is a 'shift' in the pest status of several key species in recent periods, though these shifts may not be solely attributable to climate change. In the present situation, one can expect significant changes in the growth, development and population dynamics of various insect pests. The duration of the insect life cycle is altered under increased temperature and elevated CO2 concentrations resulting in variable number of generations per year. Under elevated CO2 higher consumption of foliage by leaf chewing insects with extended larval duration is observed in many studies. Published data reveals that some pests become more serious while others may decline. Evaluating the impact of climate change on insect pests is a very complex exercise and requires greater understanding of interactive factors. A more critical database on biotic stresses and their relationship with drivers of climate change is required for evolving adaptation strategies by farmers. The impact of rising temperature and CO2 are also likely to change insect-pest dynamics. Dilution of critical nutrients in crop foliage may result in increased herbivory of insects. For example, Tobacco caterpillar (*Spodoptera litura*) consumed 39% more castor foliage under elevated CO2 conditions than ambient environment (Srinivasa Rao *et al.* 2012). Low carbon soils of mainly dryland areas of India are likely to emit more CO2 compared to high or medium carbon temperate region soils.

Table. 2: Impact of monsoonaland annual run-off in India.

Region / location	Impact
India subcontinent	• Increase in monsoonal and annual run-off in the central plains
	• No substantial change in winter run-off
	• Increase in evaporation and soil wetness during monsoon and on an annual basis
Orissa; and West Bengal	One metre sea-level rise would inundate 1700 kin^2 of prime agricultural Laud
Indian coastline	One metre sea-level rise on the Indian coastline is likely to affect a total area of 5763 km* and pul 7.1 million people at risk
All-India	Increases in potential evaporation across India
Central India	Basin located in a comparatively drier region is more sensitive to climatic changes
Kosi Basin	Decrease in discharge on Ihe Kosi River; Decrease in run off by 2
Southern and Central India	Soil moisture increases marginally by 15-20% during monsoon months
Chenah River	Increase in discliarge in the river
River basins of India	General reduction in the quantity of the available run-clll increase in Mahanadi and Bmhmini basins
Danuxlar Basin	Decreased river flow
Rajasthan	Increase in evapotranspiration

Facing aberrant weather for crop production - particularly drought

Recommendations for aberrant weather have been worked out by the ICAR System in India as back as 1979 and passed on to the implementing agencies. The classical example was the worst drought the country faced in 1979 was managed well resulting in satisfactory agricultural production throughout the country.

The following strategy may be of some help in mitigating the adverse effects of the abnormal monsoon conditions (Vayugrid, 2014)

- Even with incipient showers that may occur any time between April and the end of May, get the soil prepared and keet ready for planting with the subsequent showers. During this process it would be advisable to incorporate substantial quantum of organic matter (a minimum of 3 tons / hectare)
- Prepare the land in the form of ridges and furrows across the slope spaced with a minimum spacing of 60 cm between the rows (could be up to 90 cm in case deep-rooted and long duration crops like castor / pigeon pea that are to be grown under rain-fed conditions)

- Alongside make a provision for farm ponds in low lying unproductive land (if available) to collect run-off water in the vent of high intensity rainfall.
- Plant suitable crops depending upon the time of occurrence of next spell of showers. If the rains occur in June, one can plant drought resistant crops like soghum, groundnut, pigeon pea etc.
- If the rains are delayed beyond first week of July Castor, sunflower, ragi etc., should be helpful in ameliorating the adverse situation.
- Foliar spray of 1 to 2% urea during the drought spell helps in maintaining a healthy and productive plant growth.
- Pest and disease management under altered or unseasonal rainfall situations is of great importance. To cut down the cost of production due to ever increasing prices of chemical pesticides and to avoid the harmful residues of chemicals, it is desirable to use preventive spays of extracts of *Pongamia*, neem and custard apple leaves and neem based pesticides. Application of *Pongamia* cake to the soil following preparatory tillage not only prevents a wide range of pests and diseases; but also enhances the soil fertility.
- Allocate part of the land holding for tree planting. Allocate less fertile part of the land to plant high yielding grafts of *Pongamia* and other fruit trees viz., cashew nut. These trees are highly drought resistant and even in the years of drought that result in failure of annual crops, the above mentioned trees yield satisfactorily, thereby mitigating adverse effects of drought
- The above tree plantations provide for inter-cropping of fodder and forage crops that would be in short supply during the drought years. This strategy of intercropping perennial trees with fodders / forages ensures steady supply of green fodder to feed cattle.
- The trees such as *Pongamia* provide for bio-fuel that shall be of immense value in ensuring environmental protection. Plantations of high yielding strains of *Pongamia* grafts ensure consistent monetary returns irrespective of weather patterns. (Vayugrid, 2014)

Climate Change Impact on Livestock

India owns 57 % of the world's buffalo population and 16 % of the cattle population. It ranks first in the world in respect of cattle and buffalo population, third in sheep and second in goat population. The sector utilizes crop residues and agricultural by-products for animal feeding that are unfit for human consumption A rise of 2 to 6°C in temperature is expected to negatively impact growth, puberty and maturation of crossbred cattle and buffaloes. The low producing indigenous cattle are found to have high level of tolerance to these adverse impacts than high yielding crossbred cattle. Therefore, high producing crossbred cows and buffaloes will be affected more by climate change. A rise of 2–6 °C due to global warming (time slices 2040–2069 and 2070–2099) is likely to negatively impact growth, puberty and maturity of crossbreds and buffaloes and time to attain puberty of crossbreds and buffaloes will increase by one to two weeks due to their higher sensitivity to temperature than indigenous cattle. Lactating cows and buffaloes have higher body temperature and are unable to maintain thermal balance. Body temperature of buffaloes and cows producing milk is 1.5–2 degrees C higher than their normal temperature, therefore more efficient cooling devices are required to reduce thermal load of lactating animals as current measures are becoming ineffective (Upadhyay *et al.* 2009)

Poultry

Poultries are extremely sensitive to temperature-associated issues, specifically heat stress. Endocrinological changes caused by prolonged heat stress in broiler chickens enhance lipid accumulation, reduced lipolysis, and induced amino acid catabolism (Geraert *et al.* 1996). Due to heat stress, feed intake of poultries will be reduced (Deng *et al.*, 2012), which leads to less body weight, egg production and quality of meat, and also reduces the thickness of eggshell and increases the egg breakage (Lin *et al.*, 2004). Heat stress has negative effect on strength, weight, ash content and thickness of the eggshell (Miller and Sunde, 1975). Rising environmental temperature may cause seasonal improvement in growth and development of fishes, but increases the risks to the populations living beyond the thermal tolerance zone (Morgan *et al.*, 2001). The rise in temperature of 1° C will affect the mortality of fish and its geographical distribution (Vivekanandan *et al.*, 2009). The temperature rises of 0.37° C to 0.67° C alter the pattern of monsoon seasonal variations, eventually shifting the breeding period of Indian main carps from June to March in West Bengal and Orissa's fish hatcheries (DARE/ICAR Annual Report, 2008-09)

The analysis of mortality data from 2004 to 2009 at Project Directorate of Poultry, Hyderabad revealed that the overall mortality was increased with rise in the ambient temperature of broiler, layer and native chickens. The mortality started increasing when the temperature reached 32°C and the peak was observed at 38 to 39°C (13.5%). The mortality was highest in broiler type chickens followed by layers and native chicken. The mortality due to heat stress in broiler type birds started appearing at the ambient temperature of 30°C, while in layer and native chicken the heat stress related mortality was observed at the ambient temperature of 31°C. The deaths due to heat stress were 10 times more in broiler type chickens as compared to layer and native type chicken. The mortality due to heat stress was negligible in native (Desi type chickens) which may be due to low metabolic rate and natural heat tolerance (Wasti *et al.*, 2020).

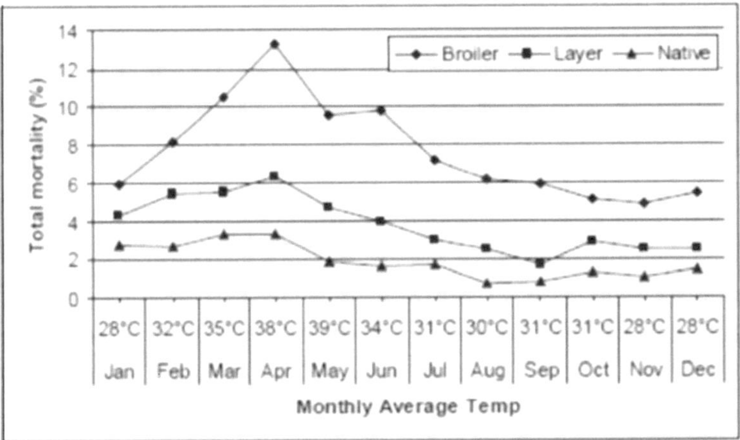

Fig. 6: Effect of ambient temperature on the survivability of meat type (broiler), egg type (layer) and native (desi) chicken. Another study was conducted to find the influence of high ambient temperature on feed intake body temperature and respiratory rate in commercial layers for 13 weeks. The consumption which was 108 g/bird-day at 28°C was reduced to 68 g/bird-day at the shed temperature 37.8°C (Wasti *et al.*, 2020).

Climate Change Impact on Fisheries

Marine fisheries

A rise in temperature as small as 1°C could have important and rapid effect on the mortality of fish and their geographical distributions. Oil sardine fishery did not exist before 1976 in the northern latitudes and along the east coast of India as the resource was not available and sea surface temperature (SST) were not congenial. With warming of sea surface, the oil sardine is able to find temperature to its preference especially in the northern latitudes and eastern longitudes, thereby extending the distributional boundaries and establishing fisheries in larger coastal areas as (Vivekanandan *et al.* 2009).

Livestock, Poultry and Fishery sectors

The climate change, whether it is global, regional or in a smaller scale, has a greater impact on biological production, or sum of those processes acts directly on individual organisms or species. The growth and development of any species with some specific characteristics is governed by their resilience and tolerance to the changes in their environment. Global climate changes affect numerous factors which are associated with production, reproduction, health and adaptability of every animal. Higher temperatures abruptly change the animal's body physiology (Pereira *et al*., 2008) such as rise in respiration rates (> 70- 80/minute), blood flow and body temperature (>102.5° F). In Bangladesh, decrease in livestock production due to diseases, lack of forage, heat stress and breeding strategies resulted in huge economic losses (Chowdhury and Monzur, 2016). The correlation between performance of cattle production and temperature-humidity index is negative (Shinde *et al*., 1990; Mandal *et al*., 2002). Erratic changes in weather conditions directly impact the production level of animal by 58 per cent and reproduction by 63.3 per cent (Singh *et al*., 2012). Dairy breeds are more vulnerable to heat stress than the meat breeds. An increase in metabolic heat production in higher milk producing breeds leads to higher susceptibility to heat stress; while the low milk producing animals are resistant (Dash *et al*., 2016). Increase in temperature and temperaturehumidity index value beyond the critical threshold level reduces the dry matter intake and milk yield. It also interrupts physiology of animal's body (West, 2003). During 2009-10, the extreme events like floods and cyclones devastated agricultural production in large range in southern and central Mozambique, consequently loss of livestock, its infrastructure and feed (Musemwa *et al*., 2012).

Poultries are extremely sensitive to temperature-associated issues, specifically heat stress. Endocrinological changes caused by prolonged heat stress in broiler chickens enhance lipid accumulation, reduced lipolysis, and induced amino acid catabolism (Geraert *et al*. 1996). Due to heat stress, feed intake of poultries will be reduced (Deng *et al*., 2012), which leads to less body weight, egg production and quality of meat, and also reduces the thickness of eggshell and increases the egg breakage (Lin *et al*., 2004). Heat stress has negative effect on strength, weight, ash content and thickness of the eggshell (Miller and Sunde, 1975). Rising environmental temperature may cause seasonal improvement in growth and development of fishes, but increases the risks to the populations living beyond the thermal tolerance zone (Morgan *et al*., 2001). The rise in temperature of 1° C will affect the mortality of fish and its geographical distribution (Vivekanandan *et al*., 2009). The temperature rises of 0.37° C to 0.67° C alter the pattern of monsoon seasonal variations, eventually shifting the breeding period of Indian main carps from June to March in West Bengal and Orissa's fish hatcheries (DARE/ICAR Annualreport2014)

CSA Frame Work

Sustaining agricultural production in the face of climate change requires a series of scientific inputs and practices which are broadly known as Climate Smart Agriculture (CSA). As defined by FAO, CSA is an integrated approach that addresses the interlinked challenges of climate change and food security with the objectives of (i) Sustainably increasing productivity to support equitable increases in farm incomes, (ii) Adapting and building resilience of food production systems to climate change at multiple levels and (iii) Reducing greenhouse gas emission from agriculture (including crops, livestock and fisheries). CSA is not a technology but an approach which relies on use of modified production technologies, new policies and investments on managing risks.

Regional and National Initiatives

CGIAR has initiated a major research programme on the application of CSA approach to reduce the vulnerability of farming in developing countries (https://ccafs.cgiar.org/), while the Indian Council of Agricultural Research has launched a mega project on Climate Resilient Agriculture in India i.e. National Initiative on Climate Resilient Agriculture (NICRA) (http://www.nicra-icar.in/). Considered as one of the few major projects in developing countries, NICRA has the twin objectives of generation of appropriate climate resilient technologies in crops, horticulture, livestock, fisheries and poultry and its demonstration on farmers' fields through more than 130 Krishi Vigyan Kendras (KVKs) to provide the farm level resilience.

Experience of the NICRA clearly indicated that by adoption of already available technologies and practices with a whole village approach, we can bring much needed resilience to agriculture against climate change. By adopting scientific water conservation methods, ground water recharge, use of drought tolerant varieties, adjusting the planting dates, modifying the fertilizer and irrigation schedules and adopting zero tillage, farmers are able to realize satisfactory yields even in deficit rainfall and warmer years. These interventions have also lead to positive carbon balance in the villages quantified by using FAO, EX-ACT model (Srinivasarao *et al*, 2016).

Internationally, extensive research is under way on evolving climate resilient technologies and practices. Varieties are being developed in different crops with multiple abiotic stress tolerance. The first such example is the identification of *sub*-1 gene in paddy by IRRI Philippines which confers sub mergence tolerance. The introduction of *sub*-1 in mega variety like *Swarna* has resulted in *Swarna sub*-1 which is now grown successfully in large parts of eastern

India and north east were paddy crops gets inundated during kharif season. In wheat, ICAR-NBPGR has screened more than 300 germplasm lines of wheat and identified several promising heat tolerant lines. Biodiversity conservation is important in fighting climate change. Several local germplasm lines may be low yielding but may contain very useful traits like heat and drought tolerance.

Other Climate Smart Practices relate to water and nutrient management. Direct seeded rice for example emits less methane compared to puddled rice. Neem coated urea and application of nitrogen fertilizers in wheat, maize and paddy based on leaf colour charts reduces the total N requirement and indirectly cuts down the N_2O emission. Crop residue burning releases tons of carbon dioxide in to the atmosphere in states like Panjab where paddy straw is burnt every year in order to vacate the field for wheat planting. Conservation agriculture practice under which wheat is planted with zero tillage offers an alternative to burning which is now being promoted. Innovations in water management including extensive use of micro irrigation is by far the most important input for CSA.

The country needs to provide robust agro advisories to farmers based on real time basis. This requires a huge effort of weather data collection, assessment of crop condition, soil moisture, pest and diseases under field conditions on real time basis and translating this entire data in to a simple advisory which the farmer can follow. Lot of investments are made on installation of automatic weather stations in different states but no serious efforts are made to actually utilize the weather data for climate resilient agriculture. With the use of dynamic models like CLIMEX and DYMEX, we can assess the spatial distribution and abundance of important pests and study how future climate change may cause emergence of new pests and biotypes.

The global result indicates that business as result of climate change by 2080s would reduce world agriculture production capacity by 16% if carbon fertilization is omitted and by about 3% if it is included (Cline 2010). The reduction by 16% would be severe and would potentially cause major price increases because of inelasticity of demand for food (Cline, 2010).

There are 21 countires and regions of larger countries in which production capacity falls by more than one-third without carbon fertilization; but there are seven, which experience severe losses of the above order despite carbon fertilization (Cline, 2010).

India would face a major loss of the order of 30% despite carbon fertilization. The region wise losses of the order of 35% was projected for the North-east and North-west zones.

While the impacts of climate change and variability are to be assessed on regional basis, the adoption of climate smart practices can best be done at village level with a goal of creating Climate Smart Villages across the country. These villages will basically adopt weather smart, water smart, energy smart, carbon smart and nitrogen smart practices. Besides technologies, policy inputs on prudent use of water, nutrients, carbon and energy are essential to promote CSA. (Venkateswarlu and Shankar 2009). To sum up, investments on generation of new production and conservation technologies, policies that support rational use of resources, sharing of global best bet practices and capacity building of farmers will go a long way towards making agriculture climate smart. While these are national initiatives, global efforts on mitigation of greenhouse gas emissions are equally important to tackle global warming on long term basis. The ongoing COP26 at Glasgow hopefully will come out with strong commitments and positive action on climate change mitigation

Agriculture of the countries located below equatorial latitudes are found to be more vulnerable to adverse effects of climate change (Cline, 2010). Among these countries, India figures prominently in terms of anticipated negative effects of climate change including global warming. Indian small farmers without any adequate monetary, technological and logistic support systems shall be worst affected leading to misery in their agricultural programs.

Decline in Forest cover and Climate Change

With the decline of forest cover, there has been increase in atmospheric CO2 to the tune of 13%, causing higher atmospheric temperatures (Brovkin *et al.*, 2004). Currently this is the problem before the entire world. The consequences of cutting down the dense forests and enhancement of atmospheric temperatures are leading to unimaginable changes in the climate (Anonymous 2012). The occurrence of severely intense floods year by year in India, Bangladesh and Nepal is attributed to the consequences of climate change due to the progressive elimination of dense forest cover for mining and other industrial activities (Narain 2014; Anonymous 2022). According to the prestigious scientific journal Lancet, in India the human deaths of the order of 1.6 million in a single year 2019 are stated to be due to air-pollution with increase in CO2 concentration due to decline in forest cover (Anonymous 2022).

Planting Trees Could Be the Best Way to Fight Climate Change

Tree-planting has become a cornerstone of many environmental campaigns in recent years. The call to plant trees is everywhere, seen as a simple and effective way to help reduce the impact of carbon emissions and restore natural ecosystems.

Perhaps the most ambitious example is the 1 trillion trees campaign launched by the World Economic Forum in Davos in January 2020 in support of the UN's Decade of Ecosystem Restoration, which aims to restore, protect, or plant 1 trillion trees by 2030. That followed a similar campaign aiming to plant 1 trillion trees by 2050 which was kicked off in 2018 by nature nonprofits including WWF. (https://www.globalcitizen.org/en/content/why-planting-trees-helps-fight-climate-change/)

How trees work: Trees are the ultimate carbon storage machines. Like all plants, they take carbon dioxide from the atmosphere for their own growth and energy, a process called photosynthesis. They produce oxygen which we breathe as a by-product of this process.

Bukhari (1998) studied tree-root influence on soil physical conditions at under three sites; a forest, a-two-year-old logged-over area and a-one-year-old abandoned farm. His results showed higher soil bulk density at the 60 to 90 cm layer in forest, low in logged over and intermediate in farm while soil moisture content varied in the reverse order. He also found the ameliorated soil texture, structure and fertility following tree felling

Woodlands and forests can lock up carbon for centuries – which is something humans and the planet desperately need them for, given the damage done to the atmosphere by carbon-emitting human activity.

According to the Woodland Trust, a UK conservation charity, 400 tons of carbon can be locked into one hectare (which is 10,000 square metres — or about two and half football pitches) of woodland alone.

It follows then that cutting down trees has serious consequences for carbon emissions. A 2018 study found that in Oregon, US, logging had been responsible for releasing 33 million tons of carbon dioxide each year since 2000, dwarfing other sources of carbon emissions such as transportation in the state.

The restoration of trees remains among the most effective strategies for climate change mitigation. The mapped the global potential tree coverage showed that 4.4 billion hectares of canopy cover could exist under the current climate. Excluding existing trees and agricultural and urban areas, it has been found that there is room for an extra 0.9 billion hectares of canopy cover, which could store 205 gigatonnes of carbon in areas that would naturally support woodlands and forests. This highlights global tree restoration as one of the most effective carbon drawdown solutions to date. However, climate change will alter this potential tree coverage. It is estimated that if we cannot deviate from the current trajectory, the global potential canopy cover may shrink by ~223 million hectares by 2050, with the vast majority of losses occurring in the tropics. The results highlight the opportunity of climate change mitigation through global tree restoration but also the urgent need for action. Ethiopia has planted more than 350 million trees in 12 hours, its government announced, claiming a world record. Unfortunately, tree planting as a sole strategy is not the silver bullet to solve multiple environmental and social problems. Instead, well-planned tree planting incorporated amongst the toolbox of strategies to protect and increase forest cover can provide many benefits to people and the millions of other species that depend on forests.

Carbon sequestration is the process of capturing, securing and storing carbon dioxide from the atmosphere. The idea is to stabilize carbon in solid and dissolved forms so that it doesn't cause the atmosphere to warm. The process shows tremendous promise for reducing the human "carbon footprint." There are two main types of carbon sequestration: biological and geological.

Biological carbon sequestration is the storage of carbon dioxide in vegetation such as grasslands or forests, as well as in soils and oceans.

Oceans absorb roughly 25 percent of carbon dioxide emitted from human activities annually. Carbon goes in both directions in the ocean. When carbon dioxide releases into the atmosphere from the ocean, it creates what is called a positive atmospheric flux. A negative flux refers to the ocean absorbing carbon dioxide. Think of these fluxes as an inhale and an exhale, where the net effect of these opposing directions determines the overall effect.

Colder and nutrient rich parts of the ocean are able to absorb more carbon dioxide than warmer parts. Therefore, the polar regions typically serve as carbon sinks. By 2100, most of the global ocean is expected to be made up of carbon dioxide, potentially altering the ocean chemistry and lowering the pH of the water, making it more acidic.

Soil: Carbon is sequestered in soil by plants through photosynthesis and can be stored as soil organic carbon (SOC). Agroecosystems can degrade and deplete the SOC levels but this carbon deficit opens up the opportunity to store carbon through new land management practices. Soil can also store carbon as carbonates. Such carbonates are created over thousands of years when carbon dioxide dissolves in water and percolates the soil, combining with calcium and magnesium minerals, forming "caliche" in desert and arid soil.

Carbonates are inorganic and have the ability to store carbon for more than 70,000 years, while soil organic matter typically stores carbon for several decades. Scientists are working on ways to accelerate the carbonate forming process by adding finely crushed silicates to the soil in order to store carbon for longer periods of time.

Forests: About 25 percent of global carbon emissions are captured by plant-rich landscapes such as forests, grasslands and rangelands. When leaves and branches fall off plants or when plants die, the carbon stored either releases into the atmosphere or is transferred into the soil. Wildfires and human activities like deforestation can contribute to the diminishment of forests as a carbon sink.

Grasslands: While forests are commonly credited as important carbon sinks, California's majestic green giants are serving more as carbon sources due to rising temperatures and impact of drought and wildfires in recent years. Grasslands and rangelands are more reliable than forests in modern-day California mainly because they don't get hit as hard as forests by droughts and wildfires, according to research from the University of California, Davis. Unlike trees, grasslands sequester most of their carbon underground. When they burn, the carbon stays fixed in the roots and soil instead of in leaves and woody biomass. Forests have the ability to store more carbon, but in unstable conditions due to climate change, grasslands stand more resilient.

Geological carbon sequestration is the process of storing carbon dioxide in underground geologic formations, or rocks. Typically, carbon dioxide is captured from an industrial source, such as steel or cement production, or an energy-related source, such as a power plant or natural gas processing facility and injected into porous rocks for long-term storage.

Carbon capture and storage can allow the use of fossil fuels until another energy source is introduced on a large scale.

All trees store carbon but it is thought that tropical rainforests are even more useful when it comes to defending against climate change. They grow rapidly and produce rainforest cloud cover that reflects the sun rays back into space, according to the Rainforest Alliance. They are also vital for the weather system – helping to create rain through the water vapour transpiring from their leaves which in turns helps prevent droughts in the region.

Estimates for the amount of carbon in the atmosphere that the Amazon rainforest has stored vary hugely. But whatever the figure, studies have shown that it has helped to mitigate the carbon emissions produced by all the nations in its surrounding area.

For all of these reasons, climate scientists have hailed the benefits of planting trees and protecting the trees that we already have.

A huge study from ETZ Zurich University, published in 2019, concluded that by restoring a very large area of forest globally, equivalent to the size of the US, it would store 205 billion tons of carbon. That's about two thirds of the 300 billion tons that has been released into the atmosphere as a result of human activity since the industrial revolution, the study found.

Forests store or sequester a lot of carbon. Nationwide, USDA Greenhouse Gas Inventories indicate that forests, urban trees, and harvested wood account for the majority of natural sinks of carbon dioxide (USDA 2016).However, trees outside of forests, including those planted in agroforestry practices, also play an important role in carbon sequestration and reducing greenhouse gas emissions.

Riparian forest buffers are areas of trees, shrubs, and other vegetation found next to stream channels and other waterways. They are modelled on natural communities such as bottomland hardwood forest, coastal scrub, and upland oak-hickory-pine forests When planning riparian forest buffers, the buffer zones closest to fields can be designed to be harvested, by selecting species that can be used on-farm or sold. Harvesting can help remove nitrogen from the site that is captured in the harvested woody materials and help maintain more actively growing plant materials, which increases future nutrient uptake. According to Virginia Government Department of Conservation USA, (https://www.dcr.virginia.gov/natural-heritage/riparian#), Riparian wetlands are characterized by plant species adapted to periodic flooding and/or saturated soils. They support a high diversity of plant and animal species. More energy and materials, born by moving water, enter, are deposited in, and pass through riparian ecosystems than any other wetland ecosystem. Drier upland forests adjacent to waterways also provide many of the same ecosystem values.

- Riparian forest buffers help control the rate and volume of water flowing in streams and rivers, greatly influencing flood levels. Water flowing through a riparian forest is slowed by the vegetation, leaf litter, and porous soils found there.
- The leaf litter acts as a filtration system by capturing sediment from upland runoff. This action also helps to filter out phosphorous bonded to sediment particles. Sediments, and any nutrient which may be bonded to them, become part of the forest soil rather than clouding our waterways.
- Chemical and biological processes of the forest remove nutrients, such as phosphorous and nitrogen, and store them in the soil or as plant

tissue. Pesticides are also converted to nontoxic compounds by various chemical and microbial activities within the forest. This helps to protect fish, which are most threatened by pesticide pollution.

- Riparian forest soils act as areas of water storage. Plants take up water into their tissues and release it into the atmosphere.
- A canopy created by riparian forest provides shade and controls water temperature, which is essential for instream organisms, including trout and the invertebrate food source on which they depend. Instream, the leaf litter and woody debris from the canopy and forest create food and habitat vital to the aquatic food web.
- Riparian forests provide food and habitat for a variety of terrestrial wildlife and serve as safe corridors for movement between habitats. Habitat conversion and fragmentation have reduced wildlife habitat and limited the ability of animals to move between existing habitats. Riparian forests provide for both these needs.
- Riparian forest buffers offer recreation to fishermen, birders, hikers, canoeists, and picnickers. The diversity of habitats and life and the scenic beauty provided by riparian forests can be enjoyed by many people in so many different ways.

These ecological functions combine to make riparian forest buffers critical investments in human and ecological health and well-being today, and for our children tomorrow.

The year 2015 emerged as a pivotal year for climate management when the Parties at COP 21 reached an agreement to limit the average global temperature increase to well below 2°C above pre-industrial levels and to pursue efforts to limit warming to 1.5°C (UNFCCC 2015). Carbon dioxide removal options have become a common feature in climate change mitigation scenarios that are consistent with the goals of the Paris Agreement (Clarke *et al.*, 2014). Given that carbon budgets are tight and rapidly being depleted (Rogelj *et al*,2016). carbon dioxide removal options are more widely used to compensate for temporary budget overshoot (UNEP, 2017).

Agroforestry practices entail the integration of trees into agricultural systems, in combination with crops, livestock, or both. Afforestation and reforestation and agroforestry projects form part of several voluntary and mandatory carbon-offset trading schemes worldwide (Diaz *et al* 2011; Miles *et al* 2015).

Agroforestry systems have proven to create environmental, economic, and social benefits as per the available data (Pandey, 2007). In the Northeast, as in other temperate regions, the most common agroforestry practices are riparian forest buffers, windbreaks, silvopasture, alley cropping, and forest farming. Most agroforestry practices are designed to be multifunctional. Here, this means that they can be designed to sequester carbon while also benefiting the farmer or forest landowner in other ways.

Agroforestry contributes to climate change mitigation in three ways

(1) Sequestering carbon in biomass and soils,

(2) Reducing greenhouse gas emissions, and

(3) Avoiding emissions through reduced fossil fuel and energy usage on farms. For a windbreak the growing trees store carbon directly in their biomass and in the soil. At the same time, the system releases fewer greenhouse gases, like nitrous oxide, because the trees take up extra nutrients and also because less area is fertilized. Finally, less fossil fuel and energy are used in the agricultural operation because some of the field is no longer cultivated.

Farmers too can mitigate climate change through agroforestry practices, they can also get other benefits. These benefits include increased yields, reduced risks, improved pollinator and wildlife habitats, or increased capacity to adapt to climate change. This makes agroforestry more appealing to farmers who are working towards multiple outcomes. Agroforestry practices can also work in cooperation with other carbon sequestration practices and make them more risk resilient. For example, windbreaks and cover cropping can work together. The reduced wind speed may make cover crop establishment easier in challenging conditions. Agroforestry can sequester carbon while leaving most of the field in agricultural production, instead of converting it to forests or other land uses. This is especially true for agroforestry practices that take place at the edge of fields, such as windbreaks and riparian forest buffers. Even if only a small percentage of farms add agroforestry practices, the potential carbon sequestration can be significant.

Specific design and management approaches can increase or decrease the climate change mitigation potential of agroforestry practices.

According to Schoeneberger *et al.*, (2017), tree species can be selected for their carbon sequestration characteristics. The amount of carbon dioxide a tree can hold is called carbon sequestration. They sequester this carbon dioxide by storing it in their trunks, branches, leaves and roots; the best trees for carbon dioxide absorption will have dense wooded canopies (Prasad, 2021). The Carbon sequestration potential of *Pongamia pinnata* during the 10 to 15 years of its growth was found to be many folds higher than that of several other tree species. *Pongamia* was found to sequester around 45 to 50 kg of C per tree per annum as against 28 to 35 kg of Neem (*Azadirachta indica),* 23 to 26kg of Mahua (*Madhuca latifolia)* and 11 to 15 kg in respect of Tendu (*Diospyros melanoxylon)* (Prasad, 2021)

Table. 3: Carbon sequestration by some popular tree species

Tree type	C sequestration / tree of 15yrs (kg)/ annum	C Sequestration (t)/ ha	Tree density (No./ ha)
Pongamia (*Pongamia pinnata)*	45-50	23.5	500
Tendu(*Diospyros melanoxylon)*	11-15	6.0	460
Mahua (*Madhuca latifolia)*	23-26	8.5	354
Neem (*Azadirachta indica)*	28-35	13.0	450

Source: Prasad, M.V.R. (2021) Pongamia for bioenergy and better environment, NIPA, New Delhi, India.pp70.

Silvipasture systems that add trees to pastures may have the greatest potential among the agroforestry practices to mitigate climate change. As before, carbon is sequestered in the trees that are planted into the pasture and in the soils. Silvopasture may also reduce methane emissions, an important contributor to greenhouse gas emissions. Silvipasture management can reduce methane emissions by using a grazing strategy of moving cattle in a rotational stocking system. Another factor leading to lower methane is more digestible feed and greater overall gain from feed efficiency due to shade-induced microclimate changes.

Table 4: Climatic Change—Implications for India's Water Resources.

Year	Season	Increase in temrerature, °C		Change in Rainfall. %	
		Lowest	Highest	Lowest	Highest
2020s	Rabi	1.08	1.54	-1.95	4.36
	Kharif	0.87	1.12	1.81	5.10
2050s	Rabi	2,54	3,18	-9.22	3.82
	Kharif	1.81	2.37	7.18	10,52
2080s	Rabi	4.14	6.31	-24,83	-4.50
	Kharif	2.91	4.62	10.10	15.18

Source: Lal (2001a)

The data presented in the table above gives the quantification of possible adverse changes India might face due to climate change in years to come. Considering the gravity of the impending adverse situation, there should be a speedy and consistent action plan to minimise the adverse effects of the climate change. In regard to agriculture, the interventions detailed in the following section would be helpful in ameliorating the adverse effects of climate change to certain extent in dryland agriculture.

Interventions that can facilitate farmers to face adverse effects of climate change

Based on the available knowledge within India, some critical technology interventions can be introduced to enable small holders cope with climate change and variability as suggested by Venkateswarlu and Shanker (2009) as described below.

1. Adoption of Climate Ready Crops and Varieties

Since weather aberrations have been quite common in India, the agriculture research programmes have historically focused on developing short duration drought tolerant varieties of various crops. Such varieties have characters of early maturity which makes them suitable for delayed planting in case delay in onset of monsoon and also maturing early to overcome the problem of early withdrawal. NARS and CG institutes have produced large number of such varieties which are part of the seed chain in different states. What is important is to strengthen the seed production of such varieties and evolve a system of decentralized storage for use in contingency planning.

The traditional cropping pattern in different agro climatic regions of India has evolved according to the soil types and rainfall pattern. However, to realize higher economic returns farmers have been shifting to high water demanding commercial crops which require more water and also increase the risk to farmers. The strategy under such situation is to enhance the economic output of the well adapted crops through value addition and processing so that the farmer can realize reasonable income from these climate ready crops or improve the water use efficiency of currently grown high water demanding crops to match with water availability in a given region. Maheswari *et al.* (2015) have listed major crop varieties tolerant to different abiotic stresses which can be deployed to cope with climate aberrations. The gene banks in India have large number of wild germplasm collections which have climate ready traits like heat, drought and submergence tolerance. We need to revisit the gene banks and identify such material for use in the regular breeding programme. Due to heavy rainfall and floods submergence of kharif crops is becoming major issue in recent times particularly in rice. The research system has responded through the development of submergence tolerant varieties like Swarna sub-1, and the farmers have been adopting this variety on fairly large scale. Crop improvement to cope with global climate change may require incorporation of traits based on adaptation to abiotic stresses and also to enhanced atmospheric

CO_2 levels. Therefore, programs to develop climate ready varieties should be accorded highest priority, so that desired varieties are available when climate change effects are experienced consistently (Sushil Kumar, 2006).

From the research point of view, to develop crop genotypes which can perform better under the predicted climate change, it is essential to understand the plant traits that are linked to adaptation. Plant adaptation is a response to a particular environmental condition. Finding and quantifying these patterns in relation to plant functioning has been the focus of research in this area. Plant traits which favour yield and also tolerance to abiotic stresses has to be considered when developing climate ready crops. These traits have the ability to control yield over a time scale influencing either water use, water use efficiency and partitioning of biomass to grain. The traits associated with stress tolerance and yield increase will not be usually the same in most crops hence the breeder is presented with a paradox of choices (Shao et.al.2007). In short, use of climate ready crops, varieties and cropping systems (e.g. inter cropping) is the first step for small farmers to practice climate resilient agriculture.

2. Building Soil Organic Matter

Adequate organic matter in the soil brings resilience from water and nutrient stresses. This is more significant in rainfed farming. Though farmers in India have traditionally managed their soils by recycling organic matter, the declining livestock population and use of crop varieties with high grain to straw ratio have resulted in non-availability of organic resources and depletion of organic carbon in the soil. This is a major challenge towards achieving climate resilient agriculture in India. However, small farmers can practice green manuring in irrigated crops and green leaf manuring/biomass recycling through vermi composting etc. in rain-fed crops to maintain organic carbon in the soil. Some of these practices are labour intensive and therefore the Government of India need to include them in programmes like MGNREGS. Conservation agriculture (CA) practices where crop residues are left on the surface is another practice small farmers can adopt to build organic carbon. Globally, CA is widely practiced in rainfed agriculture, but in India due to competitive use as fodder, adequate quantities of crop residues are not available to be retained on the surface. Unlike in the rice-wheat system of Indo-Gangetic plains, zero tillage has not been found very promising in dryland agriculture. However, more research is needed in this area.

Soil carbon sequestration is the most important strategy towards mitigation of climate change. Although, tropical regions have limitation of sequestering carbon in soil due to high temperatures, adoption of appropriate management practices helps in sequestering reasonable quantities of carbon in some cropping systems particularly in high rainfall regions. The potential of cropping systems can be divided in to that of soil carbon sequestration and sequestration in to vegetation. Tree based systems can sequester substantial quantities of carbon in to biomass in a short period. Total potential of soil C sequestration in India is 39 to 49 Tgyear-1 (Lal 2004). This is inclusive of the potential of the restoration of degraded soils and ecosystems which is estimated at 7 to 10 TgCyear-1. Biomass energy and waste recycling are other options of climate change mitigation. A large amount of energy is used in cultivation and processing of crops like sugarcane, food grains, vegetables and fruits, which can be recovered by utilizing residues for energy production. This can be a major strategy of climate change mitigation by avoiding burning of fossil fuels. Anaerobic conversion of biomass to methane is now gaining popularity as a climate friendly practice. The Government of India has recently announced a major scheme on *waste to wealth*. Bio manure produced from waste could be a sustainable source of organic matter for addition to soils. While some of these practices may require capital expenditure, most approaches are scale neutral and small & marginal farmers can adopt them with suitable incentives and awareness generation.

3. *In-Situ* and *Ex-Situ* Water Conservation

One of the major projected impact of climate change will be on extreme rainfall events which cause more soil erosion and, runoff. The only strategy under such circumstances is to conserve the soil and harvest surplus runoff in dug out structures to be used for life saving irrigation. A large number of *in-situ* rain water conservation practices suitable for different soil types and rainfall zones have been standardized. However, their adoption is not more than 25 per cent due to various operational constrains including high labour requirement and capital cost in some cases. However, this is the most critical step small farmers need to adopt to bring resilience against climate change.

The foremost intervention in managing natural resource is watershed management in drylands (Wani *et al.* 2011). Watershed management involves the use of all natural resources available such as land, water and vegetation to relieve drought impacts, moderate flooding, control soil erosion, increase water availability and thereby increase food, fodder, fuel and fibre production in sustainable way. Watershed management includes the treatment of land with biological and engineering measures in such a manner that rainfall is conserved in situ as much as possible and the surplus is safely disposed or stored in ponds or check dams to recharge ground water. Watersheds provide alternate land use systems for effective use of marginal lands. Historically, watershed development was seen as a strategy of drought proofing, but it is an ideal approach for coping with climate variability in drylands.

Small and marginal farmers can also practice *in-situ* & .*ex-situ* conservation practices to store rain water in the profile or harvest surplus runoff in dugout ponds and use the same for supplemental irrigation. Large number of in-situ conservation practices such as ridge-furrow method, broadbed-furrows (BBF), compartmental bunding are recommended for different soil types and rainfall zones in the country. The yield benefits from these practices can vary between 10-15%.

Table 5: Impact of moisture conservation technologies on crop yields in different rain-fed farming regions (Venkateswarlu *et al.*, 2009)

Practice	Target Area	Yield Benefit (%)
Broad bed furrow/ridge furrow planting	Malwa region of M P , Vidharbha and Maratwada in Maharastra	25-50
Conservation Furrow	Alfisol regions of southern plateau	10-15
Ridge and Furrow planting of upland rice and pigeonpea	Eastern U.P/Vindhyan plateau	15-25
Compartmental bunding	Vertisol regions of North Karnataka, and scarcity rainfall zone of Maharastra	10-15
Ridge and Furrow across slopes	Sandy soils of Haryana, vertisols of Maharashtra, Eastern Rajasthan etc.,	10-15

These practices extend the moisture storage in the profile and drain off excess water during heavy rainfall events. Harvesting surplus water in dugouts is more important in rainfall zones of 750-1200mm. Rao *et.al.* (2010) made an assessment of water harvesting potential of rainfed regions based on rainfall and crop water balance. Many districts of central and eastern India growing rainfed rice and soybean have considerable surplus runoff which can be harvested. The All India Coordinate Project on Dryland Agriculture (AICRPDA) has standardised the size of the ponds for different rainfall zones, lining materials and water lifting and conveyance methods. According to estimates by CRIDA, maximum rainwater harvesting potential is available in 1000-2500mm rainfall zone, but the farm pond technology is ideal for rainfall zone of 750-1000mm where crops invariably experience water stress requiring supplement irrigation (Table 2), and the cost benefit ratios are higher.

Table 6: Potential of Rainwater Storage for Water Harvesting in different rainfall zones

Rainfall zone (mm)	Area (million ha)	Harvestable runoff (million ha m)
<500	52.07	0.78
500-750	40.26	1.51
750-1000	65.86	4.03
1000-2500	137.24	14.61
>2500	32.57	3.26

(*Source*: CRIDA, Hyderabad)

4. Crop Diversification

The traditional strategy adopted by small farmers for coping with climate variability is to follow diversified cropping pattern including mixed cropping, but, low profitability restricts farmers' choice in adoption of such practices. However, recent research in different agro climatic regions has thrown up profitable and diversified crops options to farmers including maize, summer mung, oilseeds etc. With market support in terms of procurement, small farmers can be persuaded to diversify the cropping patterns and reduce risk. Several diversified cropping systems have been recommended across different agro climatic zones of the country. This strategy is relevant both in irrigated areas to save on water and maintain soil fertility and in rain-fed areas to cope with droughts and supply fodder to the livestock.

5. Protected Cultivation

Horticulture sector in India has been growing at more than 10% annually. However, horticulture crops are highly vulnerable to climatic impacts like unseasonal rainfall, hailstorm etc. Even small farmers invest heavily on horticulture crops. Protected cultivation through green houses and shade nets enables farmers to save crops from such adverse weather events. In Maharashtra, farmers have attempted to protect grape bunches from impact of hail by covering with plastic/PVC cones. Green houses and shade nets can reduce or increase the ambient temperature by few degrees and therefore can moderate the impact of warming. Protected cultivation with nets also reduces insect damage and results in quality produce. However, there are critical research gaps in this sector. The design of the greenhouses and shade nets should be based on the weather parameters of a given location including the extreme weather events like heavy rains, high wind velocity and hailstorms. Weather based insurance should be extended to all protected cultivation. Protected cultivation requires high capital investment but several state governments have introduced schemes for small and marginal farmers providing up to 75-90% subsidy.

6. Agroforestry

Adoption of location specific agroforestry practices is an important way for small farms to adapt to climate change. Trees improve microclimate, sequester carbon and provide supplementary income to farmers in the form of fodder, fuel wood, timber, etc. Different agroforestry modules have already been recommended in the country keeping the small farmers in view. Instead of looking for large scale production of marketable products, the AF strategy can be adapted more as a risk mitigation option for smallholders.

Agroforestry systems make maximum use of the land. Emphasis is placed on perennial, multiple purpose crops that are planted once and yield benefits are received over a long period. Such benefits include construction materials, food for humans and animals, fuels, fibers, and shade. Trees in agroforestry systems also have other benefits like control of soil erosion and improving soil fertility. In a system with trees and pasture, with foraging animals, the trees provide shade and/or forage while the animals provide manure. In arid regions, trees as shelter belts improve micro climate and enhance crop yields up to 20%. Thus, agroforestry systems limit the risks and increase sustainability of both small- and large-scale agriculture. Agroforestry systems are considered as the main parts of the farming system itself, which contains many other sub-systems that together define a way of life for small farmers.

7. Integrated Farming Systems (IFS) Approach

Practicing more than one enterprise reduces the risk to small farmers. Livestock component in particular buffers small farmers from vagaries of weather. Several studies have established that income optimization with risk minimization are possible only through IFS approach. Though the emergence of nuclear families even in rural areas is constraining small farmers in adopting an ideal IFS approach, incentives to family farming both internationally and nationally, besides promotion of flexible IFS modules can enhance adoption of IFS approach by small farmers.

There are two types of farming systems *viz*, watershed based farming systems and farmer/family centric farming systems. The farmer centric farming systems are again subdivided into crop based, horticulture based and livestock based. Economics and employment generation of different farming systems depend on the relative share of the components. In general, livestock based farming systems provide more income and livelihood security in rain-fed regions while horticulture based farming systems are advantageous in irrigated areas. Gopinath *et al.* (2012) reported that improving existing farming systems with critical interventions is a better option to enhance the income of small and marginal farmers in drought prone regions than introduction of a complete new IFS. The important role of IFS approach is to minimize risk and spread the income throughout the year and optimum utilization of family labour. Secondly due to the inter connectivity and flow of energy from one component to another, the water, carbon and energy use efficiencies are significantly higher as compared to sole crops. Therefore, IFS forms an important component of climate smart agriculture (CSA).

According to The Economist, in Brazil, agricultural techniques that combine crops and livestock with forestry practices can make a farm five times more productive than the current average across Brazilian agriculture. At present, however, such techniques are used on only about 5% of the country's farm land. (Sources: Bloomberg, The Economist, The Wall Street Journal and the News Item from Daily Dose Ozy May09, 2022)

8. Livelihood Diversification

Due to high risk, small and marginal farmers cannot adapt to climate change with crop diversification alone; instead, they need livelihood diversification with a significant portion of income coming from non-farm enterprises. Small farmers who have non farm income or remittances from migrant family members can cope with crop failures more effectively. Providing works through drought relief and MGNREGA programmes are best examples of helping poor and landless to cope with adverse climatic conditions like droughts and floods. Kareemuila *et.al.* (2010) studied the impact of NREGS on livelihood security and agricultural capital formation in four states (AP, Karnataka, Maharashtra, Rajasthan). It was reported that water conservation works have significantly helped the farmers in coping with droughts and reduced distress migration. Small and medium enterprises based on agro processing and value addition provide work to the rural youth when agricultural activities are negatively impacted due to adverse climatic events. However, skill development among rural youth is essential for up scaling this approach. Overall, with proper planning and skill development, livelihood diversification can be effective strategy to cope with climate variability.

9. Agromet Advisories and Use of Information Technology

Timely information on weather enables smallholders to take crop management decisions that minimize their risk. Crispino Lobo *et al.* (2017) provided an excellent review of how timely agromet advisories can make small and marginal farmers climate smart based on pilot studies in Ahmednagar district of Maharashtra. Weather based agro advisories and contingency crop planning information are being disseminated to farmers by several agencies through mass media, mobiles and other modern IT channels. Many states are disseminating contingency plans through *M-Kisan* portal. ICAR Institutes and Agricultural Universities are also effectively using information technology for dissemination of agro advisories through agromet field units and ATICs. The participation of private sector is rapidly growing in this area. IT based platforms are also being used for surveillance of pests and diseases and development of fore warming systems. With increased use of mobiles by farmers, the dissemination of weather information and agro advisories will become more wide spread in future. More value can be added to these services if market information can be embedded.

Epilogue: It is necessary to recognize the inevitability of climate change and global warming, which shall have serious implications on farming. The tropical countries situated below the Equatorial Zone including India shall be affected worse than the others. While it is not possible to wipe out all the adverse effects of this complex problem, the complexities of the issues which are highly site specific can be tackled to certain extent integrating diverse components of technologies of multidisciplinary nature as explained in the above pages. It should be recognized that focus should be on small farmer, who has no ware-withal to face the crisis. There should be efforts to resolve the problems on community basis. While the technologies have shown us the way, it is absolutely necessary that judicious choice of technological inputs combined with logistics should be implemented to beat the crisis imposed by climate change.

9

Climate Resilient Practices for Dryland Agriculture

Dryland agriculture is a more vulnerable system in view of its high dependency on monsoon rains which are of aberrant nature. The risk of crop failure and poor yields always restrain farmers from investing on new technologies and input use.

Concept of resilience

Resilience is used to describe the magnitude of a disturbance that a system can withstand without crossing threshold into a new structure or dynamic. Resilience is defined as "the ability of a system and its component parts to anticipate, absorb, accommodate, or escape from unacceptable standards of living due to the effects of a hazardous event, in a timely and efficient manner". In other words, resilience describes the capacity of communities in a given context or landscape to maintain and improve their livelihoods—despite stresses and shocks—through the sustainable management of natural resources while maintaining key ecological functions

Climate-resilient agriculture (CRA) is an approach that includes sustainably using existing natural resources for crop and livestock production systems to achieve long-term higher productivity and farm incomes under climate variabilities. These practices facilitate reduction of hunger and poverty in the face of climate change. CRA practices can alter the current situation and sustain agricultural production at the local to the global level.

Strategies and technologies for climate change adaptation

Crop based strategies

The choice of the crops and their varieties for an agro-ecosystem could further be narrowed down by matching crop requirements with prevailing location specific climatic and soil information. The analysis of long-term climatic data on onset and withdrawal of monsoon, intermittent dry spells and effective cropping period etc., which serve as a good guide to select crops and varieties. While intrinsic soil properties viz., soil depth, texture, slope and available water holding capacity etc., reflect the manner of a soil to mitigate the adverse impact of weather events, an additional advantage of greater magnitude can be achieved by combining the said soil attributes with the choice of appropriate crop varieties. Generally, the crop to be grown under rain-fed areas should be of short duration with early vigour, deep root system with ramified roots, less vegetative with erect leaves and stem, moderate tillering in case of tillering crops and varieties, resistance/tolerance to biotic stresses, lesser period between flowering and maturity so that the grain filling is least affected by adverse weather, resistance/tolerance to abiotic stresses, low rate of transpiration, less sensitive to photo-period and wider adaptability. Thus, under changing climate conditions, introduction of high yielding, drought resistant/tolerant varieties hold the promise for getting higher yields

As a general rule, rainfed/dryland crops are sown early with the onset of monsoon to realize higher yields. And any delay in monsoon beyond normal period affects sowing of many crops of longer duration or narrow sowing window. The crops with wider sowing windows can still be taken up till the cut-off date without major yield loss and only the change warranted could be the choice of short duration cultivars. Beyond the sowing window, choice of alternate crops or cultivars depends on the farming situation, soil, rainfall and cropping pattern in the location and extent of delay in the onset of monsoon. For example, pulses and oilseeds are preferred over cereals with respect to water economy and for delayed kharif sowing. Cluster bean, moth bean and horse gram are better choices for low rainfall areas receiving less than 400 mm of rain as compared to other kharif season pulses. For cultivation on conserved soil moisture during rabi season, chickpea and lentil are preferred over peas and French bean. Similarly, among oilseeds, groundnut, castor sesame and Niger perform well under rainfed conditions during kharif season. In the rapeseed-mustard group, taramira is the best choice for light textured soil with low moisture storage capacity, followed by Indian mustard. Among the kharif cereals, coarse cereals (millets, ragi and sorghum) are better choice over maize and rice. Similarly in rabi season, barley does better under conserved soil moisture than wheat. Among the millets, foxtail millet (*Seteria*) is most suited for late sown condition without any adverse effect on productivity.

Drought

The development, identification, and use of climate resilient crop varieties along with different adaptation and mitigation strategies are essential for agriculture to successfully cope with climate variability. Stress tolerance traits such as disease resistance, drought tolerance and submergence tolerance were transferred to high yielding background to make crops resilient to adverse climatic variations. The superior crop varieties identified for short duration & drought tolerance are presented in Table 1.

It is evident that short duration crops and varieties are more suited to get stable yields in dryland agriculture. Different short duration drought varieties were assessed in different AICRPDA centers over the years (Table 1)

Table 1: Short duration and drought tolerant varieties in drylands

Location	Crop	Variety	Duration
Arjia	Maize	surya	70-75
Anantapur	Groundnut	Vemana	105-110
Indore	Soybean	JS-90-41	87-98
Akola	Cotton	AKH-081	150-160
Bijapur	Sunflower	KBSH-1	90-95
Solapur	Sorghum	Mauli	105-110
Hissar	Pearl-millet	HHB-67	60-62
Bangalore	Fingermillet	GPU-26	90-105

(*Source*: Chary,C.R.*et al.*,2020)

During 2011 to 2018, under delayed onset of monsoon conditions, varieties of major rainfed crops were assessed for their suitability and best performing varieties were identified. On an average, these varieties gave about 15-35% higher yields compared to local/farmers' varieties (Table 2)

Table 2: Crops and varieties suitable for delayed onset of monsoon

Centre(State)	Delay onset by	Crop	variety
SKNagar(Gujarat)	31days	Pearlmillet	GHB558
		Clusterbean	GG2
Bhilwara(Rajasthan)	15days	Maize	PM-3
Vijayapura(Karnataka)	28days	Pearlmillet	ICTP-8203 (ICMV-211)
Hissar(Haryana)	26days	Pearlmillet	HHB-226

(Maheswari *et al* (2019) Chary C.R.*et al.*,2020)

Extreme events

Heat stress

The continuous exposure of plants to high temperatures or heat stress during crop growth cycle is a major impediment to agricultural production and cause an array of morpho-anatomical, physiological and biochemical changes in plants, which affect plant growth and development eventually reducing economic yield. Heat stress is often defined as the rise in temperature beyond a threshold level for a period of time sufficient to cause irreversible damage to plant growth and development. In general, a transient elevation in temperature, usually 10–15°C above ambient, is considered heat shock or heat stress. The adverse effects of heat stress can be mitigated by developing thermo-tolerant crop varieties through genetic improvement and when coupled with various adaptation and mitigation strategies can counter production losses. Heat tolerance is generally defined as the ability of the plant to grow and produce economic yield closest to its genetic potential under high temperatures. The heat-threshold level that the plant can withstand without adverse effects varies considerably at different developmental stages in different crops. For instance, during seed germination, high temperature may slow down or totally inhibit germination. High temperature, in general adversely affects photosynthesis, respiration, water relations and membrane stability production of ROS (Reactive Oxygen Species) and anti-oxidants, accumulation and adjustment of compatible solutes etc. In addition, plants intrinsically respond to high temperature stress by triggering a cascade of events and adapt by switching on numerous stress-responsive genes. The genetic resources, especially land races and wild relatives from areas where past climates were similar to the projected future climates for agriculturally prime areas, could serve as the starting genotypes for breeding crops for heat tolerance. Major food, vegetable and horticultural crop varieties with tolerance to heat stress released by various Institutes/ Universities(Table 3)

Table 3: Heat tolerant varieties suitable different climatic zones

S.N.	SI. Varieties No.	Zone	Sub-Zone	State	Source of seed availability
			Cereals		
Wheat					
1.	Lok-1, Vidisha, GW-173, Arpa	ACZ-II	Northern Hills	Chhattisgarh	IGKV Jabalpur/ Private sector
2.	RAJ 3777, RAJ 3765, Rad 3077	ACZ-I, IVA	North West Plain Zone-18 of Rajasthan	Rajasthan	NSC, RSSC, NSP, ARS, Durgapura

S.N.	SI. Varieties No.	Zone	Sub-Zone	State	Source of seed availability
3,	RAJ 3777, RAJ 3765, ACZ-I-B RAJ 3077, RAJ 4037, RAJ 4083	ACZ I B	North Western Plane	Rajasthan	NSC, RSSC, NSP
4,	RAJ-3777, Raj-37G5, MP-3288, HI-1500	ACZ-IV B	ARS Banswara	Rajasthan	RSSC, RSSC, MPSSC
Chickpea					
1.	JG-14, Indira Ghana, ACZ-I JG-31S, JG-11	ACZ-I	Chhattisgarh Plains Zone	Chhattisgarh	IGKV Jabalpur
2.	JAKI 9218, JG 6	ACZ-I, II, III, IV, V, VI	Vindhya Plateau Zone, Bundelkhand, Malwa Plateau Zone, Jhabua Hill Zone	Madhya Pradesh	RVSKVV, GWalior
3.	R5G 888, GNG 663	ACZ-I-B	Semi-arid Eastern Plains, Rajasthan Jainur	Rajasthan	NSC, RSSC, NSP
Moth bean					
1.	RMO-40, RMQ-22S, RMO-425, RCG1033	ACZ-I-A, l-C	Arid Western Plains, Rajasthan Hyper arid and Western Plains	Rajasthan	RSSC, NSC, SKRAU, Bikaner
Soybean					
1.	JS-335	ACZ-I	Chhattisgarh Plains Zone	Chhattisgarh	IGKV, Jabalpur
Sunflower					
2.	DRSF113	ACZ-IV, VI	Central, Southern Dry Zone	Karnataka	GKVK, UAS(B)/ KSSC / NSC

(*Source*: Maheshwari etal.,2015)

Cold stress

Major food crops, maize (*Zea mays*) and rice (*Oryza sativa*) are very sensitive to low temperatures by which the growth of these crops are severely affected in terms of their growth and development by temperatures below 10o C resulting in considerable yield loss or even crop failure. When the temperature decreases to less than 5°C for more than three consecutive days it is considered as cold wave/stress in areas where normal temperature remains 10°C or above, while in areas where normal temperature is below 10°C, if temperature goes below

3°C for more than three days it is considered as cold wave . Many plants, especially those, which are native to warm habitat, exhibit symptoms of injury when subjected to low non-freezing temperatures. These plants including maize, soybean, cotton, tomato and banana are in particularly sensitive to temperatures below10–15°C. Various symptoms in response to cold/chilling stress include reduction of leaf expansion, wilting, chlorosis and necrosis. In chilling stress, primary injury is the initial rapid response that causes a dysfunction in the plant, but is readily reversible if the temperature is raised to non-chilling conditions . Major food, vegetable and horticultural crop varieties with tolerance to cold stress released by various Institutes/ Universities (Table 4)

Table 4: Tolerant varieties for cold in different subtropical zones of the country

S.L.	Varieties	Zone	Sub-Zone	State	Source of seed availability
Cereals					
Pearl millet					
1.	GHB-53S	ACZ-III, IV, V, VI, VII & VIII	Middle Gujarat North Gujarat, South Shurashtra, North Shurashtra & Bhal costal	Gujarat	JAU, Jamnagar
Fodders					
Anjan grass/ Buffalo grass					
1.	Bun del Anjan-1	All Zone	All Zone	All over India	IGFRIJhansi
Dhaman grass/ Bird wood grass					
2.	Bundel Dinanath-2	All Zone	All Zone	All over India	IGFRI, Jhansi
Rye grass					
3.	Pb- Ryegrass No.1		North Western Plain Zone	Punjab	PAU, Ludhiana
Soybean					
3.	RGN-73	ACZ-IV B	AR5 Banswara	Rajasthan	MPSSC, MPSSC

(*Source*: Maheshwari *et al.*, 2015)

Salinity stress

Salinity is a major environmental stress and is one of the chief constraints to crop production on drylands. Dryland salinity is the accumulation of salts in the soil surface and groundwater in non-irrigated areas. It is usually the result of three broad processes: groundwater recharge (or deep drainage),groundwater movement, groundwater discharge.

Often it results from replacing deep-rooted native vegetation with shallower-rooted crops and pastures, which take up less water. Unused rainwater leaks into the ground causing groundwater to rise and dissolve salts stored deep in the soil. The salty water may: rise to the surface causing waterlogging and/or scalding, emerge at the break of a slope as seeps flow over the surface or underground into streams and rivers.

Dryland salinity may also be caused by the exposure of naturally saline soils such as hypersaline clays, and can be associated with sodic soils (soils with an exchangeable sodium percentage (ESP) of more than 6%).

Some of the most important tolerant varietiesfor salinity are: KRL-1-4, KRL-19, KRL-210 and KRL-213 of wheat and CS-52 and CS-54 of mustard. These varieties are being grown widely in salt affected areas of Punjab, Haryana, Gujarat, Maharashtra, and many other states. These varieties are contributing immensely to the food basket of salt affected areas. (Table 5 and 6)

Table 5: Salt tolerant varieties of different crops

Crop	Tolerant varieties	Adaptability		
		Sodlc pH_2	Saline ECe, dS/m	Coastal saline ECe, dS/m
Rice	CSR nr, CSR11. CSR 12. CSR13\	9.6-10,2	6-11	-
	CSR19, CSR23'. CSR27*, CSR30V CSR36'	9 4-9.0	611	-
	CSR1, CSR2, CSR3, CSR4‘, CST7-1‘, SR26B. Sumati'	-	6-9	4
Wheat	KRL 1-4*. WHI57	<9 3	6-10	-
	Raj 3077, KRL19"	<9 3	6-10	
Bariev	DL200, Ralna. BH97, DL348	8.8-9.3		
Indian	Pusa Bold, Varvina	88-9.2	6-8	
Mmustard	Kranti. CS52', CSTR330-1	8.8-9.3	6-9	
(Raya)	CST609-B 10. CS54*	8.S-9.3	6-9	
Gram	Karnal Chana 1	<9.0	<6.0	
Sugarbeet	Ramonskaaya 06, Maribo Resistapoly	9.5-10	<6.5'	
Sugarcane	Co453, C01341	<9.0	EcE-10	-

Source: Maheshwari *et al.*,2015

Table 6: Tolerant levels of different of crops

Tolerant ESP, 35-50	Moderately Tolerant ESP, 15-35	Sensitive ESP < 15
Kamal grass (Leptcchlaa fuses) Rhodes grass (Chloris gayana) Para grass (Brachiaria mutica) Bermuda grass (Cynadon dactykxi) Rice (Oryza sativs) Dhaincha (Sesbania aculeate) Sugflrbeet (Beta vulgaris) Teosinle (Euchlaena max cana l	Wheal (Trilicum aestivum) Barley (Hordeum vulgare) Oat (Avena saliva) Shaftal (Trifolbum resupinatum) Lucerne (Medicago saliva) Turnip (Brassica rapa) Sunflower (Helianthus anus) Safflower (Carthamus tinclorius) Berseen (Trifolium alexandnnum) Lindsed (Linum usitatissimum) Onion (Alliumcepa) Gallic (Allium salivum) Pearl millet (Pennisetum typhoides)	Gram (Cicer anetium) Mash (Phaseolus mungo) Chickpea (Cicer arielinum) Lentil (Lens esculenta) Soyabean (Glycine max) Groundnut (Arachis hypogea) Sesamum (Sesamum Oriental) Mung (Phaseolus aureus) Pea (Pisum salivum) Cowpea (Vigna unguiculata) Maize (Zea mays) Cotton (Gossypium hirsulum)

(*Source*: Gurung, T.R and Azad, A.2013)

In the physical method, salt scrapping, land leveling, mixing of sand, deep ploughing and sub-soiling are the main practices to be employed for the reclamation of alkali soils that change the soil surface conditions resulting in low salinity in the root-zone. Sub-soiling operation improves the permeability which helps in leaching of soluble salts

The chemical approach involves adding suitable amendments in the sodic soils to lower the exchangeable sodium percentage (ESP) and increase the concentration of calcium in the soil. The most commonly used amendments are gypsum, lime, calcium carbonate and pyrites

In Biological approach the practice of growing salt-tolerant crops, plants, grasses and trees is adopted to make the soil productive. Growing legumes improve the physical properties (i.e. Water holding capacity and infiltration rate) and the soil structure. Green manuring is being adopted for the biological reclamation of salt-affected soils.

Alkali Soil and its Reclamation

Soils in which electric conductivity of the saturation extract is less than 4desisimens per meter (ds/m), exchangeable sodium is more than 15per cent and pH value of the soil is more than 8.5 are called alkali or non-saline or sodic soils.

The reclamation of these soils involves deep ploughing to break the hardpan for improving the free-water movement and the application of chemical amendments such as gypsum, calcium carbonate, pyrites, etc.

Saline Soil and its Reclamation

Saline soils are defined as soils having electric conductivity of the saturation extract more than 4decisimens per meter (ds/m), exchangeable sodium less than 15per cent and pH value of the soilless than 8.5. These soils are also called as white alkali soils as white salts accumulated on the soil surf

Many dryland areas which are now cropped, existed under forest or grass cover. The ecosystem under these conditions was balanced - the grasses and trees utilizing all the precipitation in their respective areas and keeping the groundwater tables low. The surface drainage may be adequate for removing excess water from the recharge areas, it is often necessary to resort to subsurface drainage in the discharge areas which is a costly proposition.

Planting salt-tolerant species is an effective way of obtaining some economic returns while efforts are being made to improve the saline seeps. Investigations on the relative tolerance of plant species in several Australian states have indicated the value of growing the following in the saline seep areas (Maas, 1986 ; Bresler, 1987):

Agropyron elongatum Puccinellia ciliate, Atriplex spp. Sporobolus virginicus, Hordeum hystrix Trifolium fragiferum; Kochia brevifolia T. glomeratum, Lolium rigidum T. resupinatum, Paspalum vaginatum T. subterranum

For soils with higher salinity levels a mixture of alfalfa and salt-tolerant grasses like *Agropyron elongatum, Agropyron cristatum, Agropyron trachycaulum* and *Festuca elatior* was found to perform well. Where soil salinity levels were even higher, (more than 14.0 dS/m), annual crops or grasses were not recommended. Under these conditions *Kochia scoparia*, a prolific growing native weed, was allowed to grow and had excellent feed value.

Source: https://www.fao.org/3/x5871e/x5871e06.htm

The reclamation of saline and alkali soils on drylands that are situated in the zones of acute water scarcity need alternate technologies that do not involve lot of water as indicated above. In such cases the following procedures may hold valid depending upon the site specific situations.

- Application of soil amendments viz., as gypsum, calcium carbonate, pyrites, etc. should be carried out only with the certain onset of monsoon rains.
- Use of salt tolerant crops including their transgenic derivatives must be encouraged. Using the salt-tolerant crops is one of the most important strategies to solve the problem of salinity. Tolerance will be required for

the "de-watering" species, but also for the annual crops to follow, as salt will be left in the soil when the water table is lowered.

- Add organic matter and manure to keep moisture and reduce irrigation.
- Restrain from deep tillage/heavy machinery not to transfer soil salts to the root zone area, which induces salinization.
- Use cover crops or mulch to protect the ground surface: As the name hints, these are plants to cover soils for certain reasons. Unlike primary species, they support secondary farmer's needs rather than are grown for trade or human consumption. They improve soil health, boost yields, and feed the cattle. However, it does not mean that these plants are some exclusive species. In alternative situations, they serve as cash cultures, and you can find them on the plate as well (for example maize or corn). The difference is that in the case of late *kharif* cover crops, these species are used as grasses. Farmers plant them in different seasons, either late *kharif* or late spring/summer, uniformly or between rows. Some are winter-killed, and some require removal and residue management. They also suggest one species at a time or their mixtures. The latter method is reported to bring more prolific results. Common cover crops are legumes, grasses (forage grains), brassicas, turnips, radishes, etc. This practice is strongly welcomed in crop rotation, no-till, and organic farming. This kind of approach doesn't involve replacing regular *kharif* crops www.ncbi.nlm.nih.gov.
- Not much work has been done on breeding for salt tolerance in wheat, which is an important food crop. Australian plant breeders are aware of the need for salt tolerance, but this is only one of a number of constraints, the major one being drought, so salt tolerance is not specifically targeted. Targeted breeding has been largely confined to India and Pakistan. The most successful releases have been the Indian KRL1-4 and KRL 19, released by the Central Soil Salinity Research Institute (CSSRI) at Karnal, the Pakistani LU26S and SARC-1, released by the Saline Agriculture Research Cell (SARC) at Faisalabad, and the Egyptian Sakha 8, released by the Agricultural Research Centre at Giza. (Munns *et al.*, 2006)

Floods

Generally, the flooding in the field can be either water logging in which root and some portion of the shoot under water are complete submergence where the whole plant is under water. Lack of oxygen supply for the plant is main cause of damage in water logging conditions, because of which plant shifts its metabolism from aerobic to anaerobic mode. Aerenchyma formation, greater activity of glycolate pathway, involvement of antioxidative defense mechanism are some of the adaptive mechanisms to cope with flooding tolerance. Ethylene contributes in induction of the genes associated with the adaptive mechanism of flood tolerance (Alamgir and Uddin 2011).

Effects of flooding stress on plants are: 1. Decay and death of leaves 2. Wilting 3. Abscission 4. Epinasty 5. Lenticels formation Nutrient deficiency & Toxicity: Under the anaerobic condition Fe toxicity is high. This leads to increase the polyphenol oxidase activity, leading to the production of oxidized polyphenols. It also causes leaf bronzing and reduced root oxidation power. (Table8)

Table 7: Major crop varieties with tolerance to flooding stress released by various Institutes/ Universities

S.N.	Varieties	Zone	Sub-Zone	State	Source of seed availability
Pulses					
Chickpea					
1	DCP92 3, Pusa240r GNG16	ACM	Chhattisgarh Plains Zone	Chhattisgarh	IGKV farm/ Private sector
2	DCP92-3, Pusa240r	ACZ-I, II, III, IV, V, VI	Vindhya Plateau Zone, Gird	Madhya Pradesh	RVSKW, Gwalior
Fodder Crops					
Anjan grass/ Buffel grass					
1	Marwar Anjan (CAZRI 7S)		Northern zone	Arid and semi-arid areas in the	CAZRI, Jodhpur

(*Source*: Maheshwari *et al*., 2015)

Mitigation of flooding stress 1. Providing adequate drainage for draining excessive stagnating water around the root system. 2. Spray of growth retardant of 500 ppm cycocel for arresting apical dominance and thereby promoting growth of laterals. 3. Foliar spray of 2% DAP + 1% KCl (MOP). 4. Spray of 0.5 ppm brassinolide for increasing photosynthetic activity. 5. Foliar spray of 100 ppm salicylic acid for increasing stem reserve utilization under high moisture stress. 6. Foliar spray of 0.3 % Boric acid + 0.5 % ZnSO4 + 0.5 % FeSO4 + 1.0 % urea during critical stages of the stress . 7. Balance the use of fertilizers (NPK or NPK + lime). 8. Apply sufficient K fertilizer. Apply lime on acid soils, do not apply excessive amounts of organic matter (manure, straw) on soils containing large amounts of Fe and organic matter.

Resilient cropping systems

Right kind of crops/cropping systems which require less water is key in successful rain-fed farming. Intercropping is an efficient strategy that can be followed with desirable outcome in the present climate change scenario. Grain-legume intercrops have many potential benefits such as stable yields, better use of resources, reduction of weeds, insect-pests and diseases, and reduced N leaching as compared to sole cropping systems. Crop/cropping system based technologies need to be centered on promoting the cultivation of crops and varieties that fit into new cropping systems and seasons. Improved and novel agronomic and crop production practices like adjustment of planting dates to minimize the effect of high temperature increase-induced spikelet sterility can be used to reduce yield instability, by avoiding flowering to coincide with the hottest period.

Different intercropping and double cropping systems were demonstrated in AICRPDA-NICRA villages and agro-ecology specific best performing/risk resilient cropping systems were identified(Table 8)

Table 8: Climate Resilient cropping systems in different rainfed environments

Centre/soil type/climate	Intercropping systems	Double cropping systems
Arjia/semi-arid/Vertisols	Corn+Urd bean (2:2)	_
	Groundnut+sesame (6:2)	
	Sorghum+mungbean (2:1)	
Anantapur/arid/Alfisols	Groundnut+pigeonpea(11:1)	
	Pigeonpea+Pearlmillet(1:1)	
Agra/semi-arid/Inceptiols	Pearlmillet+clusterbean (4:4)	
Hissar/arid/Inceptisols	Pear lmillet+mungbean (6:3)	
	Pearl millet+clusterbean(4:4)	
Kovilpatti/semi-arid/Vertisols	Cotton+radish (1:2)	
	Cotton +onion(1:2)	
Rajkot/semiarid/Vertisols	Cotton+sesame (1:1)	
	Cotton+Foddermaize(1:1)	
	Cotton+soybean(1:1)	
	Groundnut+castor (3:1)	
SKNagar/semi-arid Entisols	Castor+mungbean(1:1)	
	Castor+sesame (1:1)	
Solapur/semi-arid Vertisols	Pigeonpea+sunflower (1:2)	
	Pigeonpea+soybean (1:3) Pearl-millet+Pigeonpea (2:1)	Black gram/green gram/ cowpea-rabi sorghum

Bijapur/semi-arid/ Vertisols	Pearlmillet+pigeonpea (2:1)	Greengram-rabisorghum chickpea
	Pigeonpea+groundnut(4:2)	
	Pearlmillet+castor (2:1)	
Akola/semi-arid/Vertosols	Cotton+greengram (1:1)	Soybean-chickpea
	Pigeonpea+soybean (2:4)	Greengram-safflower
	Cotton+sorghum+pigeonpea+ Sorghum (3:1;1:1)	
Bengaluru/Semi-arid/Alfisols	Pigeonpea+field bean ((1:1)	Cowpea-finger millet
	Fingermillet+pigeonpea (8:2)	Fodder pearl millet-cow-pea
	Castor+fingermillet (1:2)	
Indore/semi-arid/Vertisols	Pigeonpea+soybean (2:4)	
	Maize+pigeonpea (2:4)	
Parbhani/semi-arid/Vertisols	Soybean+pigeonpea (4:2)	
	Cotton+greengram (1:1)	
	Cotton+pigeonpea (6:1)	
Hyderabad/Semi-arid/Alfisols	Sorghum+pigeonpea (2:1)	
	Pigeonpea+greengram (1:1)	

(*Souce*: Maheshwari *et al* (2019)

Crop contingency plans

The principal source of water in rain-fed/dryland crops is rain, major portion of which is received during south west monsoon period (June-October) in India. The monsoon period is beset with breaks of rains in all most all parts of the country. Sudden bursts of rain are alternated with breaks. Normally there are five important aberrations in rainfall behavior., viz., (i)early commencement of rains (ii) delayed onset of monsoon (iii)intermittent breaks during the cropping season (iv)early cessation of rainfall (v) extended rainfall. These situations call for scientists and planners to develop contingent measures to save the rain-fed crops from varied monsoon aberrations. Further suitable crops and varieties matching the effective growing season were identified in different agro-climatic regions of the country. Drought leads to moisture stress, in which in turn affects different phenological growth stages of crops. Keeping this situation in view, Central Research Institute for Dryland Agriculture(CRIDA and AICRPDA/ Agrometeorology centers located in different parts of the country developed appropriate crop and contingency planning measures matching different weather conditions.

In dryland agriculture, contingency of growing another crop in place of normally grown crop arises due to delay in the onset of monsoon. Depending upon the date of receipt of rainfall, crops are selected. It is assumed that the rainfall for the subsequent period is normal. Depending upon the economic status of the farmer, certain amount of calculated risk is taken to get good profits if season is normal or better than normal. Contingency cropping is highly location specific due to variation in amount and distribution of rainfall. Especially in arid regions, the spatial distribution of rainfall is highly variable. It is common to observe that rainfall received varies from field to field in the same location. Temperature gradually falls from August onwards reaching minimum in November and December. Contingency plan and midterm corrections vary with the type and time of occurrence of rainfall aberration

Crops have to be selected with suitable duration of growth and maturity to match with the length of the growing season. Generally short duration pulses like green gram, black gram and cowpea may suit the situation. However, if the monsoon turns to be extraordinarily good, opportunity is lost if only short duration crops are sown. Farmers with economic strength and motivation for high profits with some amount of risk can go for crops of long duration. The long duration crops with flexibility or elasticity in yield are more suitable. For example, pearl-millet, and sorghum can be ratooned if monsoon extends. Sunflower can be introduced for higher profits followed by short duration grain legume crops like green gram.

Timely availability of quality seeds for sowing and re-sowing (in the case of loss of standing crop due to early season stress) is critical for successful farming. The strategies for availability of resilient varieties include (a) developing community seed banks to meet local seed demand as a contingency measure and facilitate the revival and distribution of traditional and stress-tolerant crops and varieties, (b) a robust decentralized seed system to provide quality seed material at affordable prices at right time, (c) ensuring conservation of the local agro-biodiversity which has inbuilt tolerance to various stresses, (d) improving seed quality of farmer-saved seeds and seed production, distribution and storage conditions at farmers' level and e) strengthening seed village concept in a coordinated effort.

Drought Management in standing crops

Normally the rain-fed crops face dry spell (drought) in different growth stages viz., vegetative, early season flowering(mid-season) and reproductive stages(terminal-drought) or in any combination of these phases. (Table 9)

Table 9: Drought Management in standing crops

Climatic aberration	RTCP measures
Delayed onset of monsoon	• Beyond the sowing window, choice of alternate crops or cultivar depends on the farming situation, sail, rainfall and cropping pattern in the location and extent of delay in the onset of monsoon.
Early season drought	• Resowing within a week In 10 days with subsequent rains for • better plant stand when germination is less than 30%. • If the plant population is 50 to 75% of the optimum population (gaps in rows and also in between rows), sow the same crop of shorter duration variety in the gaps either within or in between rows. If less than 50% optimum population, sow suitable contingent crop as per the remaining effective growing season e.g., pearlmillet in place of sorghum up to first week of July. • Thinning in small-seeded crops. • Interculture to break soil crust, remove weeds and create soil mulch • for conserving soil moisture. • Avoid top dressing of fertilizers till favourable soil moisture. • Opening conservation furrows at 10 to 15 m intervals. • Pre watering along with gap filling when the crop stand is less than 75% in crops like cotton
Mid-season drought	• If symptoms such as drying of leaves, wilting of plants, and cracking in black soils are observed, blade harrowing in rows during dry spell helps in creating dust mulch and closing of cracks the in black soils. After relief of dry spell, open conservation furrows at 1.2 m distance, spray with 2% urea, particularly in pulse crops, castor, and sunflower. Apply additional 10 kg /ha. • Plant protection. • Supplemental /protective irrigation., if available. • Repeated interculture to remove weeds and create -soil mulch to • conserve soil moisture • Avoid top-dressing of fertilizer until receipt of rains. • Opening conservation furrows for moisture conservation, foliar spray of 2% KNO3 Or 1% water soluble fertilizers like 19-19-19, 20-20-20, 21-21-21. • Opening «t furrow* in alternate crop rows • Surface mulching with crop residues.
Terminal drought	• Providing life- saving or supplemental irrigation, if available. Harvesting crop at physiological maturity with w some realizable yield or harvest for fodder. • Prepare for winter (rabi)sowing in double cropped areas. • Ratoon maize or pearlmillet or adopt relay crops such as chickpea, saf-flower, rabi sorghum and sunflower with minimum tillage after soybean in medium to deep blacks oils in Maharashtra • Prefer contingency crops (horse gram/cowpea) or dual-purpose forage crops on receipt of showers under receding soil moisture conditions.

Climatic aberration	RTCP measures
Unseasonal heavy rainfall events	• Re-sowing • Providing surface drainage • Application of hormones/nutrient sprays to prevent flower drop or promote quick flowering, / fruiting and plant-protection measures against pest/disease outbreaks with need based prophylactic /curative interventions • At crop maturity stage, prevention of seed germination and harvesting of produce. • If untimely rains occur at vegetative stage, the contingency measures include: • Draining out the excess water as early as possible • *Application of 20 kg N + 10 kg K/acre (0.4 hat after draining excess water. • Gap filling either with available nursery or by splitting the tillers from the surviving hills in rice. • Suitable plant protection measures in anticipation of pest and disease outbreaks. • Foliar spray with 1%KNO3 or water-soluble fertilizers like 19-19-19,20-20-20,21-21-21 at 1% to support nutrition • Earthing up the crop for anchorage etc.,
Floods	• In sand-deposited crop fields/fallows, ameliorative measures include early removal or ploughing of sand (depending upon the extent of deposit) for facilitating rabi crop or next Kharif crop *Draining out of stagnant water and strengthening of field bunds *Community nursery raising * Retransplanting in damaged fields and transplanting of new areas or direct-seeding depending on the seed availability * Prevention of premature germination of submerged crop at maturity or at harvested produce by spray of salt solution

(*Source*: Adopted from Srinivasa Rao *et al.*, 2016)

Rainwater Management

Rainwater management is a central issue for bringing any kind of resilience in dryland/ rain-fed farming. Utilizing every drop of rainwater becomes crucial under overall efficient rainwater management. Storing rainwater in soil by various location specific water conservation measures is priority and excess runoff collection in farm ponds and its recycling at critical crop stages is the second important strategy in-situ moisture conservation.

In-situ conservation practices

Treating land before commencing of rains through summer ploughing, broad bed furrow (BBF) raised and sunken bunds etc. facilitates the maximum intake of rainwater into soil profile thus successful crop production is possible in rain-fed regions of India. Ridge furrow, sowing across the slope and paired row sowing are important water conservation measures which have proved to be highly effective not only for water conservation but also draining out of in-situ moisture conservation through conservation furrow excess rainwater during heavy rains. The yield improvements with these technologies varied from 20 to 40%depending upon rainfall and its distribution. (Table 10)

Table 10: Soil and water conservation measures for various rainfall zones

Rainfall			
<500mm	500-750mm	750-1000mm	>1000mm
1. In-situ onservation in between rows	1. Contour cultivation	1. Vertical mulching in black soils.	1. Live bunds
	2. Li ve bunds		2. Field bunds
2. Contour cultivation	3. Field bunds	2. Contour cultivation	3. Graded bunds
3. Dead furrows	4. In-situ conservation in between rows	3. Dead furrows Live bunds	4. Vertical mulching in black soils.
4. Field bunds			
5. Tic bunds	5. Tie bunds	4. Minimum tillage	
5. Mulching	6. Mulching	5. Graded bunds	
6. Ploughing across the slope	7. Dead furrows	6. Cultivation of crops across ihc slope.	
	8. Vertical mulching in black soils		

(*Source* :Pathak *et al*., 2007)

Studies by CRIDA at Hyderabad, Bangalore and Anantapur revealed that more than 80% farmers follow in-situ conservation measures like sowing across the slope, opening of dead/conservation furrows and key line cultivation since in-situ conservation methods are comparatively easier to be adopted by farmers than ex-situ rainwater conservation. This in-situ rainwater conservation in rain-fed areas is a way to bridge gap between potential productivity of available crop varieties and existing crop yields by improving soil moisture content and reducing soil erosion (Pathak *et al*., 2007) BBF and ridge-furrow technology implementation have saved soybean and maize crops during 2013 from heavy rains in Malwa regions of Madhya Pradesh and Vidarbha region in Maharashtra and crops without these interventions completely failed (Table 11)

Table 11: Broad bed furrow protects soybean during excess rainfall in Madhya Pradesh

Particular	Broad bed and Furrow	Farmers Practice	% Increase
Grain Yield (kg/haf	1937	1152	40.5
Net Return (Rs/ha)	38305	18393	51.3
H:C Ratio	351	2.41	
Rain Water Use Efficiency (kg/ha-mm)	2.09	1.24	

(*Source*: Prasad, Y.G. *et al*., 2014)

Making of BBFs Drainage in furrows Crop stand in BBf

Fig.1. Broad Bed-Furrow System as an efficient method of moisture conservation

Most effective means of soil & water conservation. @ BBF, CF and ridge furrow are important for Maharashtra. Ridge/Furrow is for water conservation and draining out excessive moisture during heavy rain spells. Soybean and Maize saved with ridge furrow systems in black soils of MP when more than 120 mm rainfall received in a day.

Table 12: Role of different interventions to manage abiotic stresses

Module	Intervention	End User	Outcome
Moisture deficit or Draught Management	In-situ soil moisture conservation practices including land modification, raised bed and furrows, broad bed furrow, mulching	Individual	Productivity
	Farm ponds	Individual	Productivity, Enhance water availability, Diversified enterprises, Cropping intensity
	Tanks	Community	
	Check dams	Community	
	Resource Conservation Technology (RCTs), conservation agriculture	Individual / Community	Productivity. Livelihood
	Recharging wells, perdolaitian ponds	Individual	Enhanced water availability, productivity
	Contingency crop planning	Individual/ Community	Resilience, Productivity
	Planting methods	Individual	Resilience, Productivity
	Biomass / residue recycling	Individual	Resilience, Productivity
	Adjustment in planting time of rabi (winter) crops	Individual	Water saving, Productivity

Module	Intervention	End User	Outcome
Floods, cyclone, excess and unseas onal rains	Land shaping in coastal tracts prone to cyclones, excess rainfall. Sea water intrusion	Individual	Sustainable increase in cropping intensity
	Improving drainage/ conveyance efficiency in flood prone areas	Community	Productivity & resilience
	Flood tolerant varieties	Individual	Resilience
	Integrated farming system models	Individual	Resilience through year round farm income
	Planting methods (Broad bed furrow, ridgo-furrow, raised bed and sunken furrow system ete.)	Individual	Productivity and resilience

(*Source*: Alok Sikka *et al.*, 2017)

Farm ponds

Farm pond is an option for rainwater harvesting and to provide life-saving irrigation to standing crops when they are exposed to mid-term/terminal drought and also for pre-sowing irrigation in post-rainy season crops in rain-fed areas. Ponds may be lined in light textured soils, while unlined farm ponds in soils having higher seepage can be utilized to recharge. In a watershed, a series of ponds may be constructed in farm fields and across the water courses/ first- and second-order drainage channels to intercept run-off, reduce peak flow, control erosion and store water for supplemental irrigation and to recharge groundwater. Selection of crops and cropping systems based on profitability and irrigation requirement is needed to efficiently utilize the harvested water. Modern methods of irrigation like drip and sprinklers may be adopted for increasing water-use efficiency. Insufficient awareness among farmers, small farm holdings, relatively high initial investments, evaporation and seepage loss, and moderate benefits during 'normal' years are some of the constraints that impede adoption of farm ponds on a large scale in rain-fed ecosystems of India.

Lining of farm pond for seepage control with different materials were also demonstrated. Among different lining materials viz., stone dust + cement, soil + cement, brick, granite slabs which were executed as part of study indicated that, lower seepage loss of 25.61 l m was recorded under brick lining, while -3 higher seepage loss (84.6 l m) was recorded in soil + cement (8:1). The evaporation loss (3.4 l m) was lower in brick lining and higher in soil + cement (8:1) .

Supplementary Irrigation (SI)

Supplemental irrigation can help mitigate greenhouse gas (GHG) emissions in two ways: (1) Increased yields achieved with SI (compared to rain-fed production) result in higher carbon sequestration rates in plant biomass and for the build-up of soil organic carbon (SOC), especially if crop residues are returned to the soil or if cover cropping practices keep the soil protected as much as possible (Lal, 2004). This increase in SOC, in turn, has a synergistic effect on adaptation through improved soil fertility as well as reduced nutrient leaching and soil erosion. If this soil fertility improvement results in a decrease of mineral fertilizer use, additional mitigation benefits can be obtained through a reduction of GHG emissions linked to fertilizer manufacturing and use. (2) When substantial yield gains are achieved through SI compared to a relatively low increase in farming inputs, the GHG emission intensity per unit of produce decreases. Although this may imply higher GHG emissions per area, the yield gains can prevent conversion of additional areas to crop land and associated release of carbon from soils. The GHG emission intensity can be further reduced by the use of renewable energies instead of fossil fuels for SI, e.g. through solar-powered irrigation systems that lower the carbon dioxide emissions associated to the delivery of irrigation water.

Examples of successful interventions with efficient use of harvested rainwater for crop production during deficit rainfall

Table 13: Use of harvested rainwater collected in farm ponds in different villages

Village	KVK/ District	No. of farm pond	Storage capacity (m^3)	No of farmers covered	Irrigation potential created (ha)	Increase in cropping intensity
Kukurha	Bu xar	1	1800	06	1.75	100%
Affaur	Saran	2	11500	42	10	25%
Chopanadih	Koderma	1	16560	50	20	40%
Takli	Am^avati	2	6400	25	10	20%
Nacharam	Khammam	4	1460	4	9.6	40%
Sanora	Datia	3	9500	10	19	100%
Napenhalli	Tumkur	2	2500	25	65	135%
Naqerhaili	Tumkur	13	22100	25	12	100%

(*Source*: Alok Sikka *et al.*,2017)

integration of farm pond in the dryland farming systems has a great potential to increase farm productivity and income in dryland regions. The farm ponds in Maharashtra resulted in significant increase in farm productivity (12 to 32 %), income and cropping intensity. Similarly, in Andhra Pradesh farm pond water was useful for supplemental irrigation to mango tree plantation, vegetables and other crops and animals and resulted in significant increase in household income adding net returns of US$ 120 to 320 ha-1 annum

The crop and livestock productivity and farm income increased significantly in Akola and Chittoor districts due to farm ponds. Increment in productivity of different rainfed crops in these districts ranged from 8 to 45 per cent. Moreover, the gross cropped area also increased by 20 to 26 per cent. As a result of availability of supplemental irrigation using harvested rainwater, the farmers planted additional fruit plants and it also enhanced the productivity of existing fruit plants namely mango in Chittoor and coconut in Vellore. With the provision of supplemental irrigation, not only the productivity of mango increased but their fruiting was also regularized. (Table 14)

Table 14: Benefits of far pondsin relation to yield, income and employment

District	Increase in gross croppedarea	Increase in productivity of different rainfed crops (%)	Additional frnir plaids raised per household (No.)	Increase in existing fruit plant	Increase in fodder avaibility (ton)	Increase inlivestock	Additional employment generated (Mandays/ annum)	Aiidiiinnnl income per household (USS)
CtuRixir	19.4	8.35	46	31	2.7	14	212	700
Annncpur	4	5-11			14		66	317
Akola	25.8	12-15	7	24	3_5	9	196	927
Bangalore rural	-	Q-S	6		D_5	4	26	47
Vellore	4	3-13	16	41	u	7		503
Hhilivnrn	8.5	4-1 ll	12		20	7	72	307
Jodhpur			12	15	1.2	5	38	177

*Luu case was due lo supplciftritlal liflgaliuil and also due to iltlpcuvcil varitfica and package of pncttCCfl
(*Source*: Kumar Shalander, etal2013 2013)

In Bhilwara, the net benefits due to a farm pond ranged from US $ 132 to US $ 345 per annum. In Jodhpur, where soils are sandy and evaporation losses are high, the rainwater harvested in a covered concrete underground structure (Tanka) was mainly used for drinking purpose, animals and supplemental irrigation to fruit plants in the initial stages. The adoption of small Tanka by farmers was low mainly because the net benefits from perennial component raised with the help of harvested rainwater in Tanka were small.

Table 15: Potential benefits of farm ponds in different rainfall and soil situations

Rainfall	Major soil types/ Odder	State	Major production systems	Type of farm pond	Potential benefits		
					Increment in system's yield ("V|	Additional net returns (US 5 ba' annum'1)	Unit cost of structure <USS)
500 in 750	Alfiwls	Andhra Pradesh	Sorghum, castor based	250 in1 bum pond with lining + sprinkler with 3 hp pump set	15-25 (Increase in cropped area 10-15%)	84-131 per farm pond' annum	1024
			Groundnut castor based	Water harvesting nnd recycling through iarm ponds with hnuig with soil + cement (6' 1 ratio)	20-24 (+ diversification into vegetables	72-121	1024
	Inceptisols	Rajasthan	Maize based	500 m' farm pond with lining + sprinkler with 3 hp pump set	20- 25 l+divcfsitkaLkMi - bring waste-land mfn culiivationy	158-224	1397
	Vertisols	Gujarat	Groundnut based	Recharging defunct open wells through fillers (Retains 67% sediment load and enhance ground water level) flitter and deepening of the defunct well)	15-30	26-47	931
750 to 1000	Alfisol	Karnataka	Finger millet based	250 ra* Farm pond with lining • sfinnk lor with i hp pump set	15-20 (+10-15% more area under vegetables)	102-224	1024
	Vertiwl	Madhya Pradesh	Soybean based	1000 1o 4000 m^3 pond without Lining mainly for large farmers	20-40 (+diversification into lion-flori-horticulture	144-298 per pond of 1000 m^3	1.6 per m

(*Source*: Kumar Shalander, *et al* 2013)

Ground water recharge

Runoff Conservation Structures are normally multi-purpose measures, mutually complementary and conducive to soil and water conservation, afforestation and increased agricultural productivity. They are suitable in areas receiving low to moderate rainfall mostly during a single monsoon season and having little or no scope for transfer of water from other areas.. The structures commonly used are bench terracing, contour bunds, gully plugs, nalah bunds, check dams and percolation ponds.

Bench terracing involves leveling of sloping lands with surface gradients up to 8 percent and having adequate soil cover for bringing them under irrigation. It helps in soil conservation and holding runoff water on the terraced area for longer durations, leading to increased infiltration and ground water recharge. Contour bunding, which is a watershed management practice aimed at building up soil moisture storage involve construction of small embankments or bunds across the slope of the land. They derive their names from the construction of bunds along contours of equal land elevation. This technique is generally adopted in low rainfall areas (normally less than 800 mm) where gently sloping agricultural lands with very long slope lengths are available and the soils are permeable

Contour trenches are rainwater harvesting structures, which can be constructed on hill slopes as well as on degraded and barren waste lands in both high- and low- rainfall area. The trenches break the slope at intervals and reduce the velocity of surface runoff. The water retained in the trench will help in conserving the soil moisture and ground water recharge. These structures are constructed across gullies, nalahs or streams to check the flow of surface water in the stream channel and to retain water for longer durations in the pervious soil or rock surface. As compared to gully plugs, which are normally constructed across 1st order streams, nalah bunds and check dams are constructed across bigger streams and in areas having gentler slopes. These may be temporary structures such as brush wood dams, loose / dry stone masonry Checkdams, Gabion check dams and woven wire dams constructed with locally available material or permanent structures constructed using stones, brick and cement. Percolation tanks, which are based on principles similar to those of nalah bunds, are among the most common runoff harvesting structures in India. A percolation tank can be defined as an artificially created surface water body submerging a highly permeable land area so that the surface runoff is made to percolate and recharge the ground water storage. The existing village tanks, which are normally silted and damaged, can be modified to serve as recharge structures. Desilting of village tanks together with proper provision of waste weirs and cut off trenches on the upstream side can facilitate their use as recharge structures. As such tanks are available in plenty in rural India, they could be converted into cost-effective structures for augmenting ground water recharge with minor modifications. Recharge pits and shafts are artificial recharge structures commonly used for recharging shallow phreatic aquifers.

In majority of NICRA villages, Checkdams (new/desilting of existing ones), low cost temporary check dams (sand bag Checkdams), and 'Boribandhans' (poly bag Checkdams) were major interventions in areas vulnerable to droughts. These structures not only increased the water storage capacity but also increased the groundwater recharge.

Desilting of Community Tanks

Community tanks/ponds for augmenting village level water resources Community tanks, traditionally recognized as resilient systems, are silted up and have become defunct due to neglect. De-siltation of these tanks helped in increasing the surface water storage, besides increased groundwater recharge. The spin off effects of spreading rich silt deposited in these structures by farmers in their fields helped improve the soil water holding capacity. An increase in the yield of cotton and castor (15 to 18%) was observed in silt applied fields (Kurnool districts of AP) over the check due to improved fertility status. De-siltation and renovation of three community ponds with surface area of 1045, 5035 and 3605 m2 and storage capacity of 5735, 23000 and 16500 m3 in Tamil Nadu resulted in additional increase in water storage of 36617 m3 as well as recharging of groundwater. This additional water helped in cultivation of crops in 137.5 ha area at Jambumadai and Vadavathur villages of Namakkal District

Table 16: Effect of desiltations of community ponds

Name of the Pond	Dimensions (m x m x m)		Ground-water recharge	Area Irrigated (ha)	Major cultivated crops	
	Before	After			Before	After
Senguttai	36 × 26 × 3,65	38 × 28 × 5,4	18 open welts & 11 bore wet Is	13.6	Onion and sorghum	Onion, groundnut, and sorghum
Aayiramkuttai	30 × 22 × 1.65	92 × 55 × 45	45 Open wells & 112 bore wells	73 6	Onion and sorghum	Onion, Paddy and groundnut
Periyakalingikuttai	49 * 22 x 3,65	112 x32 x 4.5	45 open wells & 156 bore well	137.5	Onion and sorghum	Onion, groundnut and sorghum

(*Source*: Alok Sikka *et al*., 2012)

Fig 2: Check dam at D. Nagenahalli village (*Source*: Prasad Y.G. 2014)

High value crop

Use of Tank silt

Red and sandy soils are predominant in dryland areas and these soils have low clay content, due to the low clay content, percolation of rainwater is more. Hence, to increase the water holding capacity of these soils, tank silt can be added to the top layers of soil, through which water is stored by reducing deep percolation. Application of silt besides increasing water holding capacity is also a good source of nutrients. Moreover, desilting of tanks improves moisture availability and storage capacity of tanks. Tank silt application reduces soil erosion and prevents the loss of nutrients from soil. Studies at CRIDA have shown that application tank silt increases 10-20% of yield along with 30-40% of increased soil moisture. The quantity of tank silt to be applied to soil depends on clay content of soil. Tank silt can be applied once in every three years. This will help in proper mixing up of silt to the soil whenever rains occur. The major precaution to be taken is to avoid high pH of tank silt. The added advantage of tank silt application is desilting of tanks improves storage capacity of tanks. And these tank beds can be used to grow fodder in summer season.. Tank silt can be applied once in every three years. This will help in proper mixing up of silt to the soil.This will help in proper mixing up of silt to the soil whenever rains occur. The major precaution to be taken is to avoid high Ph tank silt.

Building resilience with better soil management

Healthy soils are essential for higher crop productivity but also to contribute towards drought mitigation particularly mid-season droughts. Improved soil organic matter storage in soil profile retains more water and provides drought proofing in rainfed agriculture during long gaps between two rains. Based on 16 long-term manurial experiments under rainfed conditions in AICRPDA network, it was showed that each ton of soil organic carbon improved productivity of rainfed crops by up to 0.15 t/ha/year. Location specific integrated nutrient management (NM) practices were identified and being promoted based on locally available organic resources. Balanced nutrition particularly optimum potassium nutrition also contributes to mitigation of water stress conditions as K controls water relations in plant growth. Foliar sprays of KNO3 and thiourea during mid-season droughts reduce the negative effects of droughts in several rainfed crops

Agroforestry

Agroforestry products such as timber, fiber, fruit, food, fodder, fuelwood, fertilizers, medicine and others are meeting the subsistence needs of farmers and providing greater opportunity for sustained productivity.Agroforestry has a particular potential role in the mitigation of atmospheric accumulation of greenhouse gases. Agroforestry systems offer opportunities for the creation of synergies between adaptation and mitigation and have a technical mitigation potential of 1.1-2.2 PgC in terrestrial ecosystems over the next 50 years. The enhancement of forest C stocks through agroforestry can be considered as one of the main options for reducing greenhouse gases in the atmosphere Agroforestry is a unique extensive action involving the integration of woody plants with crop and livestock components. Thus, the greatest role of agroforestry in relation to climate change is in mitigating the emissions of CO2 by sequestering carbon from the atmosphere. Globally, forestry. has taken a central stage as one of the options to mitigate CO2 climate change. Conversely, agriculture & plantation could also be a solution for climate change by adoption of mitigation and adaptation action . This happens with the help of well managed agroforestry practices

Crop Diversification

Crop diversification is an important risk minimizing strategy for drought proofing in the scarce rainfall zones. Inter-cropping is a feasible option to minimize risk in crop production, ensure reasonable returns and improve soil fertility with a legume intercrop in such areas. The different approaches of crop diversifications are

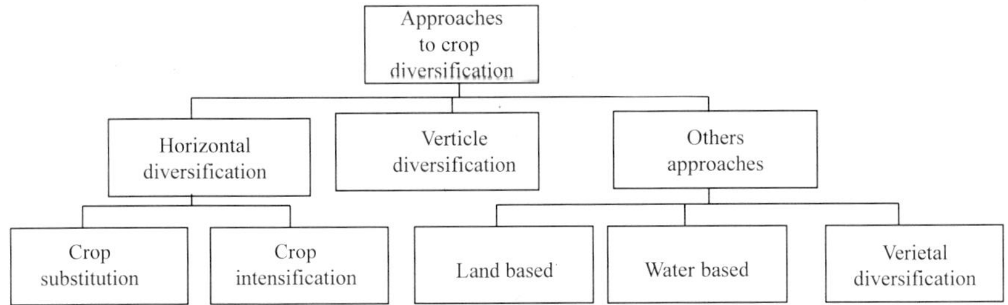

Fig. 3: Different approaches of crop diversifications.(*Source*:Barman Anamica *et al.*, 2022)

Cotton, soybean, pigeon pea and millets are the major crops in the scarce rainfall zones. Inter cropping of these crops is more profitable. In contingency situations such as delay in onset of monsoon, adoption of inter cropping for delayed plantings can be remunerative instead of sole cropping. Inter cropping systems such as Groundnut + Pigeonpea(10:2), Maize + Pigeon pea (6:1) were demonstrated in drought prone districts of Karnataka, Andhra Pradesh, Telangana, and Madhya Pradesh.

Table 17: Different measurements of crop diversification and their characterization

Measure of crop diversification	Characterization
1. Temporal crop diversification	
Crop rotation	Growing of two or more different crops by one after another in consecutive ways
Catch crop	Growing of crops to in between the space of two main crop or when no main crops are being grown
Double or multiple cropping	Growing two or more crops in one growing season
Relay cropping	In relay cropping second crop is grown in standing crop before the first crop is harvested
2. Spatial crop diversification	
Alley cropping	It is an agroforestry system in which food crops are grown in alleys formed by trees
Intercropping	Growing two or more crops simultaneously on the same Land with definite pattern
Mixed cropping	Growing two or more crops simultaneously in the same field
Variety mixture	Growing two or mote varieties of a same species
Trap	Growing commercial and non-commercial crop simultaneously in the same land

(*Source*: Barman Anamica *et al.*, 2022)

Integrated Farming System Modules

Mono-cropping is generally practiced in drought prone areas of the country. Diversification of farm enterprises can augment farming income and build resilience among the dryland farming community to tide over losses. Several integrated farming system modules with a combination of small enterprises such as crop, livestock, poultry, piggery, and duck rearing can be adopted by farmers. Integrated farming system modules minimize risk from a single enterprise in the face of natural calamities and diversified enterprises can provide better income to farmers and improve their livelihoods. Integrated farming system modules such as, Pearl millet - chick pea - cows, Maize - groundnut + cow, Pearl millet + guava + goat / cow, Pearl millet + agro forestry + goat / sheep have been demonstrated in different rainfed lands for better profit..

Management of livestock (climatic vulnerabilities)

Livestock becomes an integral part of traditional farming systems in dryland agriculture. This component will act as buffer to stabilize and enhance the income of dryland farmers for different weather aberrations Among livestock enterprises, the cattle,buffalo, and small ruminants (sheep and goat) are important source for income and livelihood improvement in dryland farming community in the country. There is need to take care of hygiene, health, and nutritional management to improve the productivity of the livestock. Animal production and animal husbandry including poultry management are important components of dryland agriculture. Appropriate technologies for enhancement of animal management and production including poultry are very important to ensure stable income levels to dryland farmers. Livestock face several climate change-related challenges in drylands. It include Droughts and floods, Thermal stress, mortality and reduced yields, Water availability and quality (drinking, for forages and feed crops).(Pankaj *et al*2013)

Dairy Animals

Drought

Feed and Fodder Resources

- During early season drought, short to medium duration cultivated fodder crops like sorghum (Pusa Chari Hybrid-106 (HC-106), CSH 14, CSH 23 (SPH-1290), CSV 17 etc) or Bajra (CO 8, TNSC 1, APFB 2, Avika Bajra Chari (AVKB 19)etc.,) or Maize (African tall, APFM 8 etc.,) which are ready for cutting in 50-60 days and can be sown immediately after rains under rainfed /drylands conditions in arable lands during Kharif season .
- In waste/degraded lands, grasses like *Cenchrus ciliaris*, *C. setigerus*, *Chloris gayana*, *Panicum maximum*, *Desmanthus virgatus* and *Stylosanthes scabra* can be taken up to increase forage production

- Harvest and use biomass of dried up crops (soybean, wheat, green gram, black gram, sorghum, bajra, maize, chick pea) material as fodder
- Use of unconventional and locally available cheap feed ingredients especially soya meal waste for feeding of livestock during drought
- Harvest all the top fodder available (Subabul, Glyricidia, Pipol, Prosopis etc) and feed the LS during drought
- Concentrate ingredients such as Grains, brans, chunnies & oilseed cakes, low grade grains etc. unfit for human consumption should be procured from Govt. Go downs for feeding as supplement for high productive animals during drought
- Promotion of Horse gram as contingent crop and harvesting it at vegetative stage as fodder
- All the hay should be enriched with 2% Urea molasses solution or 1% common salt solution and fed to Long season
- Continuous supplementation of minerals to prevent infertility. Encourage mixing available kitchen waste with dry fodder while feeding to the milch animals

Drinking water

- Adequate supply of drinking water.
- Restrict wallowing of animals in water bodies/resources
- Ddalum in stagnated water bodies

Health and disease management

- Carryout deworming to all animals entering relief camps
- Identification and quarantine of ck animals
- Constitution of Rapid Action Veterinary Force
- Performing ring vaccination (8 km radius) in case of any outbreak
- Restricting movement of livestock in case of any epidemic
- Tick control measures be undertaken to prevent tick borne diseases in animals
- Rescue of sick and injured animals and their treatment
- Organize with community, daily lifting of dung from relief camps

Heat and cold wave

- Allow the animals early in the morning or late in the evening for grazing during heat waves
- Feed green fodder/silage / concentrates during day time and roughages / hay during night time in case of heat waves
- Put on the foggers / sprinklers during heat weaves
- In severe cases, vitamin 'C' and electrolytes should be added in H2O during heat waves.
- Apply / sprinkle lime powder in the animal shed during cold waves to neutralize ammonia accumulation

Poultry

Drought

1. Feed and fodder sources
2. Supplementation only for productive birds with house hold grain
3. Supplementation of shell grit (calcium) for laying birds
4. Culling of weak birds

Drinking water

Mixing of Vit. A,D,E, K and B-complexi ncluding vitamin C in drinking water (5ml in one litre water)

Heat Wave

Shelter Management

- In severe cases, foggers/water sprinklers/wetting of hanged gunny bags should be arranged
- Don't allow for scavenging during mid day

Health and disease management

- Supplementation of house hold grain
- Provide cool and clean drinking water with electrolytes and vitamin C
- In hot summer, add anti-stress probiotics in drinking water or feed

Climate smart livestock villages

Clusters programme may be taken up to sustainably increase productivity of livestock production, income of farmers, build resilience to climate change and reduce GHG emissions in different bioclimatic zones. Researchers, local government partners, farmers and private sector may collaborate to efficiently implement prioritized interventions, monitor and evaluate progress and disseminate outcome.

- Promoting stress tolerant species and breeds of livestock suitable and acceptable in different bioclimatic zones.
- Scientifically designed shelters and improved management of animals including health management. Community animal housing clusters in grazing areas in arid, semi-arid and coastal areas. Community livestock housing and management connect with value chain for small holder livestock reared in intensive and semi-intensive systems.
- Balanced / strategic feeding of livestock as per requirement of climatic zone, species and farming system.
- Weather information on near term and long term and timely warning for extreme climatic events.
- Integrated land use planning incorporating fodder crops, food feed crops, through cross visits to analogue sites and to other areas practising climate smart agriculture, agroforestry and pasture management as per local area requirements and acceptability.
- Installation of community biogas plants for efficient use of dung for clean fuel and manure as well as waste to compost programme.
- Crop residue management, fodder conservation and socioeconomically viable community processing centres and fodder bank. v Educating and facilitating farmers to cope with climate change and its impact

Weather based agro-advisories

Weather based agro meteorological advisory services (AAS) help in cultivar selection based on seasonal rainfall forecast choosing windows for sowing/ harvesting operations; mitigation from adverse weather events; nutrient management; fertilizer application; feed, health and shelter management for livestock (optimal temperature for dairy/ hatchery). There is an increased role of weather based AAS in farming activities for access to real time weather information, timely agricultural operations, improved crop yields, reduced cost of cultivation, need based changes in cropping patterns and finally improved livelihoods.

This weather based agro-information can be made available to farmers through audio and visual media and also effectively through mobile phone networks. Weather forecasting and early warning systems will be very useful in minimizing risks of climatic adversaries. Information and communication technologies could greatly help the researchers and administrators develop contingency plans

Table: Economic Impacts of Agro meteorological advisory services

Name of the farmer	Date of issuing AAS	Crop	Rainfall forcast	Advisory given	Observed Rainfall	Action Taken by the farmer	Benefit/loss
Anantapur							
Dastagiri (Yaganti-palle	27/02/15	Summer Greengram	02.03.15 4.0 mm	Postpone pre- sowing irrigation	02/03/15 25.6 mm	Postponed pre sowing irrigation	Undertook sowing with the rain-fall received. Saved Rs.1000/- per acre towards the cost of pre-sowing irrigation
Maddileti Reddy (Ya-gantipalle)	15/05/15	Black gram	16,17 May 14 mm	Postpone irrigation	16/05/15 11.0 mm	Postponed irrigation	Farmer saved Rs. 400/- per acre towards pre-sowing irrigation
Mahesh Reddy (Ya-gantipalle)	02/06/15	Paddy	05.06.15 06.06.15 9mm	Land prepa-ration for nursery utiliz-ing forecasted rainfall	05.06.15 8.2 mm	Postponed irrigation for land preparation	Farmer saved Rs. 400/- per acre towards pre sowing irrigation.
Shiva Satyam (Ya-gantipalle)	19/06/15	Maize	20-22nd June 29-65 mm	Keep ready all inputs for sowing and sow the crop	0.0 mm	Kept all inputs ready for sowing	Farmer kept all inputs ready for sowing but unable to take up sowing as there was no rainfall up to 15/07/15. Hence the sow-ing was delayed.
Bandi Manmadha Reddy (Yer-raguda)	07/08/15	Pigeon pea	6mm rainfall during next 5 days	Wait for protective irrigation	25 mm in next 5 days & 46 mm on 13 .08.15	Postponed irrigation	Because of no rainfall from 23/07/2015, he planned to irrigate the crop. Based on the advisory he postponed irrigation and saved Rs.400/- per acre.

Name of the farmer	Date of issuing AAS	Crop	Rainfall forcast	Advisory given	Observed Rainfall	Action Taken by the farmer	Benefit/loss
Bijapur							
	2nd & 3rd week of Dec, 2015	Tomato, Cucumber Vegetative stage	Dry weather	Spray water in the early morning hours to wash dew		Complied	Got good yield (480 trays) and price (RS. 120/tray) and got net income of Rs. 45,000
July to Sep, 2015	Soybean	Less rainfall	Timely advise for irrigation to the crop, sprays to control diseases		Decided date of harvest based on the forecast.	Got more net income of Rs. 10,200/- compared to other farmers.	
Udaipur							
Ram Singh		Mustard	Possibility of rainfall (3mm)	Avoid pesticide spraying in mustard (to control aphid)	7.1mm on 23 Jan & 22.9 mm on 24 Jan		Saved labour cost and insecticide cost of Rs 100.
Manohar Lal		Wheat (Harvest stage)	Possiblility of light rain on March 27	Harvest the crop and ensure safe storage	7.1mm on 27 March	Kept harvested produce in field for sun drying	Seed quality affected due to rainfall for those who did not follow AAS. Wheat grain fetched Rs. 1200/- per quintal as against Rs. 1500/- per quintal from those who adopted AAS.
Manohar Lal		Wheat (Harvest stage)	Possiblility of light rain on March 27	Harvest the crop and ensure safe storage	7.1mm on 27 March	Kept harvested produce in field for sun drying	Seed quality affected due to rainfall for those who did not follow AAS. Wheat grain fetched Rs. 1200/- per quintal as against Rs. 1500/- per quintal from those who adopted AAS.

Name of the farmer	Date of issuing AAS	Crop	Rainfall forcast	Advisory given	Observed Rainfall	Action Taken by the farmer	Benefit/loss
Manohar Lal		Wheat (Harvest stage)	Possiblility of light rain on March 27	Harvest the crop and ensure safe storage	7.1mm on 27 March	Kept harvested produce in field for sun drying	Seed quality affected due to rainfall for those who did not follow AAS. Wheat grain fetched Rs. 1200/- per quintal as against Rs. 1500/- per quintal from those who adopted AAS.
Bhawar Lal	10.7.2015	Capsicum (infested with mosaic disease)	No rainfall in next four days	Spray imidacloprid	7 mm rainfall was received on 13.7.2015	Sprayed imidacloprid@1 ml /3 lit water in capsicum in 2500 m^2 area on 13.7.2015	Due to rain insecticide was washed out. Loss of Rs.100/- and labour charge.
Manohar Lal		Wheat (Harvest stage)	Possiblility of light rain on March 27	Harvest the crop and ensure safe storage	7.1mm on 27 March	Kept harvested produce in field for sun drying	Seed quality affected due to rainfall for those who did not follow AAS. Wheat grain fetched Rs. 1200/- per quintal as against Rs. 1500/- per quintal from those who adopted AAS.
Bhawar Lal	10.7.2015	Capsicum (infested with mosaic disease)	No rainfall in next four days	Spray imidacloprid	7 mm rainfall was received on 13.7.2015	Sprayed imidacloprid@1 ml /3 lit water in capsicum in 2500 m^2 area on 13.7.2015	Due to rain insecticide was washed out. Loss of Rs.100/- and labour charge.

(*Source*: U.M.B.Rao *et al* 2015)

Institutional interventions

Institutional interventions either by strengthen existing ones or initiating new ones relating to seed bank, fodder bank, commodity groups, custom hiring centre) collective marketing, introduction of weather index based insurance and climate literacy through a village level weather station are introduced to ensure effective adoption of all other interventions and promote community ownership of the entire programme. Under NICRA, in each of the village clusters, a Village Climate Risk Management Committee (VCRMC) was formed to effectively co-ordinate with farmer groups on climate variability/ change related issues. VCRMC representing all categories of farmers in the village is formed with the approval of Gram Sabha and committee involved in programme implementations and need based decisions depending on weather aberrations. A village-level weather station and custom-hiring centers were also established to promote weather literary and enable farmers i" ti-.ly completion of farm operations during delayed monsoon.

Real time contingency planning is considered as "Any contingency measure, either technology related (land, soil, water, crop) or institutional and policy based, which is implemented based on real time weather pattern (including extreme events) in any crop growing season" The real-time contingency measures aim to (i) to establish a crop with optimum plant population during the delayed onset of monsoon; (ii) to ensure better performance of crops during seasonal drought (early/mid and terminal drought) and extreme events, enhance performance, improve productivity and income; (iii) to minimize damage to horticultural crops/produce; (iv) to minimize physical damage to livestock, poultry and fisheries sector and ensure better performance) to ensure food security at village level and (vi) to enhance the adaptive capacity and livelihoods of the farmers. Some of the methods/measures to be adopted as real-time contingency plan implementation during various weather aberrations

Policy framework

Implementation of agriculture contingencies in real time needs a stronger policy support. India has a very comprehensive framework of legal and institutional mechanisms in the region to respond to the tremendous challenges to the environment it is facing, owing to population growth, poverty and illiteracy augmented by urbanization, and industrial development. The country has had, over the last six decades, major programs addressing climate variability concerns. These include cyclone warning and protection, coastal protection, floods and drought control and relief, major and minor irrigation projects, food security measures, research on climate resilient agriculture, and several others. A sound policy framework should address the issues of redesigning the social sector with focus on vulnerable areas/populations, introduction of new credit instruments with deferred repayment, liabilities during extreme weather events, and weather insurance as a major vehicle to transfer risk .Better seed systems to be established to ensure quality seed availability during delayed monsoon conditions and in the conditions of complete crop failures in severe droughts or flood conditions monsoon conditions and in the conditions of complete crop failures in severe droughts or flood conditions.

Availability of seed material of contingent crops such as legumes, millets and oilseed crops need to be ensured. In many dryland regions, often the window for crop sowing operation is limed due to less number of rainy days. Hence, it is necessary to ensure farm implement availability in the village to complete sowing of crops. However, farm machinery supply should be for both bullock and motor operated mode as large numbers of farm holdings are small and marginal. Similarly, ensuring the availability of bullock/tractor drawn implements of sowing of different intercropping systems is important for promoting adoption of resilient intercropping systems. Every part of India experiences some weather aberration like droughts, floods/cyclones, heat wave and hailstorm, frost etc. often resulting in grain damage during harvesting stage. Hence, strong implementable interventions are needed to ensure the purchase of damaged grain at local level. A good convergence among research organizations and various government programs such as national/ state action plans, NMSA, NFSM, Pradhan Mantri Krishi Sinchayee Yojana (PMKSY), soil health schemes, water mission, green climate fund etc. will further contribute to scaling up of resilient crops and cropping systems to cope with weather aberrations in rainfed regions of India.

A good convergence among research organizations and various government programs such as national/ state action plans, NMSA, NFSM, Pradhan Mantri Krishi Sinchayee Yojana (PMKSY), soil health schemes, water mission, green climate fund etc. will further contribute to scaling up of resilient crops and cropping systems to cope with weather aberrations in rainfed regions of India.

10

Alternate Land Use Systems in Drylands

Marginal and sub marginal lands are being brought under cultivation to meet the demand of human and livestock in terms of rising demand of 6F viz., food, fodder, fuel wood, fibre and fertilizer. These in lands are unable to sustain productivity. Cultivation on such lands leads to imbalance in the ecosystem. In order to meet the above said demands and to conserve the natural resources, a diversified land use system needs to be adopted in different agro-ecological regions of the country as an alternative to conventional cropping systems. Through this approach, the biological productivity and quality of resource base of degraded lands can be significantly enhanced. Land use systems which are alternatives to crop production are called by the term 'Alternate Land Use Systems'. The system involves the addition of perennial component which has drought tolerance, can withstand the aberrations of monsoon and imparts stability to production (Korwar,1992). When a land is put under an alternative production system in relation to its capability more appropriately to the new land use to achieve more sustainable biological and economic productivity on a long term, it is known as alternate land use. Depending on the components of the alternative production system, various types of alternate land uses are recognized. (Prasad 2011)

Network research carried out in India revealed that alternate land use involving perennials (tree/crop, grass shrub or a combination of both) has advantages and can conserve natural resources and increase productivity.

Some of the advantages of perennial trees/ grasses/shrubs are:

- They can thrive in relatively resource-poor soils,
- Provides vegetative cover to the soil round the year, thus, substantially control erosion caused by both wind and water. The net result would be amelioration of microclimate.
- Provides good quality green fodder, which is in short supply to support livestock.

- Improves soil quality through nutrient cycling by mining of deeper layers, litter fall and root turnover.
- Reduce surface evaporation and weed growth when pruned material is applied as mulch at surface, thus, improve water use efficiency
- Provide fuel, timber and minor forest products (e.g. gum, honey) and thus will lessen the dependence of farmers on forest.
- Supplement the diet of poor farm families and ensure their nutritional security.
- Generate much needed cash when aromatic and industrial value plants are included.
- Supports development of flora and fauna particularly soil microbial and earthworm activity.
- Generates employment throughout the year, which substantially increases the cash flow and reduces migration

Land Capability-based Land use systems

As per land capability classification (LCC), there are eight classes and arable farming can be practiced in LCC I to IV, and beyond that, arable farming is considered as a risky enterprise. Lands worse than LCC IV are devoted for forestry and grass land management for safer side. Alternate land use options integrating farming with other land uses like horticulture, silviculture and pasture management are preferred options. Suitable Alternate land use systems based on rainwater availability and land capability are presented in Table 1.

Table 1: Alternate land use options for different land capability classes

Land capability class	Sub-class	Suitability for cultivation or not	Special precautions	Preferable alternate land use and other options
1.very good cultivable land	Deep, nearly level	Intensive cultivation of crops depending upon soil and climate	-	–
II.Good cultivable land	IIe Good soil with minor erosion problem	Cultivation with precaution	-	Agri horticulture
	IIs land with minor soil problem	Selection of proper crop	-	Agrisilviculture
III Moderately good Cultivable land	IIIe Good soil on moderate slope	Cultivation with precaution against permeant land degradation	Special attention to erosion control	Agri horticulture
IV Fairly good land Suited for occasional occasion	IVe.Moderately step land subjected to severe water erosion	Occasional cultivation in rotation with pasture or orchards protected by permanent cover crops	Intensive erosion control when in cultivation	SilviPasture
	IV w Bottom land that is very wet	Cultivation with special crops	Intensive drainage	Silviculture with proper drainage
	IVs Fairly good land with limitations due to shallowness, gravel stone or strong alkali	Occasional cultivation in rotation with pasture	Very intensive treatment to over-come soil limitation, careful selection of trees and tree crops	Hortipastoral
Land capability class	Sub-class	Suitability for cultivation or not	Special precautions	Preferable alternate land use and other options
	IV c Good with just enough rainfall in favourable years	Cultivation during wet years	Conserve all rain	Hardy fruit and fodder trees

Land capability class	Sub-class	Suitability for cultivation or not	Special precautions	Preferable alternate land use and other options
V nearly level land not suitable for cultivation because of stoneiness,wetness etc.,	_	Not suitable	To be developed as range land with regulated grazing	Hortipasture/Silvipasture after providing micro-catchment like cresecent shape bunding
VI Steep slopes, highly erosion prone with shallow soil	_	Not suitable	Careful management if developed as forest wood lot orgrazing land	Silvipasture, continues
Contour trenches(CCT) or staggered trenches with provision ofbio-fence or Cattle proof trench(CPT) or fencing				
VII Steep slope with severe soil erosion resulting in eroded stony and rough soil surface with shallow soil depth	_	Not suitable	Contour trench,gully control works and controlled grazing	Cultivation of improved grasses after land treatment

Note: e: Susceptible for erosion, s: shallowness to rooting zone and stoniness: w:excess water, poor drainage and high water table, c: moisture limitation with poor rainfall

(Venkateswarlu *et al.* 2013)

Alternate Land use systems for Arable lands

Alternate land use systems can be classified as Alley cropping, Agri-horticulture, Silvipasture, Agri silviculture and horticulture etc., depending upon the nature of the component in the system. Tree farming, social forestry are other alternative land uses, which are meant to improve the degraded natural base, besides providing economic products to the community. Alternate land use systems are aimed at the optimizing the use of resources through the principle of recycling, internalize input production, reducing the risk and conserving the natural resources. It reduces the erosivity of the rainfall and erodability of the soil through dissipation of energy of rain drops by canopy at low heights, surface litter, obstructing the run-off, root binding and improve the organic matter, physicochemical and biological properties.

Among the various agro-forestry systems, agri-horticulture system is the most important in terms of economic returns to the farmers and their preferences. For example, based on a long term economic analysis of different alternate land use systems evaluated in semi -arid alfisol region of Andhra Pradesh, agri horticulture was found to be the most profitable (Reddy and Sudha, 1988) giving a CB ratio of 1:5.53, followed by silvipasture and agri- silviculture (Table 2).Among different land use systemssystems,Agri-horti system(Ber+arable crops)gave highest net returns folles by hortipature systems (Tanwar *et al.* 2016).

Table 2: Benefit cost ratio of different alternate land use systems in semi-arid Alfisols

Agroforestry system	Period (seasons)	Benefit cost ratio
Arable farming (crops)	1	1.34
Agroforestry	10	1.65
Agri-horticulture	30	5.53
Silvi-agriculture	10	1.99
Silvi-pastoral	10	2.43

(Reddy and Sudha, 1988)

Table 3: Economics of Alternate land use systems

System	Component	Cost of cultivation (Rs10³/ha)	Net retums (Rs10³/ha)	Benefit: cost ratio		
				2012 -13 Rs	2013 -14	2014 -15
Arabic ramnnj;	Crops alone	14.46	11.9	1.40	1.86	2.0
Ajiruluresty	*P cineraria* + crops	19.4	24.3	2.02	2.34	2.4
Agri-silviculture	*Hardwickia binata* + crops	19.5	19.0	1.52	1.82	2.3
Afri-horti	*Ziziphus mauritiana* + crops	47 4	66.1	2.42	2.39	13
Horti-pasture	*Ziziphus mauritiana* + grass	29.2	19.8	1.56	2.54	11
Silvi-pasture	Colophospernum mopone + grass	16,4	39.6	2.94	3.75	3.4
Agri-pasture	Grass alome	16.3	32.2	2.9	3.19	2.6

Alley cropping is a system in which food crops are grown in alleys formed by hedgerows of trees or shrubs. The essential feature of this system is that hedgerows are cut back and kept pruned during cropping to prevent shading and to reduce competition with food crops. *Leucaena leucocephala* has been mostly used as a hedge-row species in the system. Likewise, other species like *Gliricidia* and *Sesbania* could be used as hedge rows. The main criteria for the species should be that it must be amenable to lopping besides being multipurpose and fast growing. The alley width varies with the choice of the species and watershed location in a climatic region. By and large, alley width has been found varying from 2 to 10 m; wider the alley, the lesser the adverse effect on crop performance. Alley cropping with *Leucaena/gliricidia* hedges and grass barriers have been found effective in controlling erosion upto 30% slope under humid sub-humid and subtropical climate conditions. Contour paired row of *Leucaena* as a hedge, *Leucaena* and eucalyptus trees and 0.75m wide grass barriers at 1.0m vertical interval in maize at 4% slope re duced the runoff from 40 to 30% and soil loss from 21 to 8 t./ha/year (Table 4)

Table 4: Effect of paired rows of barrier hedges, grass strips and trees on runoff and soil loss in maize at 4% slope

Treatment	Runoff (%)	Soil loss (t ha^{-1} yr^{-1})
Maize on contour	40.0	21.0
Leucaena hedge	21.3	12.1
Panicum grass (0.75 m wide)	36.7	7.0
Bhabar grass (0.75 m wide)	42.7	10.0
Vetiveria (0.75 m wide)	39.6	8.1
Leucaena trees (6-8yrs)	20.4	8.4
Eucalyptus trees (6-8 yrs)	16.3	5.8
Agroforestry land uses (mean)	30.0	8.7

(*Source*: Sharda V.N. and Venkateswarlu, B.V. 2009)

Trees such as species of *Leucaena, Gliricidia, Sesbania, Cajanus*, and *Cassia* have been widely tested in alley cropping system. Short duration rainy crops such as pearl millet and sorghum were found to be compatible with *Leucaena leucocephala* and *Gliricidia sepium* than long duration species such as castor *(Ricinus communis*) and pigeon pea (*Cajanus cajan*). Wider alleys and low cutting heights were found to give higher intercrop yields in *Leucaena* system in semi- arid condition (Korwar 1992; Rao *et al.*1991). Alley cropping of *Leucaena* with sorghum (*Sorghum bicolor*) and cowpea (*Vignaunguiculata)* was studied at Hyderabad to quantify the effect of shoot pruning, fertilization, and root barriers around *Leucaena leucocephala* trees on crop production under rainfed conditions. Crops grown with pruned trees attained higher dry matter and leaf area index than did those with unpruned trees. Alley cropping of *Leucaena leucocephala* with pearl millet (*Pennisetum typhoides*) and pigeon pea (*Cajanus cajan*) reveals that reduction of yields with the increase of alley widths (10, 15, and 20 m) might be due to dense shade of the tree (Patil *et al.* 1999). Mittal and Singh(1983) reported that hedge row plantation of *Leucaena leucocephala* with maize, black gram and cluster bean reduced the yields of maize*(Zea mays*),black gram (Vigna mungo), and cluster bean (*Cyamopsis tetragonoloba*) by 38,34, and 29 %, respectively. This reduction in yield was compensated by relatively higher fodder and fuel production of Leucaena. Maximum returns were obtained with intercropping compared to growing of pure trees or crops.

Ley Farming

The dry lands are not only thirsty but are also hungry because of very low nutrient content. The dry land farmers are reluctant to apply fertilizers, due to risk of lack of matching returns. Farmers apprehend fertilizer wastage if the monsoon fails. Although research results have been able to dispel these fears, fact remains that due to poor economic conditions, dryland farmers must be burdened less with bought inputs. In this situation, ley farming, a crop rotation including a period of pasture (a ley), may prove more appropriate to enhance soil fertility. The modern concept of ley farming includes grass-legume mixture or ley as a farm crop. The grass, in general, has the quality of improving soil structure while the legume enriches the soil nitrogen. It was observed that *Stylosanthes hamata* raised for a period of 2 to 3 years in a four year rotation helped in building soil nitrogen to the extent of 35 kg/ha. With this, a seasonal crop productivity can be sustained without fertilizer – N application. Acceptance of ley farming by farmers needs to be evaluated.

Agro-silviculture systems

The agro silviculture includes agricultural crops as major component and trees are combined with them. In this system, tress can be chosen that shed the leaves while the crop is being taken up. (*Acacia albida, poplar, pawlonia Sp*) or avoid shade effects by lopping(eg.*Prosopis cineraria, Acacia tortilis),*pollarding (eg. *Subabul, Acacia albida*). The system is recommended for land capability class IV with annual rainfall of 750 mm. Short duration dryland crops such as perarlmillet, black gram and green gram, combined with widely spaced tree rows of *Faiderbia albida* and *Hardwickia binata*, have been found compatible in semi-arid tropical areas.(Korwar,1992).

Table 5: Crop yields in *Faidherbia albida* agroforestry system at various ages and densities under rainfed conditions at CRIDA, Hyderabad

Tree age (years)	Tree density (per ba)	Crop(s)*	% increase (+) reduction (-)
5	156 ami 625	Sorghum. groundnut	+57 and +32 % higher sorghum and groundnut yield (156 trees ha^{-1}) in comparison to higher tree density or625 trees ha^{-1}
A	156	Sorghum, groundnut	Sorghum -6 %, Groundnut + 14% over sole crop control
6	625	Sorghum, groundnut	Sorghum -46% and groundnut -24% over sole crop
8-9	625	Con pea	- 9% in grain sield and -1(1 ti iu dry mailer producilou
8-9	625	Sorghum	-12 % in grain yield and -13% in dry matter production
10	156	Sorghum, cmvpea	Sorghum -11%, coirpea -8% over sole crop control
11-12	625	Green gram, black gram, cmvpea	Green gram -26-29 %, cotvpea -59-64 lit. and black gram -91 96 In grain yield in comparison to control
11-12	156	Green gram, black groin	Green gram +12-20% ti, hi ark gram 4 2-27% grain yield in comparison to control

(Korwar and Pratibha 1999)

Results of the experiments conducted under rainfed conditions at CRIDA on yield of intercrops with two densities of trees (156 and 625 trees ha^{-1}) of Faidherbia albida from 5 to12 years of tree age are presented (Table 5)

At 5 years of tree age, sorghum (*Sorghum bicolor*) and groundnut (*Arachis hypogea*) yields were higher in lower tree density by 32– 57 % in comparison to higher tree density. Tree population did not affect the dry matter production in both the groundnut and sorghum crops at 5 years of tree age. At 6 years of tree growth, though, groundnut yields improved when compared to the lower tree density, there was a marginal reduction in yields of sorghum in comparison to the sole crop. The crop yields were significantly affected under higher tree densities of 625 trees ha^{-1} during the sixth year. Seed and dry fodder yields of sorghum were reduced by 12 and by 13 %, respectively, when intercropped with 8 to 9-year-old *Faidherbia albida* at a density of 625 trees ha^{-1} in comparison with the sole crop. However, the extent of reduction in *Faidherbia albida* is lowest in comparison to other N-fixing tree species such as *Acacia ferruginea* and *Albizia lebbeck*. At 10 years of age, the yield reduction in sorghum and cowpea (*Vigna unguiculata* cv. C 152) was to the extent of 8–11 %, respectively, in com- parison to the sole crops. However, in 11 to 12- year-old trees, the seed yields of green gram(*Vigna radiata* cv. ML 267) and black gram (*Vigna mungo* cv. T 9) were higher under the lower tree density (156 plants ha^{-1}, spacing 8 m x 8 m) than when monocropped or grown under the higher tree density (625 plants ha^{-1}), whereas cowpea yields were lower under the trees at both densities than when cultivated as monocrop (Korwar and Pratibha 1999).

During the years of low rainfall, the yield reduction is more with the higher tree density. Under no input conditions, viz., under no nitrogen application situation, the *Faidherbia albida* + sorghum and *Faidherbia albida* + cowpea system recorded higher grain yields in comparison to sole crop controls and the extent of yield increase was 54 and 32 % in sorghum and cowpea, respectively (CRIDA 1996).

Agri-horticultural system

Agro-horticulture system is one form of Alternate land use systems adopted in arable lands having fruit trees as component. The system provides higher income per unit area. and its distribution decide largely the type of fruit species that can be successfully grown. From climatic point of view, temperate sub-tropical regions are ideally suited for fruit cultivation. Fruit crops such as mango(*Mangifera indica*), guava (*Psidium guajava*), pomegranate*(Punica granatum*), goose berry (*Emblica officinalis*),jamun *(Syzygium cuminii*), wood apple(*Feronia limonia*), and tamarind (*Tamarindus Indica*) can be grown in areas where the rainfall is more than 600 mm. Tree species like custard apple, Ber (*Ziziphus* spp), Phalsa (*Grewia ubinaequalis*), Karonda (*Carissa carandas*), Lasura(*Cordia* spp.), and Bel (*Aegle marmelos)* can be grown in areas where the rainfall is less than 500 mm under dryland situations.

Leguminous crop sown under ber (*Ziziphus mauritiana* cv. Seb) plantation produced 0.2 t /ha of grain and 0.8 t /ha quality *ber* fruits from same land unit even when seasonal rainfall is 200 mm, thus imparting a drought proofing mechanism to the system. The density of *ber* plants were kept 400 individuals/ ha (Gupta 1997). The economics of this improved system indicated that in case of sole leguminous crop (mung bean) farming, the net profit per hectare was Rs. 4800/-, however, in case of Ber intercropping, the profit was to a tune of Rs. 8000/- per ha. (Table 6)

Table 6: Improved agroforestry practices: agri-horticultural (Ber+ Mung bean)

Treatment	Annual rainfall (mm)	Fruit yield (kg ha1)	Grain yield (kg ha1)	Net profit (Rsha1)
Sole crop	200	-	520	4800
Intercropped with Ber	200	800	200	8000

(*Source*: Tiwari *et al*., 2014)

Experiments at CRIDA have shown that in *Emblica officinalis* fruit yields were significantly higher (5,277 kg /ha) in sole trees than intercropped ones (3,478 kg/ ha) due to competition for moisture. During low rainfall years, the arable crops like Ragi (*Eleusine coracana*), common millet (*Panicum miliaceum*), and species of *Echinochloa* and Seteria failed whereas some fruit yield in goose berry was obtained which shows the resilience of the tree-based systems.

Patil *et al*. (1999) revealed that the fruit trees like ber (*Ziziphus mauritiana*) and goose berry (*Emblica officinalis*) had no adverse effect on yield of arable crops. Under rainfed condition on a light soil, the cropping system with ber, planted in 10 m alley width was more remunerative than other alley width.

At Parbhani, in drumstick based Agri-horticultural system, green gram as an intercrop yielded seed yield of 238 kg/ha while black gram and soybean yielded 197 and 426 kg/ha, respectively. The sole drumstick crop yielded 22200 kg/ha. At Bijapur, among aonla, henna and custard apple based Agri Horti system intercropped with chickpea and safflower., aonla + custard apple + henna + chickpea system was superior with maximum chickpea equivalent yield of 2270 kg/ha. Aonla + custard apple + henna + chickpea + safflower was the second best with chickpea equivalent yield of 2162 kg/ha.

At Bangalore, in aonla *(Emblicaofficinalis*) based agri-horti system, significantly higher aonla equivalent yield was recorded in aonla + cowpea system (1903 kg/ha) and was on par with aonla + finger millet system (1849 kg/ha). Finger millet, cowpea and horse gram proved to be better intercrops in aonla which registered higher B:C ratio and RWUE.

In a custard apple based Agri Horti system, significantly higher custard apple equivalent yields were obtained with fodder maize (1468 kg/ha) and finger millet (1317 kg/ha) with net returns of Rs. 59633/ha and 53398/ha, B:C ratios of 3.09 and 3.08, respectively and RWUE (86.85 and5.58 kg/ha-mm). In various agroforestry systems evaluated in Andhra Pradesh, agri-horticulture was found to be most profitable giving CB ratio of 1:5.53 followed by silvi -pasture and agri- silviculture. (Table 7)

Table 7: Compatible intercrops with ber and custard apple in different AICRPDA centres

Centre	Tree-crop	Intercrops	Yields (kg/ ha) of intercrops
Solapur	*Ber*	Pearl millet + pigeonpea	246 + 12
	Custard apple	Pearl millet + pigeonpea	220 + 13
Rewa	*Ber*	Blackgram	450
	Custard apple	Blackgram	380
	Ber	Pigeonpea	139
	Custard apple	Pigeonpea	176
Dantiwada	Sole ber		320
	Ber	Castor	259 + 80
		Clusterbean	249 + 79
		Pearl millet	231 + 54
		Greengram	2.9 + 12
Agra	*Ber*	Greengram	370
		Cowpea	285
		Clusterbean	645
Hisar	*Ber*	Greengram	295
		Cowpea	290
		Clusterbean	302
Jhansi	*Ber*	Blackgram	374
		Greengram	.67

(*Source*: AICRPDA, 2000 to 2006)

In Ber based agri-horti system, Pearl millet + pigeon pea (Solapur), Pigeon pea + black gram (Rewa), Castor (Dantiwada) and Cluster bean (Hyderabad) showed promising results in rainfed environment (AICRPDA,2006). Ber on an average gave 40 kg fruits tree^{-1} along with the 100 kg of horse gram and 450 kg of cowpea cultivated in interspaces (Osman *et al.* 1989) of a hectare area

In Hyderabad, the yields of sorghum, groundnut and mung-bean grown with pomegranate and custard apple were reduced by 23-26% compared to the respective sole crops (CRIDA, 1999). The yield reduction was higher in association with custard apple. Among the systems, groundnut grown in interspaces of either pomegranate or custard apple gave the highest gross income (Rs. 19,540-19,770 /ha). Custard apple + mungbean system recorded the highest yield advantage (54%) compared to respective sole crops.

Guava is another fruit tree with which crops or fodders can be grown for 3-4 years. Besides green gram, cowpea and cluster bean, *Stylosanthes hamata,* and *Cenchrus ciliaris* can be cultivated successfully in the interspaces. However, keeping the basins clean at least 3-4 times a year is essential. A dry fodder yield of 2-3 t ha/ yr of Cenchrus and 4-5 t ha-1 yr-1 of stylo can be harvested apart from yield from guava fruit (Rao, 1999). Among different intercrops (groundnut, green gram and cowpea) tried in four year old mango orchard at Hyderabad, groundnut proved to be a successful intercrop (CRIDA, 1995). However, there was considerable reduction (53 – 56%) in pod yield of groundnut (506-539 kg ha-1) in agri-horticultural system over sole groundnut (1148 kg ha-1). Leguminous intercrops can be taken in mango orchard up to at least 8 years after planting with minimum reduction in yields of annual crops. In young mango plantations in Karnataka, ragi and groundnut intercropping is very common (Rao and Sujatha, 2003).

Awasthi *et al.* (2009) studied intercropping under arid conditions of Bikaner in newly established 'NA7' aonla *(Emblica officinals* Gaertn). Moth bean was grown during kharif season as a common crop in rotation with rabi crops, ie fenugreek, chickpea, mustard and cumin. Growth parameters in terms of plant height stem girth, canopy spread and canopy volume of aonla was recorded to be significantly more with intercrops compared with its sole plantation. Higher grain and straw yield were recorded in Moth bean-chickpea (497, 1250 kg/ha) and moth bean-fenugreek (465,1161 kg/ha) crop sequence. Amongst the rabi crops, grain yield of fenugreek, chickpea, mustard and cumin were higher by 28.05, 38.11, 19.96 and 36.50%, respectively, when grown in association with aonla compared to its sole crops. The highest net profit (Rs 28,260/ha) was obtained from moth bean–cumin cropping system, followed by moth bean–chickpea (Rs. 25024/ha) cropping system. Moth bean–chickpea intercropping with aonla supplemented 22.01, 5.00and 27.90 kg/ha nitrogen, phosphorus and potassium through crop residues, followed by Moth bean-fenugreek crop sequence (Table 8)

Table 8: The growth and yield of component crops as influenced by sole and intercrops with Anole based

Crop sequence	Growth para meters		Grain yield (kg/ha)		Economics	
	Stem girth (cm)	Canopy spread (m)	Kharif	Rabi	NR (Rs/ba)	B:C Ratio
With Aonla						
Mothbean Fenugreek	0.16	1.07	465	986	l ISOS	0.96
Mothbean-Chickpea	0.18	1.13	497	1493	25024	2.00
Mothbean-mustard	0.14	0.90	443	1232	17499	1.47
Mothbean-cumin	0.13	0.93	451	602	2S260	1.85
Sole Aonla	0.29	0.65				
Snlr cropping						
Molhbean-fenugreek	-	-	406	770	9590	0.94
Mofhbean-chickpea	-	..	420	10S1	162S3	1.43
Mothbean-muslard	—	-	407	1027	14S22	1.41
Mothbean-aunin	—	-	40S	441	18035	1.32
CD (5%)	0.03	0.11	29.9	44.3		

Source: Awasthi *et al.* (2009)

Horti silvipasture

Horti silvipasture involves integration of fruit trees with pasture. Guava - glyricidia - maize (African tall) based hortisilvipastoral system was found to be the best for semi-arid regions of Tamil Nadu by producing higher output in terms of fodder and fruit yield . Grasses like *Cenchrus ciliaris* and *Cenchrus glaucus* and trees *like Prosopis cineraria* and *Acacia senegal* were suited for this system. Stylosanthes and *Cenchrus* were found to be compatible fodder crops with guava, custard apple and mango (Ramasamy *et al.* 2007)

Agri silviculture

Intercropping with Nitrogen Fixing Tree Species (NFT's) :Nitrogen continues to be one of the primary constraints for increased and sustained productivity of crops in drylands. Among various nitrogen fixing tree species (NFT's), Prosopis cineraria (Khejri) was found to be the most compatible with arable crops followed by Acacia ferruginea, Albizia lebbeck and Leucaena leucocephala. Pollarding of Leucaena leucocephala at 2.0 m height and frequent pruning during cropping period was found to reduce the competition between tree and crop and enabled a good harvest of sorghum and cowpea (Osman *et al.* 1998)

Alternate land use systems in non-arable lands

Horti-pastoral systems

Horti- pastoral system is a combination of fruit trees and pasture or legume. It is ideal land use option for degraded lands. In Guava based Horti-pastoral system at CRIDA, yield reduction of stylo was less under widely spaced trees (8x5 m) compared to closer spacing (5x5m), indicating the necessity of wider spacing of fruit trees which grown with stylo. Buffer grass out yielded stylo and took less time for establishment (Table 9)

Table 9: Fresh yield of forage and fruit in 8 years old guava based horti-pastroal system

Spacing (m)	Forage yield (t ha^{-1})		Fruit yield (kg $plant^{-1}$)
	Stylo legume	Buffel grass	
5 x 5	5.22 (40.4)	2.45 (3.9)	95.4
8 x 5	6.56 (25.1)	2.14 (16.1)	99.7
Control	8.76	2.55	-
Mean	6.84	2.38	97.5
CD (0.05)	0.88	NS	NS

Figures in parentheses indicate percent reduction in forage as compared to control (sole forage) (*Source*: CRIDA)

At CRIDA, the horti-pastoral system of *Cenchrus / Stylo* in rainfed guava and custard apple, *Cenchrus* yielded dry forage 7 t/ha with 17.5% of crude protein during the first year while *stylo* recorded 5.6 tons of dry fodder during the second year of plantation.). Higher fodder yield in P. cineraria-based production system (Prosopis in association with pearl millet – Brassica tournefortii) than sole cropping has been reported from arid regions of Haryana (Kaushik and Kumar 2003.) (Table 10)

Table 10: List of forage grass types suitable for different soils and rainfall situations

Region/Grass	Rainfall (mm)	Soil type	Dry forage yield (t/ha)	Crude protein content (%)
Semi arid				
Schima ttw*vosuM	600-1000	Mixed red and black	3-5	5-8
LUcanlhium annulatum	500-1000	Sandy loam .Clay silly loam	2.5	4-7
Hetewpo&on comorats	600-1000	Mixed red uiul black, red soils	3.0	2-3
(Shrysopogon Jutvus	600-1000	Hilly areas and crevices af rocks	3.5	4-7
heitema lax tun	700-1000	J.mv lying, clayey black soils.	3.0	4-6

Region/Grass	Rainfall (mm)	Soil type	Dry forage yield (t/ha)	Crude protein content (%)
Arid				
Lmtunis sindicus	100-150	Sandy	3.5	8-14
Cenchms cHiaris	150-300	Sandy	4.0	8-9
Cenchna aetigems	150-300	Sandy	3,0	8-9
Panicum anlidotale	200-600	Sandy	3.0	9-14

(*Source*: Kaushik and Kumar 2003)

Results of a study conducted in hot arid region at Jodhpur demonstrated that integrated production system comprising *Cenchrus ciliaris* and *Hardwickia binata* gave 17% higher fodder yields than sole pasture (Patidar and Mathur 2017). Kumar *et al.* (2009) assessed fodder production of sole Dichanthium and Dichanthium + aonla in semi-arid environment at Jhansi and reported that Dichanthium + aonla production system produced 0.56 t ha-1 higher fodder than sole Dichanthium along with 12.1 t/ha of fruit yield. Higher income:

Yadav *et al.* (2011) assessed the effect of trees (*P. cineraria, Dalbergia sissoo, Acacia leucophloea, Acacia nilotica*) on soil biological characteristics in semiarid regions and reported that agroforestry system enhanced soil biological activity (soil microbial biomass C, N and P), and amongst the trees, P. cineraria based system brought maximum enhancement of soil biological properties. The IFS ensures recycling of the leftover by-product of one enterprise as an input for other enterprises (Gill *et al.* 2009).

Silvi- pasture systems

Silvipasture is an ideal land use for the degraded lands in the peninsular region. The tree component for this system should be fast growing, high palatability, good coppicing ability, tolerant to drought, and to have ability to withstand browsing. The system should essentially consist of a top feed tree species along with grass or a legume as understory crop. *Leucaena leucocephala* + *Stylosanthes hamata*/*Cenchrus ciliaris* system was found very productive and profitable in these situations.

In an improved silvi-pasture, *Hardwickia binata* was taken as tree component at 3 m × 3 m spacing with Cenchrus ciliaris grass. Results of 9 years study) revealed that average carrying capacity of the system was4.1 sheep ha^{-1} $year^{-1}$ against 3.7 for sole C. cil- iaris pasture and 1.6 for sole H. binata plantation. In this type of improved silvi-pastoral system, in addition to grass + top feed production to the tune of 3.06 t ha^{-1} $year^{-1}$, a biomass of 0.26 t ha^{-1} $year^{-1}$ of fuel wood was also obtained.(Table 11)

Table 11: Production potential of *Hardwickia binata* + *Cenchrus ciliaris* based improved system

Year after planting	Grass yield	Leaf fodder yield	Total fodder yield (t ha^{-1})	Carrying capacity
	(t ha^{-1})	(t ha^{-1})		(sheep ha^{-1} yr^{-1})
1	0.82	–	0.82	1.9
2	1.23	–	1.23	2.9
3	1.64	–	1.64	3.9
4	2.05	–	2.05	4.9
5	1.64	–	1.64	3.9
6	1.64	–	1.64	3.9
7	1.24	2.31	3.55	8.5
8	0.84	0.66	1.50	3.6
9	0.84	0.66	1.50	3.6
Average	1.33	0.40	1.73	4.1

Source: Korwar *et al.*, 2014

Energy Security and Biofuel Production

The main biomass energy sources in rural areas include wood (from forest, croplands and homesteads), cow dung and crop biomass. Among the sources 70-80 per cent energy comes through biomass from trees and shrubs. Due to the agroforestry/alternate land crops initiatives large amount of woods are now being produced from outside the conventional forestlands. Small landholdings and marginal farmers, through short rotation forestry and agroforestry practices are now providing the bulk of country's domestically produced timber products. Ravindranathan *et al.* (1997) reported that Karnataka villages get 79 per cent of all the energy used mainly from trees and shrubs. *Prosopis juliflora* is the major source of fuel for the boilers of the power generation plants in Andhra Pradesh, due to its high calorific value of over 5000 Kcal. The market price offered *for Prosopis juliflora* wood at the factory gate is around IRs.700-1300 /ton, depending on the season and wood moisture content. Promoting bioenergy through *Prosopis juliflora* also encourages tremendous employment generation to the tune of 6.34-million-man days and 7.03-million-woman days for fuel making in Tamil Nadu alone. The fuel wood potential of indigenous tree species *(Acacia nilotica, Azadirachta indica, Casuarina equisetifolia, Dalbergia sissoo, Prosopis cineraria and Ziziphus mauritiana*) and exotic trees (*Acacia auriculiformis, Acacia tortilis, Eucalyptus camaldulensis and Eucalyptus tereticornis*) was studied by Puri *et al.* (1994). The calorific value of indigenous tree species ranges from 18.7 to 20.8 MJ/kg and the corresponding values of 17.3 to19.3 MJ/kg are reported for the exotic ones. Pathak (2002) opined that species such as *Casuarina equisetifolia, Prosopis juliflora, Leucaena leucocephala and Calliandra calothyrsus* have become prominent due to their potential for providing wood energy at the highest efficiency, shorter rotation and also their high adaptability to diverse habitats and climates.

Biofuels

Biofuels are important renewable and alternate source to minimize the use of fossil fuels. India has a rich potential, with more than 100 diverse tree species producing seed oil suitable for biodiesel. These oil tree species can be grown on non-agricultural lands and hence do not compete for food and fodder. Some of the important TBOs (Tree born oil seeds) like *Jatropha*, *Karanj*, *Mahua*, *Simarouba* etc., have been included in the agroforestry systems and resulted in beneficial results, thus providing a large source of non-edible tree borne oilseeds.

The country has enormous potential of oilseed bearing trees like *Mahua (Madhuca indica*), Neem (*Azadirachta indica), Simarouba* (*Simarouba glauca*), *Karanja* (*Pongamia pinnata*), Ratanjyot *(Jatropha curcas*), Jojoba (*Simmondsia chinesis*), Cheura (*Diploknema butyracea*), Kokum (*Garcinia indica*), wild apricot *(Prunus armeniaca*), wild walnut (*Aleurites molucana*), Kusum (*Schleichera oleosa*), Tung (*Vernicia fordii*) etc., which can be grown and established in the wasteland and varied agro-climatic conditions. These have domestic and industrial utility like agriculture, cosmetic, pharmaceutical, diesel and substitute, cocoa-butter substitute etc. Besides, lack of proper facility for storage, seed collection, long gestation period, fruiting/maturity season coinciding with rain are main constraints limiting collection and utilization of above TBOs.

Studies conducted at the Indian Institute of Oilseeds Research, Hyderabad, over a period of six- to eight-years with *Melia* and Kapok plantations, revealed that horse gram and cowpea were compatible as intercrops, which provide immediate return but need regular cultivation. Perennial fodder legumes such as *Stylosanthus hamata* provide leguminous green fodder in three to four cuttings per rainy season (9 to 12 t/ha) and provide good soil cover to resist wind and rain erosion, weed control and enrich soil fertility (Sudhakara Babu *et al*, 2006). The productivity of tree-oilseed crops also improves with intercropping (Mukta *et al*, 2000). Ayyasamy (2004) reported that intercropping black gram, green gram, cowpea and red gram in *Simarouba* plantations recorded similar yields to sole cropping besides improving soil fertility in terms of pH, EC, and soil NPK content. Intercropping two rows of pigeon pea with Jatropha (3 x 3 m spacing) increased income in the early periods of establishing plantations (Rao *et al*, 2006).

(1) *Pongamia* is a drought tolerant, semi-deciduous medium sized tree with short bole and spreading crown. It is widely grown from tropical dry to subtropical dry forest zones. It is a good shade bearer, suitable for planting in pastures, for afforestation in watershed areas and provider part of the country. It grows under a wide range of climate and soil conditions and can grow even in dry areas with poor, marginal, sandy and rocky soils. In addition to drought, it can tolerate saline conditions. *Pongamia pinnata* is a preferred species for controlling soil erosion and binding sand dunes because of its dense network of lateral roots. Root, bark, leaves, flower and seeds of this plant also have medicinal properties and traditionally used as medicinal plants. The seed cake, oil and leaf extracts of Pongamia are known to possess pesticidal properties.

Fig. 2: Planting material of Pongamia (*Source*:Dhayani SK *et al.*, 2015)

Table 12: Improved varieties/accessions in different agroclimatic zones

Agroclimatic /zone	Improved varklieVarrasion*
Eastern Plateaus A: Hills Region	Baramunda, Mancheswar
Southern Plateau A Hilts Region	KNR M 21 II •/,). KSK I (2? 34%) and ICR-MD<27 2 TCR-CJ-1, PT I and PT 2 ol Dhanvad.
Central Plateau A Hills Region	NRCP-13(earlv flowering). NRCP 24, NRCP 26 (high yielding). 1C 527152 (34 %)
Eastern Plateaus A Hills Region	KKVPP-02(31 7%). KKVPP-17(42%)
Central Plateau A Hills Region	RPK 14
Western Plateau A Hills Region	RHRRAK 50. RHRAK 44

Source: (Dhyani, S. K. *et al.*, 2015)

Pongamia produces natural chemical compounds, such as karanjin and pongamol, which impart bitter taste to the foliage and other parts of the tree. These compounds deter consumption of leaves by herbivores and insect infestations, resulting in lower management costs and enhanced sustainability. In addition, these compounds are valued for use in crop protection sprays, cosmetics, and sunscreen products. Pongamol and karanjin isolated from the pods of Pongamia pinnata possess significant antihyperglycemic activity in streptozotocin-induced diabetic rats and type 2 diabetic db/db mice and protein tyrosine phosphatase-1B may be the possible target for their activity (Prasad, 2021A).

(2) Simarouba *(Simarouba glauca)* or the paradise tree, is an ever-green multi utility tree of medium size. It is both a source of edible oil and also biofuels. The fat can be used as substitute to cocoa butter and extenders in confectionary and bakery industry. The cake is valued organic manure. The plants are polygamodioecious with varying percentage of staminate, pistillate, male and few bisexual flowers in the population. It can grow well in tropical climate and can withstand scanty to high rainfall. All types of well-drained soils are suitable for simarouba cultivation. Origin: Florida (United States) Family : Simroubaceae Species : *Simarouba glauca* Vernacular name : Simarouba, Paradise tree, Aceituno, Bitterwood (Table 13)

Table 13: Improved varieties/accessions simarouba

Agroclimatic zone	Improved varieties/accessions
Southern Plateau A Kills Region	Palem-1,2,3,4
Central Plateau A, Hills Region	HAUP-09,13,22, 28,
Western Plateau A Mills Region	PDKV SG-23,25,27,30

Source: (Dhyani, S. K. *et al.*, 2015)

(3) Mahua (*Madhuca indica*) Mahua is one of the important tree species in central India, as it produces abundant delicious and nutritive flowers. This is used for edible purpose either fresh or dried and stored for indefinite period. It is fast growing with 20 m height, ever green tree cultivated in warm regions for its oleaginous seeds. Its oil is used by tribal as vegetable butter in addition it is used in skin care products, soaps, detergents, etc. It serves as an important fuel oil, hence a good source of biodiesel. The seed cake is used as manure. The flowers are used to produce an alcoholic drink in several parts of India.

Improved varieties/accessions Central Plateau & Hills Region NDMC9, NDMC 7, NDMC 10 & NDMC 3

Fig. 2: Planting Material of Mahua (*Source* Dhyani, S. K. *et al.*, 2015)

(4) **Kokum** is an important minor fruit, besides seed oil source for edible and non-edible applications. It is reported to be imported from Zanzibar to India. It is found to be grown widely in tropical rain forests of Western ghats in Konkan, Goa, Southern Karnataka and Kerala. It is a slender ever green tree with dropping branches. Kokum fruit is a promising industrial raw material for commercial exploitation in view of its interesting chemical constituents. The seed yield a valuable edible fat known as kokum butter, used in chocolate and confectionary preparations and also in manufacture of soap, candle and ointments. It is an important source for biodiesel.

(5) Jojoba is a potential oilseed tree, which can be commercially cultivated in arid and semi-arid areas having low rainfall, extreme cold and hot temperature. It was introduced in India around 1965, but it was given importance only during 80s. It is an ever-green slow growing dioecious desert shrub, which can live up to 150 years. It can tolerate extremely high and low temperature. It can grow on wide range of soils but not suitable for waterlogged, heavy soils and soils prone to flooding. The seed oil is used as lubrication as a substitute for sperm whale oil, in heavy machinery, as it can be used with little or no refining at high temperatures and pressures. It is a good source for biodiesel, also in pharmaceuticals, cosmetics, food related and other chemical industry.

(6) Neem is one of the most valuable tree species found in India. It can grow on wide range of soils up to pH 10, which makes it suitable for every agro-climatic zones except in high and cold regions. It has deeper root system, thus does not compete with annual crops for soil moisture. It is a multipurpose tree which provides all the requirements of rural areas viz., timber, fuelwood, fodder, oil, fertilizers, pest repellant, etc. Neem is known for its Azadirichtin content and it has been commercially exploited in many products like medicine, manure, pesticide, insecticide, etc.Neem is regarded for its environmental value.(Fig 4)

Fig. 3: Planting material of Neem
(*Source*: Dhyani, S. K., *et al*. 2015)

(7) **Olive** (*Olea europaea*) Olive is a commercial plant of the Mediterranean since ages and in India it was imported for use in salads and other culinary preparations. In 2007, with an Indo-Israel collaboration, its cultivation was tried in Rajasthan and first fruits were harvested in 2011-12. The oil content of the seeds from Indian cultivation ranged from 9-14 % in comparison to 12-16 % of the other olive growing countries. At present, olives are grown in an area of 182 ha in several parts of Rajasthan with 14000 MT. It has global demand for culinary preparation, edible oil and also for biodiesel.

Agroclimatic zone Improved varieties/accessions Arid zone:

Arbequina Olive; Barnea Olive; Coratina Olive; Frantoio Olive; Koroneiki Olive; Picholine Olive and Picual Olive (Fig 4)

Fig. 4: Planting material of Olieve (*Source*: Dhayani *et al*. 2015)

Bush Farming

Long term observations revealed that crops are compatible with short growing shrubs than with trees in drylands. Shrubs are the perennials with smaller canopy compared to trees. These are well suited to the rainfed regions and they offer least competition with the associated crops / vegetation. Plantation of economic shrubs and intercropping with arable crops would improve the production and income from rainfed areas apart from imparting the required stability. Some of the shrubs that have shown the potential under rainfed conditions are henna (*Lawsonia inermis*), curry leaf (*Murraya koenigii),* annatto (*Bixa orellana*) and jatropha (*Jatropha curcas*). Henna has a potential of yielding 2500 kg dry leaf ha/yr from second year onwards. Curry leaf yields about 10,000 kg fresh leaf ha/yr from second year onwards. It can yield an additional yield by 2500 to 3000 kg ha/yr if the rain water is harvested and recycled to irrigate the plants during the off-season. This system is profitable in places where there is demand for fresh curry leaf.Annatto yields food grade dye and yields of 600-700 kg seed/ha from third year onwards. A stable market and price would make this plant popular. Further, short duration legumes such as black gram and green gram can be intercropped with curry leaf and annatto; they would yield 70 per cent of the sole crop. Cultivation of these species on a commercial scale will requires assured forward linkages with market for obtaining higher returns.

Medicinal and aromatic plants

The plants which are rich in secondary metabolites and, are potential sources of drugs are called medicinal plants. These secondary metabolites include alkaloids, glycosides, coumarins, flavonoids, steroids etc., Aromatic plants are those which possess essential, and these essential oils are the odoriferous steam volatile constituents of the aromatic plants. These essential oils are used in perfumery, cosmetic and pharmaceutical industries whereas the essential oils obtained from spices and condiments which impart the flavors and improved the taste of the food are used are several flavor industries. *Withnia somnifera* Suitable Medicinal and Aromatic Plants for Dry Lands are Ashwagandha(*Withnia somnifera)* , *Solanum nigrum*, Senna(*Senna alexandrina)*, Periwinkle(*Vinca rosea* or *Perwinca*) Glory lily(*Gloriosa superba)*, *Phyllanthus niruri*, *Aloe vera*, Tulsi (*Ocimum tenuiflorum)*, *Stevia*, and Mint(*Mentha alaica)* are suitable to cultivate under dry land conditions (Narashima Reddy, 2006). Pareek and Gupta (1993) suggested that *Psyllium*, senna, periwinkle, lemongrass (*Cymbopogan citratus)*, palmarosa (*Cymbopogon martini),* vetiver (*Chrysopogon zizanioides*) indigo, mesta, henna can be grown over low fertility soil in dry and warmer tracts in the country. Studies at CRIDA suggested that in rainfed areas low water-requiring aromatic grasses like lemongrass, palmarosa etc., medicinal plants like senna, and rographis (*Andrographis paniculata)* etc., can be successfully grown.

The drought tolerance of these crops helped in cultivation of these crops economically in drylands. They serve as alternatives to traditional crops and help in crop diversification in semiarid regions. (Tables14 and 15)

Table 14: Economics of some of the medicinal and aromatic plants that are grown in dryland and rainfed regions

S.1	Crops	Cost of Cultivation (Rs/ha)	Gross Income (Rs/ha)	Net Income Rs/ ha
1	Lemon grass	22500	42000	10500
2	Java citronclla	19500	33750	14250
3	Palma rosa	22JO0	40500	18000
4	Tulsi (basil)	11500	20000	8500

(*Source*: Hanumanthappa *et al*., 2018)

SI.N.	Crops	Cost of cultivation (Rs/ ha)	Grass income (Rs/ha)	Netincome Rs/ha/y
1	Lemon grass	22500	420(10	19500
2	Java Lilmriclla	19500	33750	14250
3	Mentha	20500	36000	15500
4	Palma rosa	22500	40500	mono
5	Tulsi (basil)	11500	20000	8500
6	Jama rosa/CN	25000	50375	25775
7	Kalmegh (ituirQg'ftjphxs Lthtlu}	5SO00	05000	7000
8	Sarpuandha (JtiirnYvIfiu serpentina)	30000	64000	34000
9	Shatawar (*Rauwolfia serpentina*)	25000	50000	25000

(*Source* Hanumanthappa *et al*. 2018)

Table 15: List of Medicinal and Aromatic plants suitable in drylands

S.N	Corp	Variety	Developed at	Year of Release
Medicinal plant*				
1	*Chlorophytum boriviiianum* (Safed inusli)	JS405	Mandsaur	2004
2	*Cassia angustifolia* (Senna)	Anund Late Selection	Anund	1089
3	*Diascoria floribunda*	sunn	Bangalore	1974
4	*Diascoria floribunda*	Arka Upukur	Bangalore	1980
5	*Titvcvrrhtea glabra* fLiquorice)	Haryana Mulhatti-1	Hisar	1989
6	*Ih'oscyamus milieu*s (Egyptian Henbane)	IIMI-80-1	Indore	-
7	*Lepidium sativum* (Cress)	GA-1	Anand	1998
8	*Rauxolfia serpentina* (Sarpagandha)	RI-I	Indore	
9	*Papaver somniferum* (Opium poppy)	Jawahar Aphim 16	Mandsaur	1984

S.N	Corp	Variety	Developed at	Year of Release
10	*Papaver somniferum* (Opium poppyI	Kiflimun	Fuizabad	1990
11	*Papaver somniferum* (Opium poppy)	Jawahar Aphrm 16	Mandsaur	1997
12	*Papaver somniferum* (Opium poppy)	Jawahar Aphim 16	Mandsaur	1998
13	*Papaver somniferum* (Opium poppy)	C'hctak Aphim	Udaipur	1994
14	*Papaver somniferum* (Opium poppyt	Trisna	Delhi	-
15	*Plantago ovata* (Isabgoi)	Guiarut Isabgoi- 1	Anand	1976
16	*Plantago ovata* (Isabgoi)	Gujarat Isabjsol- 2	Anund	1983
17	*Plantago ovata* (Isabgoi)	Haryana Isabgol-5	Hisar	1989
18	*Plantago ovata* (Isabgoi)	Jawahar Isubgol4	Mandsaur	1996
19	*Solatium riantm* (Khasi Hateri)	Arka Sanjccvam	Bangalore	1989
20	*Solarium \ianim* (Khasi Kateri)	Arka Mahima	Bangalore	1992
21	*Withania somnifera* (Ashwagandha)	Jawahar Asgand-20	Mandsaur	1989
22	*Muhattuj suitirhfettf* (Awmgatuih/i)	Jawahar Asgand-114	Mandsaur	1998
Aromatic plants				
1	*Cymbopogon* (lemon grass)	NI.G-R4	Faimhad	1994
2	*C. martiniilerr. Afolia* (Palmartaa)	Rosha Grass-49	Hisar	1989
3	*C martinit Far Kfotiu* (Palmarota)	CI-80-68	Indore	-
4	*Jasminum grandiflorum* (Jasmine)	Arka Surabhi	Bangalore	1991
5	*Vetiwta zizantoides* (Vytnvr)	Hyli-8	Mandsaur	-

(*Source:* Hanumanthappa *et al.* 2018)

Shelter-Belts and Windbreaks

Shelterbelts and windbreaks are important components of agroforestry and alternate land use systems in rainfed, dry, temperate, and desert areas. Windbreaks are located around the field mostly on bunds, but shelterbelts are integrated with cropping systems in the fields. Windbreaks and shelterbelts reduce wind velocity, reduce evaporative water loss from surface downwind, and thus conserve soil moisture and decrease temperature and provide shelter against direct sunlight. Therefore, it is considered as good adaptive strategies of climate change (Dixon *et al.* 1994).

Many of the boundary plantations also help as shelterbelts and windbreaks, particularly in fruit orchards. In Bihar, *Dalbergia sissoo* and *Wendlandia exserta* are most common boundary plantations. In northern parts of India, particularly in Haryana and Punjab, Eucalypts and Populus are commonly grown along the field boundaries or bunds of paddy fields; other trees which are grown as boundary plantations or live hedge include *Dalbergia sissoo* and *Prosopis juliflora*. Farmers of Sikkim, grow bamboo (*Dendrocalamus, Bambusa*) all along the irrigation channels. In coastal areas of Andhra Pradesh, Borassus is the most frequent palm. In Andamans, farmers grow *Gliricidia sepium*, *Jatropha* spp, Ficus, *Ceiba pentandra*, *Vitex trifolia* and *Erythrina variegate* as live hedges

Alternate land use systems for problem soils/areas

Large area in the country suffers from site specific environmental constraints with significant costs for rehabilitation and loss in productivity. Mechanical and engineering methods for rehabilating such lands are expensive and time consuming. Bio engineering or vegetation-based rehabilitation approaches are better alternatives for such problem areas. The degraded riverbeds can be best utilized for silvipasture systems of *Dalbergia sisso* and *Chrysopogon fulvus* in places like Doon valley. Bare erodible slopes can be protected by planting deep rooted shrubs and grasses such *as Pennisetum purpureum, Ipomea cornea, vitex negundo* and *Pueraria hirsute* in places like Himalaya region. Shifting cultivation, locally called as jhuming is extensively practiced in NEH region, Andhra Pradesh, Madhya Pradesh, and Bihar. As an alternative for shifting cultivation, a silvi agriculture model (35% hill top to forestry, middle 30% to horticulture, lower 35% to agriculture and farm pond showed promising results. The technologies were identified by CSWCRIT for rehabilitation of Degraded Ravine lands for the places like Agra, Kota, Vasan, and Chambal regions of Rajasthan. Silvi-pasture of *Leucaena leucocephala+ stylosanthes hamata/Cenchrus ciliaris* system was found productive and profitable for degraded lands of southern peninsula. Agroforestry has emerged as an ideal option for saline and alkaline soils (salt affected areas). Promising trees for alkali lands are *prosopis julifera, Acacia nilotica, Casuarina equisatifolia, Terminalia arjuna* and *Tamarix articulata* and *grasses are Leptocholoa fusca, Chloris gayana* and *Brachiaria mutica* (Table: 16)

The plant species recommended for agroforestry management for problem soils are listed in Table 16 presented below

Table16: Recommended tree species suitable for problematic soils

Problematic Areas	Kecommended tree species
Sodic soils (pH 8.62-10)	*Acacia nilotica, Acacia auriculiformis, Casualina equisetifolia, Prosopis cineraria, Prosopis juliflora, Tamrix articulate, Salvador a per Ac a, Pithecellobium dulce. Pongamia pi? mar a. Sesbania sesban, Tennina/ia cirjima, Poplars. Eucalyptus tereticornis, Butea monosperma. Azadirachta indica*
Wetland	*Alims trabeculosa, Alnus creniastogyne, Sain xuchonensts, Sain babylomca, Paulowma toiuenfosa, Tax odium distichum, Taxodium scandens and Morns alba*
Waterlogged area	*liambusa arundinacea, Eucalyptus* spp., *Coiymbia tessellaris, C'asuarina equisetifolia. Terminalia arjmia, Tamar Lx, Lucerne, Syzygium cum ini, Sairingtonia acutangula. Ficus glomareta and Psidium gujava.*
Sand dimes stabilization	*Prosopis cineraria, Prosopis juliflora, Acacia tor til is. Acacia nilotica, Azadirachta indica. Acacia senega/, Ziziphus* spp., *Tecomella inidulata, C'apparis deciduas Ailanthus exceisa and Parkinsonia Acu/eta*
Coastal Area	*Anacardntm accident ale, Ad ant bus malabanca, Casuanna equisetifolia, Pongamia glabra, Terminaha catappa, Calophyllum tnophyllurn, species of Pandanus, Nypa frutleans, Saltcornia btgeioni, Sesbania aculeate, Salvador a persica, S. o/eoides and Borassusflabeflifer*
Ravine area	*Acacia catechu. Acacia niloticaf Dalhergia sissoo, Moms alba, Dendiocalamus sir ictus, Azadirachta indica, Albizia* spp *and Holoptelia integrifolia*
Acid soils	*Pin us roxburghii, P va/lichiana. Ce/tis australis, Pinus, Cassia, Acacia meamsii, A. dec wrens, A. dealbata, A. pycnantha. Maniikara hexandra, Phoenix spp. A. auriculiformis, Pterocarpus mars upturn, Shorea robusta. Haidwickia binnata. Terminalia* spp.. *Tectona grandis, Gmelina arborea, Xyliaxy/ocarpa. Bamboo* spp., *Alnus* spp.

Source: Sharada and Venkateswarlu (2009)

Grasslands in dryland areas

The grasslands normally considered to be the cheapest source of animal feed are in a degraded and denuded state because of overgrazing and misuse over the years. Therefore, the improvement and conservation of natural grasslands and pastures deserves special and careful attention and priority in our drive for increasing forage production. Management systems that have potential for improving production and stability of the pasture are applied for assuring higher productivity and also its quality.

The sequence of activities for assuring a managed grassland system are

- Protection from unauthorized grazing,
- Soil and water conservation,
- Bush cleaning,
- Re-seeding,
- Nutrient Management
- Legume introduction and cutting, and
- Grazing management.

Both forage production and quality of grasses in terms of per cent crude protein can be improved considerably by application of nitrogen. Research studies have revealed that application of 40-60 kg N/hectare and 20-30 kg P20/hectare have increased pasture production by 50 to 100% in majority of grasses besides increasing crude protein content considerably. Although many grasses have been found to respond to very high dose of nitrogen also, it is suggested to adhere to the recommended doses, which are found to be highly economical.

Grass-legume mixtures are always desirable because of their complementary functions in providing nutritive, succulent, palatable forage for the grazing animals. The mixtures also improve the physical conditions of the soil, check soil erosion, resist the encroachment of weeds and withstand the vagaries of weather better than pure stands. They also help to check the spread of certain diseases and insect pests. Legumes usually maintain their quality better than grasses even at maturity, and being rich in protein, enhance the forage value, and also add substantially the much needed nitrogen to the soil. In the diet of animals, grasses and legumes have also greater beneficial associative effect through rumen than when they are fed separately. Further, the adopted legumes in the mixture provide a simple and practical means of meeting the nitrogen need of the associated grasses as most of the legumes fix atmospheric nitrogen.

Research studies have shown the nitrogen equivalence of legume to the tune of 40-60 kg N/ha when introduced in natural grasslands. This means that 40-60 kg N/ha can be added to the grassland soil as microbial fertilizer manufactured by legumes. Besides being a substitute and a cheap source of nitrogenous fertilizers, legume also influence total dry matter production and crude protein yield which is so much vital for the livestock. Such mixtures have shown higher grazing value of the pastures. The legumes that make better association are Caribbean Stylo (Pencil Flower), Caatinga Stylo, Shriibby Stylo, Sticky Stylo (poor man's friend), Siratro, Giant Mimosa, Wild Kulthi, etc. In the first and second year of the pasture growth, it is advisable to adopt cut and carry practice after the seeding has been completed. The fallen seeds will assure a better regeneration during the subsequent years. Although, when cutting is done after the seed maturity, the biomass is coarse with low digestibility and crude protein levels. In the subsequent years, harvest at 50% flowering level assures high nutritive value in the grass. The surplus biomass can be dried as hay and stored for the lean periods. In the subsequent years, in-situ grazing under carrying capacity can be practiced. The word 'carrying capacity' is defined as the level up to which the pasture can be grazed without appreciable damage and decline in the biomass productivity in the subsequent years.

The greatest single factor, which causes deterioration of grasslands, is overgrazing. During grazing, certain grasses are preferred while others are avoided. On account of this, selective grazing, desirable species tend to get depleted in grasslands much faster than undesirable species. In most perennial grasses, utilizing the reserve food material that is stored in the underground parts produces new shoots. Due to overgrazing, the reserve food material is lost faster, and perennial grasses are unable to re-generate due to continuous drain on food reserve. Therefore, certain period of rest is essential for the perennial grasses to recoup and rejuvenate. Based on these considerations, the following types of grazing systems can are practiced:

1. Controlled Continuous Grazing: means either limiting the number of animals or limiting the duration (season) of grazing on a given area of grassland on a continuous basis. In this system, the grassland is not divided into compartments or paddocks and animals move in the whole area. Long period of continuous grazing often leads to deterioration in composition and production of good forage grasses and increase in unpalatable ones. It also affects soil fertility levels and exposes habitat to rain leading to runoff and soil loss

2. Deferred Grazing: Here, the grazing area is divided into compartments and at least one of the compartments is rested until seed setting. This means, stopping grazing in the most vulnerable season (summer) in India. But actually, grazing takes place in all the seasons in India, which is responsible for degeneration of their quality.

3. Rotational Grazing: In this system, pasture is dividing into separate lots, fenced and the animals are allowed to graze into each lot in rotation. For example, if there are three lots, all the animals are let into lot A for a month, then into lot B for a month, and finally the animals are allowed to graze in lot C for a month. After this, all the animals are again allowed to graze in lot A for a month and repeat the process

Table 17: Rotational Grazing Method

Month	Lot A	Lot B	Lot C
1st Month	Grazing	Rested	Rested
2nd Month	Rested	Grazing	Rested
3rd Month	Rested	Rested	Grazing
4th Month	Graying	Rested	Rested
5th Month	Rested	Graying	Rested
6th Month	Rested	Rested	Graying

Deferred Rotational Grazing: is rotational grazing practiced in restricted period, that is, not continuously. This system is a mixture of above two grazing systems and is considered the best system of grazing because of the followmg benefits: (a) more number of grazing days from the same grassland used otherwise; (b) maintenance of proper vegetation composition through self-seeding; (c) health of sward is maintained as the optimum utilization of biomass takes place and a period of rest is available to grasses; (d) soil fertility level is maintained; and (e) erosion hazards are avoided.

Hohenheim system involves dividing the pasture into lots and also divide the animals into groups production-wise i.e., a) early lactation cows (high yielder), b) mid lactation cows (lower yielder), c) dry cows, and d) growing animals. Into each grazing lot, group 'a' is let in first for a week, followed by cows of group 'b', then by group 'c' and family allowing group 'd' for the same periods. Then, the animals are let into each other's lot also in order of their productive utility. The idea is to let production-wise the most valuable animals graze first, followed by second best and then the third production category.

Zero grazing or 'Cut & Carry ': is the system practiced in most parts of India. In this system, animal owner's cut grass and edible weeds from the common grazing lands and carry the same home for feeding the animals in the stall. Mostly women folks perform this task. There is a risk of damaging the grassland, if this is overdone, especially during dry season. The practice of scratching out grass along with root system from already denuded grasslands can harm the herbage permanently.

Carrying Capacity

Carrying capacity of a pasture depends upon the available biomass and its rate of production. The stocking rate (number of animals that can be allowed per ha) should be such that it balances the livestock needs with available forage supply. It is found that heavy stocking rate lowers animal liveweight production compared to moderate and light grazing. In pastures with seasonal growth pattern, stocking rate variation can be adopted with peak rate in monsoon season compared to the summers and winters. Thus, carrying capacity needs to be properly determined for grazing management to maintain a sustainable pasture production. Depending upon the seasonal productivity, stocking rate should match the carrying capacity. The successional stages show the symptoms of deterioration / development and thus, can be related in the grazing management directions. There may be further correction if several species of livestock with different body weight and food requirement are introduced together. In such cases, the number is converted into adult cow unit (acu). Carefully determined carrying capacity with proper grazing management is key to the sustainable pasture management(Table 18)

Table 18: Pasture Land Degradation (34.5% of the Land is Under Open Pastures) and Carrying Capacity

Class	Cover %	Productivity (t/ha DM)	Carry ing Capacity (acu/100 ha)
Good	13	>1.5	20-25
Moderately degraded	19	0.7-1.5	10-15
Severely degraded	44	>02-0.7	0-9
Desertified	24	<Q2	3-5

Pasture Management

The native pastures are inhabited by low productivity and less palatable species. These pastures may lack legume component, thus, making the pasture lands nutritionally deficient. Artificial renovation of such pastures is likely to provide forage of good quality as well as sufficient quantity. In drylands, different legumes from the genera *Dolichos, Desmanthus, Clitoria, Cassia,* and *Stylosanthes* were found to do well with or without grasses like *Cenchrus ciliaris.* Among these *Stylosanthes* was found to be excellent in most situations in respect of persistence, nutritive value, and palatability. Different grasses from the genera *Dichanthium, Cenchrus, Chloris, Urochloa, Panicum, Pennisetum, etc*., are performing well.

These pastures are easily established if they are seeded at the beginning of rainy season. Since the seed rates vary widely from one species to the other, seeds of *Cenchrus ciliaris @1.0kg. stylosanthes hamata* @4kgand *Stylosanthes scabra* @1kg/ha may be used as seed mixture. These seed mixtures may be broadcasted followed by light raking of the soil to improve the germination. The seed mixture may be broadcasted on a drizzling day (Table 19).

Table 19: Yield potentials of *Stylosanthes* species and grass

Species	Protein yield (kg/ha)			Crude protein
	Stylo	Grass	Total	
A.Legumes				
1.S.hamata	3550	7800	4300	-
2.S.scabra(40205)	3380	1960	5340	-
3.S.viscosa(34904)	1810	3020	4830	_
4.S.humulis(av..of 5sps.)	1580	1980	3560	
5.S.guyansis(av.of 2 sps)	620	2730	3560	-
B.Grass				
1.C.ciliaris(Biloela)				610
2.C.ciliaris(Molopo)				590
3C.ciliaris(Gynandeh)				460
4S.nervosam				560

Source: Reddy, N.V. and R.Hampaiah,1982)

Tree farming

Trees play an important role in maintenance of environment and sustenance of mankind. Trees can yield abundantly where arable crops are likely to yield poor or even fail to grow. Areas having under class V and above are put together under tree farming. Closer spacing of 2x1m or 4x2m is usually adopted in tree monoculture. In dryland areas planting should be done in early rainy season. The tree canopy should be managed by pruning, lopping, pollarding and coppicing depending on the need for fuel, fodder, poles or timber.

Chauhan *et al.* (2013) studied that fodder crop productivity in *Leucaena* (Subabul) alleys was not affected and the aboveground biomass of the system was more when compared tosole fodder production of annual (maize + cowpea followedby berseem + ryegrass) and perennial Napier Bajra hybrid. The fodder productivity of perennial and annual crops was140.2 and 110.3 t/ha in *Leucaena* alleys against the 128.7 and 98.7 t/ha in open, respectively. The Leucaena based silvi-pastoral model was found highly productive, which yielded fuel wood (3.52 t/ha on dry weight basis) in addition to fodder from inter-crops and Leucaena. Biomass yield of one hectareof inter-cropping was found equivalent to 1.45 ha by solefodder crops. Moreover, the Subabul based fodder inter-cropping produced 15 per cent higher protein than the sole fodder crop on unit area basis (Tables)

Among varieties of Leaucaena K-8 source produced the maximum green leaf biomass (3.74 kg/plant per harvest), which was significantly higher than K-156 (2.6 kg/plant) and K-743A (2.46 kg/plant). These values calculated on hectare basis reflects 14.02, 9.75 and 9.22 t green leaves/ha, respectively for above mentioned sources. The K-8 yielded 4.39 t/ha of dry leaf biomass, which was significantly more than K-156 (3.15 t/ha) and K-743A (3.04 t/ha). Irrespective of Leucaena sources, 2.82 kg/plant dry fuelwood was obtained, which is equivalent to 3.5 t/ha in addition to the leaf biomass. The value of protein, mimosine and tannin content were found positively correlated among themselves, whereas, ash and carbohydrate content were negatively related with other parameters (Table 20 and 21)

Table 20: Yield and quality assessment of annual and perennial fodder intercrops in Leucaena based silvi-pastoral model under sub-tropical conditions of Punjab Leucaena

Leucaena Sources	Fresh biomass (l.'lia)				Dev bin mass (t/lia)			
	Leaves	Shoots/ Fuel wood	Perennial (Napier bnjra hybrid)	Annual (cotvpea-maizc + bcrseem-rvegrass)	Leaves	Shoots/ Fuel tv Wood	Perennial (Napier bajra hybrid)	Annuj
K-8	14.03	12.08	137.5	L 08.3	4.650	4.387		2S 2*
K-156	9.75	8.85	127.5	96.3	3.525	3.150		23.25
K-743A	9.23	S.44	122.5	93.3	3,187	3.037		22.25
Average	11.00	7.13	129.2	99.3	3.787	3.525		23.58
Control	-	12S.7	98.7	-	21.25			

Source: (Chawvan *et al.* 2014)

Table 21: Nutritional value of feed mixed in different proportions of *Leucaena*: ryegrass + berseem ratio under alley cropping of Leucaena and fodder

Leurncnn Sources	Crude Protein (%)				Mimostne (%)				Taunlus (%		
	1:1	1:2	1:3	Mean	1:1	1:2	1:3	Mean	1:1	1:2	1:3
K-8	20.63	20.08	19.81	20.17	1.1	0.55	0.38	0.69	1.80	0.90	0.40
K-156	21.50	20.60	20.25	20,78	0.9	0.44	0.27	0.53	2.20	1.10	0.50
K-743A	20.92	20,2S	20.29	20.49	0.7	0.33	0.19	0.40	1.60	O.SO	0.30
Average	21.01	20.32	20.11		0.9	0.44	0.28		1.86	0.93	0.40
CD (5%)	0.08				0.04				0.08		

Source: Chawvan *et al*. (2014)

Carbon sequestration potential of agroforestry/Alternate land use in India

Agroforestry has gained more importance as a land use system due to its positive impact on environment as they have a huge capacity to sequester carbon (C) since Kyoto protocol. Carbon sequestration is the removal of CO2 from atmosphere and its storage in long-lived pools. Reports state that these systems have a capacity to store C that range from 0.29 to 15.21 Mg/ha/yr aboveground, and 30 to 300 MgC/ha up to 1 m depth in the soil (Nair *et al.* 2010). They can store nearly 12 per cent of the world terrestrial C in them. The sink potential of agroforestry is high compared to field crops although its sequestration potential depends on number of factors such as age, soil type, tree species and their management. Recent studies showed that tree-based agricultural systems, stored more C in deeper soil layers compared to treeless systems; and C3 plants (trees) contributed to more C in the silt- + clay-sized (< 53 µm diameter) fractions than C4 plants that constitute more stable C in deeper soil profiles. Hence, agroforestry can be a GHG (Green House Gases) mitigating activity with technical mitigation potential of 1.1-2.2 PgC (IPCC, 2007).

Measurement of Carbon Sequestered in Agroforestry/alternate crop Systems

In Agroforestry systems, biomass is the one of the elements to assess the carbon flux, further, above and below the ground biomass is to be estimated for computing carbon sequestered by them.

Above Biomass Measurements

Earlier the above biomass was estimated by cutting down sample trees, separating various parts (stem, leaves, inflorescence, etc.), and determining their dry weights followed by C content. It is known as destructive method. This method cannot be used to estimate in trees covering large area. Another method is non-destructive method where Allometric equations can be used to predict the above ground biomass in agroforestry systems (Tumwebaze *et al.* 2013). It is most commonly used to estimate above ground biomass in tree species. In these equations, the diameter at breast height and height of the tree species to be used as an independent variables and total dry weight as dependent variable. There are two different types of allometric equations- linear and non-linear that helps in calculating the above ground biomass in tree species. Such allometric equations are developed based on biophysical properties of trees. The drawbacks of allometric equations are, (i) its applicability to dicotyledonous trees because of their complex branching nature and applicable only for tree species with diameter at breast height between 60 and 105 cm. There exist huge and excellent literatures regarding use of allometric equations in estimating above ground biomass. However, there is a need to reducc thc unccrtainty in their use.

Ketterings *et al* (2001) reported that B=aDb is the common allometric equation where biomass is B, diameter is D and a and b are parameters which vary between sites and it is source of uncertainty. The site-specific relationship between height (H) and diameter, H = kDc as b=2+c can be used to estimate parameter b and average wood density (ρ) at the site as a=rρ, where r is expected to be relatively stable across sites to find out parameter a. There by an algometric equation B = rρD2+c was proposed.

Several biomass-based models were also developed through several studies in different parts of the world for different tree species. These models also use diameter at breast height as independent variable. The biomass-based models also exist for tropical species (Cole and Ewel, 2006).

In tropical areas, there exist several tree species in small areas and in such cases, and it is difficult to compute the biomass. For such situations, a pooled equation with logarithmic model was developed by Knut (1993). The equation is as follows: ln (WSUM) = ln (DBH) + ln (CW)+ ln (HT)+ e

Where, WSUM = whole dry fresh biomass (kg) DBH = diameter at breast height (cm) CW = Crown width (cm) HT- = Tree height in meters

By applying this pooled equation different tree species grown in an area can be found out.

Remote Sensing

Biomass estimation can also be done using remote sensed data as the primary source for biomass estimation. In remote sensing, for forest stand structure than the sites with complex biophysical environments, either optical sensor data or radar data are more suitable. And the same can be used for estimating biomass in agroforestry. A combination of spectral responses and image textures improves biomass estimation performance (Lu, 2006). Carbon sink estimation using remote sensing has a high degree of uncertainty due to regrowth of tree species. In such cases, approaches needed to be developed for estimating biomass changes. However, remote sensing techniques have recently gained importance in estimating above ground biomass (Zheng *et al.* 2004; Lu 2005). Estimation of remote sensing was found easy in coniferous forests due to simple forest structure and tree species composition as compared to moist tropical forests (Zheng *et al.* 2004; Lu *et al.* 2005).

Lu *et al.* (2014) reported that airborne LIDAR (Light Detection and Ranging) is one of the most accurate technologies to determine aboveground biomass.

It can be used where optical sensor or radar fails to estimate the above ground biomass. Its use in several studies were found to find out the above ground biomass in various forests as it easily detects vertical structure and extracts ground elevation.

GIS (Geographic Information System) method **:** It is one of the geospatial technologies that can be used to estimate C stock in agroforestry systems. It was used to estimate the C stock in natural forests of Eastern Ghats in Tamil Nadu. GIS helps in developing thematic maps for agroforestry adopted in larger areas. This helps in capturing different imageries as phenology plays an important role in using satellite data for estimating qualitative and quantitative characters especially in vigorous growing vegetation. To have tree-based stratification data, this technology plays an important role. GIS is gaining importance as traditional methods of estimation are tedious and time-consuming particularly where information is needed on the spatial dynamics of SOC stocks.

Below-ground (Soils) organic C

The determination of below-ground organic C dynamics in Agro forestry system is crucial for understanding the impact of the system on C sequestration, but it is more difficult than that for above ground Carbon. Organic Carbon occurs in soils in several different forms including living root and hyphal biomass, microbial biomass, and as soil organic matter (SOM) in labile and more recalcitrant forms. Difficulties of separating these different forms and their complex interactions make measurement, estimation, and prediction of soil C sequestration a daunting task. The most common method of estimating the amount of C sequestered in soils is based on soil analysis, whereby the C content in a sample of soil is determined (mass per unit mass of soil, such as g C/100 g soil) and expressed usually in mega grams (Mg = 106 g or tons)/hectare. Below ground Living Biomass in addition to SOM, belowground biomass is a major C pool (Nadelhoffer and Raich, 1992). However, belowground biomass is difficult to measure.

Modeling

In order to understand global carbon cycling, models that incorporate rates of terrestrial C- cycling are used. Such models are based on a set of assumptions that are formed from our understanding of ecological factors ; 95 processes including tree growth, and decomposition processes in the soil. The CENTURY and Roth C models are the most widely used soil C models. Mixed-effects models for biomass modeling and mapping can be used to estimate the stocks in agroforestry systems (Chen *et al*. 2016).

Agroforestry in relation to Carbon Sequestration Potential (CSP)

Carbon sequestration in different agroforestry systems occurs both belowground, in the form of cnhancement of soil carbon plus root biomass and aboveground as carbon stored in standing biomass. Some of the earliest studies of potential carbon storage in agroforestry systems and alternative land use systems for India had estimated a sequestration potential of 68-228 MgC/ha 25tC/ha over 96 Mha of land Ravindranath and Sukumar 1998. But this value varies in different regions depending on the biomass production (Pandey 2007). Studies by Jha *et al.* (2001) showed that agroforestry could store nearly 83.6 tC/ha up to 30 cm soil depth, 26% more carbon compared to crop cultivation in Haryana plains. However, the magnitude of carbon sequestration from forestry activities would depend on the scale of operation and the final use of wood (Table 22)

Table 22: Carbon storage of different Agroforestry models

Agrnforestry model	Carbon storage capactiy	Region	Author
AnrisilvLeulture system (ayed 11 years)	26.0 tC/ha	Scmiand remer	NRCAF (2005)
Block plantation raged 6 yeans l	24.1 31.1 tC/ha	Central India	Swamy S.L. *et al*; (2003)
PopitfuS deitaides "G-48" + wheal	18.53 tC/ha	Himachal Pradesh	Verma K.S. *et al*; (2008)
Silvopasturc	31.71 tC/ha		
Agrisilviculturc	13.37 tC/ha		
Acti-harticulture	12.28 tC/ha		
Silvopastofalis.il) (auod 5 years)	6.55 ME ha^{-1} y^{-1}	Kerala, India	Kumar B.M. *et al*; (1998a)[13]
Indonesian homeijurdens (aged 13.4 years)	8.00 Mg ha^{-1} y^{-1}	Sumatra	Roshetko M. *et al*; (2002)[23]

(Pratap Toppo and Abhishek Raj 2018)

In India, an average CSP in agroforestry has been estimated to be 25 Mg C ha−1 over 96 million ha (Sathaye and Ravindranath 1998), but there is a considerable variation in different regions depending upon the biomass production (Dhyani *et al.* 2009) and method of estimation. Based on global estimates of the area suitable for agroforestry, 1.1–1.2 Pg C could be stored in the terrestrial ecosystems over the next 50 yr (Albrecht and Kandji 2003). Average carbon storage by agroforestry practices has been estimated as 9, 20, 50 and 63 Mg C ha−1 in semi-arid, sub-humid, humid and temperate regions, respectively. For small holder AFS in tropics, potential carbon sequestration rate ranges from 1.5 to 3.5 Mg C ha−1 yr−1 (Montagnini and Nair 2004).

Zomer *et al.* (2016) reported that in 2010, 43% of all agricultural land globally had at least 10% tree cover. With this tree cover analysis, they estimated 45.3 Pg C on agricultural land globally with trees contributing more than 75%. On an average globally, biomass carbon increased from 20.4 to 21.4 Mg C ha−1. According to IPCC Annual Report 5 (2014), the agricultural production and ongoing land use change contribute significantly to GHG emissions accounting for 24% globally. GHG emissions in India were of the order of 2008.67 Tg (tera gram) of CO2 equivalent without emissions from land use-land use change and forestry (LULUCF). Whereas with LULUCF, the emissions were about 1831.65 Tg of CO2

Carbon sequestration by trees

Nair *et al.* (2010) have also reported world scenario of carbon stored in AFSs ranged from 0.29 to 15.21 Mg C/ha/yr in aboveground, and 30–300 Mg C/ha up to a depth of 1 m in the soil (the age varied from 4 to 35 years). Thus, the trees in agroforestry are not only improving livelihood of small and marginal farmers, but also helping in mitigating global warming by enhancing carbon sequestration potential of Indian agriculture (Ajit *et al.* 2013) (Table 23)

Table 23: Carbon sequestration potential (CSP) of trees in India

Region	Agroforestry system	Tree species	No.of trees/ha	Age (year)	CSP (Mg C/ha/yr)	References
Arid and semi-arid	Block plantation		4.32	20	3.93	Rai el at, (2000)
		Albizia procera	312	10	1.79	
		Albizia amara	312	10	1.00	
		Acacia peudttla	312	10	0.95	
		Dalbergia sissoo	312	10	2 55	
		Dichmstachys cinerea	312	10	1.05	
		Emblica officinalis	312	10	1.55	
		Hanhvickia binata	312	10	0.58	
		Afella azaderach	312	10	0.49	
		Lencaeiia leucocephala	2500	9	10.32	Rao *et al*. (2000)
		Eucalyptus comandulensis	2500	9	8.01	
		Dalbergia sissoo	2500	9	11.47	
		Albizia lebherk	625	9	0.62	
		Acacia albida	1111	9	0.82	Pragason and Karrliik 0(110
		Acacia tortiiis	1111	9	0.39	
		Acacia awiculfformis	2500	9	8.64	Newaj anti Dliyani (2008)
		Eucalyptus teretlcomis	320	2	13.86	
		A cacia farnes tana	170	2	2.42	Rai *et al*. (2002) Rao el at. (1991)
		Cassia montana	154	2	1.84	
		Prosopis juljflora	138	2	1.16	
		Acac ia nilolica	106	2	5.70	

Region	Agroforestry system	Tree species	No.of trees/ha	Age (year)	CSP (Mg C/ha/yr)	References
	Agri silviculture	Albizia procera	312	7	3.70	Prasad et a!. (2012)
		Acacia pendula	1666	5.3	0.43	
		Leucaena: leucocephala	111111	4	2 77	
			6666	4	1.00	Viswanath el al. (2004)
			4444	4	14.42	
			10000	4	15.51	
		Castiarina eqitisetifolia	833	4	1.57	NRCAF (2007)
		Delhergia sissoo		11	1.47	Swamy el ah (2003)
		Emblica officinalis		8	1.58-1.62	
		Harduickia biimata		8	1.07-1 10	
		Colophospermuni inapane		0.59-0.66		
		Acacia nilotica + Natural pasture		5	1.9-5.4	Rai ei al. (2001)
	Silvipaslurc	Acacia nilotica + Established pasture		5	5.9	
		Dalbergia sissoo-		5	2.5	

(*Source*: Ajitet *et al*. 2013)

The Central Agroforestry Research Institute (CAFRI), Jhansi has been working on carbon sequestration potential of various agroforestry systems since 2000 through in house and externally aided projects. CAFRI as a partner in ICAR scheme "National Innovations on Climate Resilient Agriculture" (NICRA) is assessing the carbon sequestration potential of selected AFS the country. The research under the NICRA scheme indicated that in agroforestry, the tree has a capacity for biomass production at least as great as that of natural vegetation. The survey (in 32 districts of 12 states) for existing agroforestry systems in the farmer's field revealed that agroforestry is practiced in all parts of India and is recognized as having high potential for carbon sequestration. The observed number of trees on farmers' fields in these districts varied from 2 to 204/ha. The baseline standing biomass in the tree component varied from 0.58 to 113.12 Mg DM/ha, whereas the total biomass (tree and crop) ranged from 4.96 to 123.58 Mg DM/ha. The soil organic carbon in the baseline ranged from 4.28 to 24.13 Mg C/ha. The CSP of existing AFS at district level has been estimated to range from 0.05to 2.78 Mg C/ha/yr. Thus, the trees in existing agroforestry systems on farmers 'fields are estimated to mitigate more than 33% of the total emissions from agriculture sector annually at the country level. (NRCAF. 2007)

Carbon sequestration by crops

In agroforestry systems, although trees sequester more carbon, crops also fix and store carbon in considerable amounts. Crop improves the organic matter in soil, which is a significant component of the terrestrial C pool . An increment in carbon pool in soil can be achieved through adoption of appropriate crop rotations integrated soil fertility management, Srinivasarao *et al.* 2012), precise use of fertilizers and organic Carbon sequestration potential of eight recommended land-use systems of arid western Rajasthan was compared. Biomass C stock was maximum in farm forestry of *Acacia tortilis* (31.4 Mg C ha) followed by *Prosopis cineraria* and *Hardwickia binata* based silvo-arable systems (8.8 and 10.6 Mg C ha). Soil C stock was also maximum in farm forestry (47.6 Mg C ha) followed by Ziziphus based systems (32.5–33.9 Mg C ha). About 50–78% of additional soil C stock was in the form of soil inorganic carbon. (fig7)The total C sequestered (biomass + soil) over a period of nineteen years was in the order: farm forestry (49.80) > silvoarable systems (11.0–13.3) > hortipasture system (8.3) > agri-horti (5.5), silvopasture (5.4) and sole pasture (5.3) compared to –1.0 Mg C ha in sole cropping (Tiwari etal., 2010)

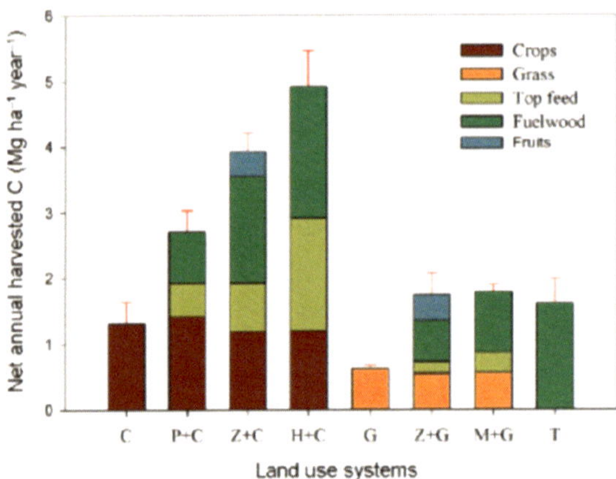

Fig. 5: Net annual harvested carbon in different land use systems (average of five years)

Carbon stocks in Agri-silvi cultural systems

a) Carbon sequestration in tree biomass

Maikhuri *et al.* (2000]) estimated species wise annual carbon sequestration potential of planted tree species on abandoned agricultural land (3.9 t/ha/yr) and degraded forest land (1.79 t/ha/yr). The highest carbon sequestration was found for Alnus nepaliensis 0.256 tC/ha/yr and Dalbergia sissoo 0.141 tC/ha/yr intercropped with wheat and paddy. In a 6-year-old Gmelina arborea based agri-silvicultural system 31.37 tC/ha was sequestered.

It was observed that the carbon sequestration in monocropping of trees and food crops were 40% and 84% less than agri-silviculture indicating that agroforestry systems have more potential to sequester carbon. In an agri-silvicultural system, *Dalbergia sissoo* at age 11 years was able to accumulate 48-52 t/ha of biomass. Carbon dynamics involving different pruning treatments were studied in an agri-silvicultural system (Dhayani *et al.* 2009) where tree biomass was 23.61 to 34.49 tC/ha with black gram-mustard. In a study (Chauhan *et al.* 2010) on poplar based agri-silvicultural system, total biomass in the system was 25.2 t/ha, 113.6% higher than sole wheat cultivation, where net carbon storage was 34.61 tC/ha compared to 18.74 tC/ha in sole wheat cultivation. In a system (Sharma *et al.*, 1995) comprising *Albizia* and mixed tree species like Mandarin, accumulated 1.3 t biomass/ha storing 6939 kg/ha in tree and crop biomass was reported.

Carbon sequestration by crops

In agroforestry systems, although trees sequester more carbon, but crops also fix and store carbon in considerable amounts. Crop improves the organic matter in soil, which is a significant component of the terrestrial C pool (Ciampitti *et al.* 2011). An increment in carbon pool in soil can be achieved through adoption of appropriate crop rotations (Wright and Hons 2005), integrated soil fertility management (Lal 2010, Srinivasarao *et al.* 2012), 8)

b) Soil organic carbon enhancement (SOC)

In a study conducted on intercropping of trees with crops, Singh *et al.* (1989]) reported an enhancement in SOC by 33.3 to 83.3% for *Populus deltoides* and *Eucalyptus* hybrid together with *Cymbopogon* sp.. In a Poplar based agroforestry system, it was observed (Gupta *et al.* 2009), that trees could sequester higher soil organic carbon up to 30cm depth during the first year of plantation (6.07 t/ha/yr) than in subsequent years (1.95-2.63 t/ha/yr). The soil carbon storage was found to be higher in sandy clays, as compared to that of loamy sand [Gupta *et al.* 2009). Traditional *Prosopis cineraria*-based systems, there was 50% increase in SOC largely due to leaf litter (Venkateswarlu 2010). (Samra and Charan 2000) observed an increase in soil organic carbon status of surface soil from 0.39% to 0.52% under *Acacia nilotica+Sacchram munja* system and an increase from 0.44% to 0.55% under *Acacia nilotica+Eulaliopsis binata* system after 5 years.

Silvipastoral systems

(a) Carbon sequestration in tree biomass

Comparative studies conducted by NRCAF [2007] on biomass production from natural grassland and silvi-pastoral system comprising *Albizia amara, Dichrostachys cinerea* and *Leucaena leucocephala* the woody perennials, with *Chrysopogan fulvus* the grass and *Stylosanthes hamata* and *S. scabra*, the legume components revealed that in 8 years, the carbon stored in the biomass of silvi-pastoral system was 6.72 tC/ha/yr, which was two times more than that of the natural grassland, which exhibited a value of 3.14 tC/ ha/yr. Singh (2002) estimated the total carbon sequestered in farm forestry with species such as *Eucalyptus* sp., Populus deltoides, Tectona grandis, Anthocephalus chinensis trees to be around 16,400 t/yr. Rai *et al.*(2009) studied the effect of introducing a silvi-pastoral system in a natural grassland in semi-arid Uttar Pradesh, where the introduced species viz., *Albizia procera, Eucalyptus tereticornis, Albizia lebbeck, Embilica officinalis* and *Dalbergia sissoo* were reported to have accumulated 8.6, 6.92, 6.52, 6.25 and 5.41 t/ha of bio mass per year. In this case, the carbon storage was 1.89-3.45 t C/ha in silvipastural system and 3.94 tC/ha in pure pasture. Saharan *et al.* [(1989) reported an increase in organic carbon from 1.7 to 2.3 times in a silvi-pastoral system involving Leucaenea leucocephala, Cenchrus ciliaris and *Stylosanthes hamata* as compared to the control. Similar studies have reported higher organic carbon levels in dry sub-humid and arid ecosystems when grass species were intercropped with annual crops. In various silvi-pasturesysystems recorded the Leucaena based system recordedthe highest carbon sequestration (3.6mg/ha/year) followed by Eucalyptus system (3.41 Mg/ha/year) (Prasad and Ramakrishna, 2018) (Table 24)

Table 24: Carbon sequestration in different silvi-pasture systems

Location	System	Carbon sequestration (\lg/ha/year)
Kamal (Kaur *et al.* 2002)	Prosopis based system	2.36
	Acacia based system	1.29
	Dalbergia xissoo based system	1.68
Himalayan foot hills (Narain *et al.* 1998)	Eucalyptus based system	3.41
	Leucaena based system	3.60
Jhansi (Rai *et al.* 2000)	Leucaena based system	1.82
	Terminalia based system	1.11
	Neem based system	0.80
	Albizia procera based system	2.01
	ttnthprvi/j xixxnn hnsed svstem	7 90

(*Source*: Prasad and Ramakrishna, 2018))

(c) Soil organic carbon enhancement

Narain *et al.* (2008) reported that planting trees and grasses in degraded lands in arid areas can help increase soil carbon stock from 24.3Pg to 34.9 Pg. Studies conducted at RRS, Kukma have reported a total belowground carbon stock of 1.6 t/ ha, which represents an increase of 23.4% of the total carbon stock under silvi-pastoral system involving *Acacia tortilis* and *Cenchrus setegerus* (Kumar *et al.* 2010).

(d) Carbon stored in block and boundary plantations

Boundary plantations are generally taken up to protect fields from wind damage, sea encroachment, floods etc. Usually a row of straight growing trees and occasionally few rows of trees are managed around farms or fields as part of crop or livestock operation to protect crops, animals and soil from wind hazards. These systems reduce the turbulence of wind at the crop level particularly under arid conditions. The plantation is generally planted on the windward direction. The tree population in these systems is normally low. The carbon sequestered in these systems can be up to 0.8 t/ha/ year in case of leucaena system at Bijapur (Table 5). The boundary plantation of poplar can sequester up to 4.56 t C/ ha/ year under irrigated conditions (Rizvi *et al.* 2011). (Tables 25 and 26)

Table 25: Carbon storage (Mg/ha/ year) in different boundary plantations Location System Carbon

Location	System	Carbon sequestration (Mg/ha/vcar)
Bijapur (Devaranavadgi *et al*l 999)	Leucaena based system	0.71
	Albizzia based	0.44
Saharanpur (Rizvi el al 2011)	Poplar based system	4.56
Yamunanagar (Rizvi *et al* 2011)	Poplar based system	3.86

Source: Rizvi *et al* (2011).

Table 26::Agroforestry models suitable for different soils and rainfall situations

ACZ / (ACIts)	Cciltrc/Domaiil Districts	Rain fall (mm)	Soil type	Agroforestry models for*			
				SamlVVC	Fruit based systems	Biomass production	1 BOs / MPTs for wastelands
Scarce rain¬fall (6.1)	Ahmednagar, eastern parts of Nusik, Pune. Saturn, Sangli, Dhulc and Nandur- har. western pans of Reed, Osmanahad, Aurangabad, some parts of Jalgaon and Buldana districts	723	Soils very shallow to shallow, medium and deep black soils	Live bunding + grasses	Pomegranate! custard apple / Bcr / Aonla + pulses / millets	l.eucaena + Marvel grass	Neern / Mel in
Scarce rain¬fall hot arid ecosystem (3 0)	Kurnool, Chittoor districts of Andhra Pradesh,	520	Mixed red and black soils	Agave + groundnut Custard apple / ber i tamarind + groundnut		Styfosanthes lut mala	Ncem
Malwa pla¬teau (5.2)	Mandsaur, Rajgarh, tJjjain, Indore, Dewas. Shajapur, Ratlam. part ofDhar, Jhabua, part of Sehorc dis¬trict of Madhya Pradesh.	ROO	M ediuin to deep black soils	Henna + grasses	Aonla / her + drumstick + soybean	Guinea grass	Teak / aon¬la' Necm

Note: * All the tree based systems should have micro catchments on mild sloping lands and contour trenches on sloping lands graded bunds.
ACZ: Agro-climatic zone; AF.Rs: Agro-ecological regions
Source: Balouli *et al.* (2018)

Conclusions

Alternate land use systems are of immense importance in the tropics especially on drylands and watershed development areas. In dry and inhospitable climates, tree growth conserves soil moisture, improves soil fertility, protects field crops against scorching and desiccating effects of wind and generally modifies the climate to be more equable and pleasant, thereby stepping up agricultural yields. Although these beneficial effects have been fully demonstrated in some advanced countries, no systematic effort has been made in India. Further, the integration of fodder trees would provide the much-needed top feed for sustenance of livestock during the lean period. In other words, it is necessary to device a land management and farming system which would produce food, fruit, fodder and fuel and conserve the ecosystem at the same time.

During the last more than three decades many agroforestry technologies have been developed and demonstrated by various research organizations. But most of them have not reached to farmer's field for want of awareness, inadequate infrastructure and lack of policy support. Therefore, the desired impact has not been observed in terms of adoption of technology. The major reason for this is that agroforestry sector is disadvantaged by adverse policies, legal constraints and lack of coordination between the government sectors to which it contributes viz., agriculture, forestry, rural development, environment and trade. Absence of appropriate policy and institutional arrangement to provide farmers with clear incentives to plant and protect trees that contribute to both ecosystem function and rural livelihoods is a great hurdle. Existing rules and regulations are greatest obstacles in felling, transport and marketing of agroforestry products and they vary from state to state. Agroforestry has to be declared at par with agriculture in availing credit and other subsidies by poor farmers (Dhyani *et al.* 2013). The agroforestry technologies have been successful in the areas where farmers got incentive in terms of quality planting material and assured market for example in case of Poplar, Leucaena and Eucalyptus. Availability of quality planting material of trees is one of the major concerns and need to be addressed urgently.

11

Biotic Stresses in Dryland Agriculture

Stress is an intrinsic part of biological systems, and successfully adapting to stimuli that induce stress is essential for the survival of dryland crops and farm animals kept in a complex and ever-changing environment. Here the biotic stresses of crop plants are treated and not that of farm animals to avoid the expansion of the volume of the chapter.

Biotic stress in plants is caused by living organisms, specifically viruses, bacteria, fungi, nematodes, insects, arachnids, and weeds. In contrast to abiotic stress caused by environmental factors such as drought and heat. Biotic stress agents directly deprive their host of its nutrients leading to reduced plant vigour and, in extreme cases, death of the host plant. In agriculture, biotic stress is a major cause of pre- and postharvest losses.

Plants respond to biotic stress through a defence system (Atkinson and Urwin, 2012).

In contrast to vertebrates, plants lack an adaptive immune system, or the ability to adapt to new diseases and memorize past infections. Though lacking an adaptive immune system, plants have evolved a plethora of sophisticated strategies to counteract biotic stresses. The genetic basis of these defence mechanisms is stored in the plant's genetic code. Plant genomes encode hundreds of biotic stress resistance genes. With the completion of several plant genome sequences during the past decade – among them are important agricultural crops such as maize, sorghum, and rice, offering the first glimpse into the wealth of biotic stress resistance genes encoded within plant genomes.

Plants respond to biotic stress through a defence system. The defence mechanism is classified as an innate and systemic response. After infection, reactive oxygen species (ROS) are generated and oxidative bursts limit pathogen spread (Atkinson and Urwin, 2012). Also, in response to pathogen attack, plants increase cell lignification. This mechanism blocks invasion of parasites and reduces host susceptibility.

The types of biotic stresses imposed on an organism depend on the climate where it grows as well as the species' ability to resist particular stresses. Biotic stress remains a broadly defined term and those who study it, face many challenges, such as the greater difficulty in controlling biotic stresses in an experimental context compared to abiotic stress.

Plants have developed various mechanisms in order to overcome these threats of biotic and abiotic stresses. They sense the external stress environment, get stimulated and then generate appropriate cellular responses. They do this by stimuli received from the sensors located on the cell surface or cytoplasm and transferred to the transcriptional machinery situated in the nucleus, with the help of various signal transduction pathways. This leads to differential transcriptional changes making the plant tolerant against the stress. The signalling pathways act as a connecting link and play an important role between sensing the stress environment and generating an appropriate biochemical and physiological response. The intensity of biotic stress varies depending on the weather, cropping system, cultivation practices, type of crops, crop varieties, and their resistance levels. Generally, hot and humid weather, input-rich intensive cultivation, and poor crop-management practices make the crop vulnerable to these stresses (Atkinson and Urwin, 2012).

Biotic Stresses in Dryland Crops

Instead of going through each and every biotic stress whether it be an insect pest or a disease or weed infestation, an attempt is made to give general principles of managing the biotic stresses in a comprehensive and integrated manner conducive to the sustainability of dryland agriculture.

Millets are mostly grown in dry climates, the adverse effect of biotic stresses in millets is less compared to other crops. There are huge areas under millet cultivation in sub-Saharan Africa, where traditional landraces of millets are grown in poor soil with negligible input, and pest and disease problems are relatively lower. In spite of that, the total losses in millets due to biotic stresses are enormous since the acreage under millet cultivation across the globe is high. The estimated loss of grain sorghum production due to biotic stresses (diseases, pests, *striga,* weeds, and birds) in nine countries of eastern and southern Africa is about 5.88 million tonnes per year as compared to 2.11 million tonnes per year due to water deficits or drought (Wortmann *et al.*, 2009). Pests and disease problems in millets are relatively higher in areas where intensive cultivation with high yielding varieties is the common practice.

In general, millets suffer more from fungal diseases than bacterial, viral, and nematode diseases. The important diseases of millets are grain mold (sorghum), downy mildew (pearl millet, sorghum, foxtail millet, proso millet), blast (finger millet, foxtail millet, pearl millet), leaf blight (sorghum, finger millet), smut (sorghum, foxtail millet, tef), rust (sorghum, foxtail millet, tef), ergot (sorghum, pearl millet), anthracnose (sorghum), and charcoal rot (sorghum) (Strange and Scott, 2005; Das, 2016). A few viral diseases (maize stripe virus, maize mosaic virus, etc.) are important in sorghum but not in other millets. Among the bacterial diseases, leaf streak, leaf stripe, and leaf spot are observed on sorghum in tropical or temperate humid environments. Recently, the bacterial soft rot caused by *Erwinia chrysanthemi* has been reported to occur in destructive forms in the Tarai region of India (Kharayat and Singh, 2013). A few species of plant parasitic nematodes have been reported to cause disease in sorghum and pearl millet, especially under poor soil and water environments. However, their damaging potential, supported by actual yield loss data are lacking. Important insect pests of millets include stem borer, shootfly, aphids, midges, headbug, etc. Besides these, grubs, armyworms, cutworms, locust, termites, black ant, and rodents also assume the dimension of important pests in some parts of the world. *Striga*, a parasitic weed, is one of the most serious constraints to cereal production in Africa (Ethiopia, Zimbabwe, Uganda, Rwanda, and Kenya), causing extensive yield losses on millets. It attacks the roots of young crops and starves them of nutrients, leading to low grain yields. Yield losses due to *Striga* are higher on pearl millet and sorghum than other millets. An estimated 100 million ha of the African savannah zones is infested with *Striga* (Ejeta, 2007).

The weeds are a global problem in agriculture and they are a major deterrent to increasing the productivity of millets, especially during the rainy season due to the weather conditions being congenial for their growth. Since millets are grown under a rainfed condition, soil moisture and nutrients are the most limiting factors. Weeds compete with millets for light, soil moisture, and nutrients and reduce their grain yield. The reduction of grain yield in sorghum may vary from 15% to 83%, depending on the crop, nature and intensity of weeds, duration of weed infestation, and environmental conditions (Stahlman and Wicks, 2000).

Bird damage of millet-grains is now considered a potential threat to millet growers. Generally, the small-grain crops are more prone to severe damage by birds than large-grain crops. The extent of damage varies depending on growing conditions and the crop and the loss may go up to 100% in isolated and unprotected conditions. Although it is a common problem in many countries, it is severe in some African countries like Ethiopia, Kenya, and Rwanda. It is estimated that *Quelea* and other birds can cause yield loss of about 1.6 million tonnes year^{-1} in eastern and southern Africa(https://www.thenewhumanitarian.org/news/2009/08/19)

Currently, climate has changed all around the globe, by continuously increase in temperature and atmospheric CO_2 levels. The distribution of rainfall is uneven due to the change in climate, which acts as an important stress as drought. The soil water available to plants is steadily decreased due severe drought conditions and cause death of plants prematurely. After drought is imposed on crop plants, growth arrest is the first response subjected on the plants. Plants reduce their growth of shoots under drought conditions and reduce their metabolic demands. After that protective compounds are synthesized by plants under drought by mobilizing metabolites required for their osmotic adjustment.

The biotic communities may respond to the climate changes in a variety of ways. The diversity and abundance of insect pests in agroecosystems are very likely to be affected negatively under abiotically stressful environment in changing climate. Vast majority of studies have documented these effects on inhabitant biota, e.g. shift in range of geographical distributions and abundance, changes in phenology and species interactions, etc. Majority of the studies dealing with the impact of climate change on crop-insect pest interactions revealed the predominance of negative impacts of climate change on crop yields over the positive impacts. In other words, there could be proliferation of new kind of biotic stresses. This may affect seriously the agricultural production and the livelihood of farming communities. This situation is expected to be more pronounced in tropical and subtropical countries of the world, where larger proportion of work force is directly depending on climate-sensitive sectors such as agriculture, livestock and forestry (FAO 2015)

Plants struggle with many kinds of biotic stresses caused by different living organisms like fungi, virus, bacteria, nematodes, insects etc. These biotic stress agents cause various types of diseases, infections and damage to crop plants and ultimately affect the crop productivity. However, different mechanisms have been developed through research approaches to overcome biotic stresses. The biotic stresses in plants can be overcome by studying the genetic mechanism of the agents causing these stresses. Genetically modified plants have proven to be the great effort against biotic stresses in plants by developing resistant varieties of crop plants (Parmar *et al.*, 2017).

Cross tolerance with abiotic stress

- Evidence shows that a plant undergoing multiple stresses, both abiotic and biotic (usually pathogen or herbivore attack), can produce a positive effect on plant performance, by reducing their susceptibility to biotic stress compared to how they respond to individual stresses. The interaction leads to a crosstalk between their respective hormone signalling pathways which will either induce or antagonize another restructuring genes machinery to increase tolerance of defence reactions (Rejeb *et al.*, 2014)

- Reactive oxygen species (ROS) are key signalling molecules produced in response to biotic and abiotic stress cross tolerance. ROS are produced in response to biotic stresses during the oxidative burst (Perez and Brown, 2014)
- Dual stress imposed by ozone (O3) and pathogen affects tolerance of crop and leads to altered host pathogen interaction (Lee *et al.*,2022). Alteration in pathogenesis potential of pest due to O3 exposure is of ecological and economic importance (Raju *et al.*, 2015)
- Without preventive measures of using pesticides, natural enemies, host plant resistance, and other nonchemical controls, 70% of crops could have been lost to pests. Weeds produce the highest potential loss (30%), compared to animal pests (23%) and pathogens (17%). However, the control measures for pathogens and animal pests show efficacy of 32 and 39%, respectively, compared to almost 74 % for weed control.
- Tolerance to both biotic and abiotic stresses has been achieved. In maize, breeding programmes have led to plants which are tolerant to drought and have additional resistance to the parasitic weed *Striga hermonthica*[1]
- Research shows that since 1960, crop pests and diseases have been moving at an average of 3 km a year in the direction of the earth's north and south poles as temperatures increase. Recent technological tools like the suitcase-sized mobile lab MARPLE, which tests pathogens such as wheat rust in near real-time and gives results within 48 hours, allow for early detection. Early warning systems are also crucial tools to warn farmers, researchers and policy makers of potential outbreaks.
- Breeding pest- and disease-resistant varieties is another environmentally friendly solution, since it reduces the need for pesticides and fungicides. Collaborating with scientists worldwide, CIMMYT works on developing wheat and maize varieties resistant to diseases, including *Fusarium* Head Blight (FHB), wheat rust, wheat blast for wheat and maize lethal necrosis (MLN) for maize.

(https://www.cimmyt.org/news/pests-and-diseases-and-climate-change-is-there-a-connection/)

Leaf eating caterpillars and sap-sucking aphids in dryland agriculture:

Correlation of weather parameters with *Spodoptera. litura* infestation on *capsicum* revealed that maximum temperature and morning relative humidity had positive correlation with moth population and fruit infestation by this pest. On the contrary minimum temperature and evening relative humidity had negative correlation with the pest; while larval population of *S. litura* had positively significant correlation with maximum temperature and evening relative humidity (Tompe *et al.*, 2020)

Incidence of aphid (*Aphis gossypii*), and leaf eating caterpillar *(Diaphania. indica*) was studied under different management modules of muskmelon crop in arid Rajasthan (Haldhar, 2014). The organic IPM system proved to be the most effective and economical approach (B: C ratio 8.80:1) against melon aphid (*Aphis gossypii*), leaf eating caterpillar (*Diaphania indica*), hadda beetle (*Epilachna vigintiopunctata*) and cucurbit fruit fly (*Bactrocera cucurbitae*) in which the lowest incidence was recorded as compared to other modules. The organic IPM module-III comprised growing resistant genotype (RM-50), spray of neem oil at 20 DAS, installation of pheromone trap (in 10/ hectares) at 42 DAS, spray of *tumba* fruit extract (TFE 5%) at 50 DAS and spray of *Spinosad* 46 SC at 60 DAS was the most effective. The conventional module-I (farmer's practices) was the second most effective system against major pests during both years. The benefit-cost ratio of the tested muskmelon production systems in the control of insect-pests decreased in the following order: module-III (B: C ratio 8.80:1)> Module-I (B: C ratio 7.74:1)> Module-IV (B: C ratio 6.60:1)> Module-II (B: C ratio 3.56:1) (Table 1).(Haldhar, 2014).

Table 1: Managing aphid and leaf eating caterpillar (Haldhar, 2014)

	2012	2013	Pooled	2012	2013	Pooled
Mndule-I	17.451 (4.30)	18.15 (4.37)"	17.80 (4.33)*	16.65 (4.20)*	17.08 (4.25)*	16.86 (4.22)'
Module-11	25.95 (5.19)*	26,58 (5.25)*	26.26 (5.22)*	20.00 (4.58)*	20.80 (4.66)**	20.40 (4.62)*
Module-Ill	11.10 (3.47)*	11.63 (3.53)*	11.36 (3.50)*	7.60 (2.93)*	8.23 (3.03)*	7,91 (2.98)
Module-1V	20.55 (4.64)'	21.30 (4.72)'	20.93 (4.681*	11.75 (3.57)"	12,25 (3.63)*	12,00 (3.60)"
Coni ml	30 JO (5.61)'	31.05 (5.66)*	30.78 (5.63)*	23.15 (4.91)*	24.03 (5.00)'	23.59 (4.95)*

Values in paranthesis are square roots transformed Value folkwing different teller .ire significantly different using Tukey's USD lest.

Fig. 1: *Lipaphis erysimi* nymphs

The investigations conducted on the incidence of leaf miner and leaf eating caterpillar on the dryland crop groundnut by Nigude *et al*., (2018) revealed that groundnut leaf miner (*Aproarema modicella* Deventer) infestation was negatively non-significant with temperature (-0.41) and positively significant with relative humidity (0.74) and rainfall (0.05).

Incidence of *Spodoptera litura,* the leaf eating caterpillar, was noticed first in the the 3rd week of August with mean population of 3.82 larvae/mrl. The population started increasing slowly and reached to its peak in the 35th MW corresponding to September 1st week (3.93 larvae/mrl) when the maximum temperature was 28.1°C, morning relative humidity 90 per cent and 1.01 mm rainfall. The population of the *S. litura* declined steadily thereafter from 3.67 at 36th MW corresponding to September 2 nd. week. The correlation coefficient was compared between the *S. litura* population and weather parameters. The analysis indicated that *S. litura* population was negatively non-significant with rainfall (-0.33) and positively significant with temperature (0.00) and relative humidity (0.52)

Table 2. Play of temperature and humidity in relation to the incidence of *Spodoptera* and leaf-miner.

Date	Temperature	Humnidity (%)	Rainfall (mm)	*S.litura* (mrl)	Leaf miner (Per cent infestation)
30 July -5 Aug	25.00	89.00	166,7	0.00	6.50
6 Aug-12 Aug	25.80	92.00	100.4	0.00	10.50
13 Aug— 19 Aug	26,70	88,00	16.7	3.82	18.50
20 Aug-26 Aug	27.20	89.00	27.1	3.84	19.00
27 Aug-2 Sept	28.10	90.00	01.1	3.93	16.00
3 Sept -9 Sept	28.50	83.00	01.9	3.67	16.50
10 Sept-16 Sept	28.30	85.00	17.9	2.50	14.00
17 Sept-23	26,00	86.00	44.0	0.00	13.00
24 Sept— 30 Sept	29.10	83.00	01.5	0.00	0,00
1 Oct - 7 Oct	28.50	82.00	02.7	0.66	0,00
8 Get-14 Oct	30.70	82.00	29.9	0.33	0.00
15 Oct-21 Oct	31.90	81.00	0.0	0.00	0.00
22 Del- 28 Oei	31.70	76.00	0.0	0.0	0.00

Source: Nigude *et al*., (2018)

Weeds, Important Biotic Stress: Weed causes harm to crops in many ways and this happens due to the unusual adaptation characteristics of the weeds and their regeneration ability**.** The yield loses caused by weeds range from 30 to 50% in pigeon pea to 78% in chickpea. The yield losses in all other crops range between the two extremes cited. The Directorate of weed Research under ICAR has played a pioneering role in conducting weed survey and surveillance, development of weed management technologies for diversified cropping systems, herbicide resistance in weeds, biology and management of problem weeds in cropped and non-cropped areas, environmental impact of herbicides and utilization of weeds. Selection of suitable weed management method depends on environmental concerns, desired management intensity, labour availability, weed pressure and crop. In general, for short duration pulses the critical period of crop weed competition is up to 30 days after sowing and for long duration pulses it is up to 60 days after sowing (Kumar *et al*.,2016). Adoption of weed management technologies has been promoted on large areas through on-farm research and demonstrations, which has raised agricultural productivity and livelihood security of the farmers.

Taking the example of wheat crop, the adverse effect of one of the prominent biotic stresses viz., weeds such as *Chenopodium album, Melilotus indica, Avena ludovicinia, Lathyrus toberosus* and *Medicago* spp. resulted in the yield losses of the order of 24 to 40%. (Oad *et al*., 2007).

Similar yield losses caused by weeds are reported in several other crops.

Diseases common to dryland agriculture: Infections affect plants at different stages of agricultural production. Depending on weather conditions and the phytosanitary condition of crops, the prevalence of diseases can reach 70–80% of the total plant population, and the yield can decrease in some cases down to 80–98%. Plants have innate cellular immunity, but specific phytopathogens have an ability to evade that immunity. Infectious agents can spread through the air, with water, be transmitted by animals, humans, and remain infectious for many months or years. The natural reservoirs of infectious agents are soil, water, and animals: especially insects (Nazarov *et al.*, 2020)

Irrigation occasionally facilitates avoiding plant diseases by growing crops in the season unfavourable for the proliferation of certain diseases. In the case of dryland agriculture, it may not be possible to adopt such a strategy. Irrigation renders the host plant more resistant or tolerant to disease by mainly increasing the plant vigour. (Rotemm and Patil. 1969)

Scabs are fungal diseases that attack leaves, fruits and tubers, resulting in hardening of the tissues. Other diseases such as Smuts, leaf spots, powdery mildew, damping off at the seedling stage do occur under conditions of dryland agriculture. Among them powdery mildew is more common. In general, higher humidity coupled with increase in atmospheric temperature are the important environmental factors promoting the incidence of the above plant diseases.

Fig. 2: Leaf spot disease

Additionally, the diseases caused mainly by fungi (*Fusarium, Gaeumannomyces, Pythium*, and *Rhizoctonia* spp.) nematodes, and to a lesser extent bacterium that infect through roots and complete part or all of their life cycle in root tissues or in soil surrounding the roots (the rhizosphere). This includes the vascular diseases (wilts) caused by fungal (*Cephalosporium, Fusarium, Verticillium* spp.) and bacterial pathogens that gain entry into the vascular system of plants through root infections, but does not include the leaf spots and other above-ground plant diseases caused by pathogens such as *Ascochyta* and *Septoria* spp. that use crop residue on the soil surface as a springboard for attack of the leaves (Cook 1990)

Management of biotic stresses in dryland agriculture

The biotic stresses of dryland agriculture may be of smaller magnitude as compared to those of irrigated agriculture; but might cause perceptible yield losses for two reasons viz., (i) The inherent physiological deficiency resulting in less vigour of the dryland crops can't withstand the on-slaught of pests and diseases (ii) The fast changing weather patterns due to climate change might promote the virulent and more devastating biotic stresses.

Therefore, the management of the biotic stresses even in dryland agriculture is a priority to save the yield losses in this deficient agro-ecological scenario.

***Chemical control*:** Firstly, there is tendency to bank upon chemical control using effective pesticides and fungicides to combat the pests and diseases. Nevertheless, the tendency to bank upon the chemical control should be last priority since the increase in the cost of the treatment might not commensurate with the yield increment.

Secondly the residues left over by the chemicals may be detrimental to the ecosystem and as well as to the consumers of the products.

Thirdly the chemicals may result in the destruction of the beneficial biocontrol agents which normally check the proliferation of pests and diseases. More importantly, it has been clearly established that chemicals used in this context are harmful to pollinating agents such as honey bees consequently resulting enormous detriment to agricultural production at large (Sanchez-Bayo and Goka 2016).

Cultural Management: The most cost effective and efficient strategy for management of biotic stresses is cultural management.

Adoption of the right kind of crop management as indicated below would save the cost factor at the same time plugging the yield losses by avoiding the incidence of biotic stresses'

- Seed Treatment with *Trichoderma harzianum* alleviates biotic, abiotic, and physiological stresses in germinating seeds and seedlings (Mastouri *et al*., 2010)

- Manipulation of sowing date in relation to the incidence of pests and diseases; more particularly adhering to timely sowing,
- Intercropping can reduce insect pest infestation and crop damage by disrupting visual or olfactory host plant location, reducing host plant quality or increasing natural enemy activity (Root, 1973; Trenbath, 1993; Hooks and Johnson, 2003; Finch and Collier, 2000). Mustard intercropped with onion and coriander significantly reduced branch and flower infestation of inset pests and increased pod formation per plant. These were found to be suitable intercropping systems as a potential multi-functional agricultural practices for pest management inin *Brassica* spp (Emery *et al*., 2021)
- When pearl millet was intercropped with cowpea under different configurations, activity of flower thrips, *Megalurothrips sjostedti* (Tryb.) and *Frankliniella schultze* (Tryb.), in cowpea was affected. Also, stem borer population and spike worm infestation rate in pearl millet were reduced by planting 1 row of millet with either 10 or 30 cowpea rows. (Gahukar, 1989)
- Intercropping of sorghum with cowpea reduced the numbers of sorghum stem borer, *Chilo partellus* as well as thrips, *Megalurothrips sjostedti* in cowpea (Nyarko *et al*., 1994).
- Intercropping French bean with other crops compromises on French bean yield but reduces damage by thrips to the French bean pods, thereby enhancing marketable yield in Kenya (Nyasani *et al*, 2012)
- Adoption of proven intercropping practices which inherently deter the incidence of pests and diseases as in the case of Sunflower grown as intercrop with pigeon pea in 2:1 proportion and also with *ragi* (*Eleucine coracana*) and soybean to check the incidence of *Helicoverpa armigenra*. (Seetharam and Virupakshappa, 1989)
- Several mechanisms combine to allow multilines and variety mixtures to provide durable protection of cereal crops against specialised, fungal pathogens. Similar approaches may control insects, insect-borne viral pathogens and nematodes. Greater activity of natural enemies of pests in intercrops can control some insect pests and possibly at least one soil pathogen (Trenbath, 1993)
- The adoption of conservation agriculture based on continuous minimum mechanical soil disturbance, permanent organic soil-cover, and diversification of crop species grown in sequences and/or associations should be applied to the fullest extent feasible in all dryland farming

systems (Thapa and Stewart, 2016) to minimize the incidence of biotic stresses.

- Conservation agriculture (CA) and integrated pest management (IPM) practices provide options to increase yields and minimize the use of chemical pesticides in dryland agriculture. The comparison of integration of CA and IPM practices (improved alternative system) with farmers' traditional practice (conventional system) under replicated on-farm tests in four different locations (Lalitpur, Banke, Surkhet, and Dadeldhura) in Nepal has shown very encouraging and positive results in combating biotic stresses (Paudel *et al.*, 2020)

Biological control of insect pests

(1) The phenomenon of natural parasitization has been observed in the case of semilooper of castor by a Hymenopteran insect *Microplitis maculipennis* which is very effective. It is advisable avoid insecticidal sprays during the play of natural parasites and predators (Prasad, 2021).

(2) Parasitization of groundnut leaf-miner of the tune of 90% was reported by Murty (1989) and Srinivasan and Rao (1987).

(3) *Telenomus manolus* has been reported as effective egg parasite on *Amsecta albistiga* (Rabindra and Balsubrahmanian, 1980, Sundaramoorthy *et al.*, 1976, Tuhan *et al* 1987)

***Biological control of plant diseases*:** This is more relevant in the case of soil-borne diseases since the pathogens of these diseases are not amenable to regular methods of disease control including chemical control. However, there is a vast reservoir of soil micro-organisms viz., fungi and bacteria which act antagonistically against the disease causing plant pathogens (Cook, 2000). Among these antagonistic agents only few have been commercialised and are currently marked as EPA registered bio-pesticides such as *Agrobcterium, Bacillus, Pseudomonos and Streptomyces* which are bacterial pesticides. Among fungi those belonging to the genera *Ampleomyces, Candida, Coniothyrum* and *Trichoderma* are important. Nevertheless, a few commercially available biocontrol products viz., (i) Deny (*Burkholderia cepia*), (ii)Intercept (*B.cepacia*) and (iii) Kodiac *(Bacillus subtilis* GBO3) have been released in the USA (Mayee,2003). In India only *Trichoderma* is being exploited. Effective strains of *T.viridae* are found to be effective in controlling *Rhizoctonia,* and *Sclerotium* by seed treatment (Mathivanan *et al.*, 2000)

The Central Institute of Cotton Research(CICR) Nagpur has developed cost effective farmer friendly *B.t Bucket* technology involving the toxin of *Bacillus thuringiensis* for control of boll worms in many crops (Mayee, 2003). There is a felt need to develop such technologies for use of small farmers involving *Tricoderma,Gliocladium, Pseudomonos, Bacillus subtilis* and *verticillium lecanii.*

Biocontrol mechanisms of Trichoderma

Seed coating with *T. harzianum* is a promising tool for crop production and protection under field conditions due to both direct antagonist activity and the indirect growth promotion. *Trichoderma* may suppress the growth of the pathogen population in the rhizosphere through competition and thus reduce disease development. It produces antibiotics and toxins such as trichothecin and a sesquiterpine, Trichodermin, which have a direct effect on other organisms. The antagonist (*Trichoderma*) hyphae either grow along the host hyphae or coil around it and secrete different lytic enzymes such as chitinase, glucanase and pectinase that are involved in the process of mycoparasitism. Examples of such interactions are *T. harzianum* acting against *Fusarium oxyporum, F. roseum, F. solani, Phytophthara colocaciae* and *Sclerotium rolfsii*. In addition, *Trichoderma* enhances yield along with quality of produce. Boost germination rate. Increase in shoot & Root length Solubilizing various insoluble forms of Phosphates Augment Nitrogen fixing. Promote healthy growth in early stages of crop. Increase dry matter production substantially. Provide natural long term immunity to crops and soil.

Methods of application

1. **Seed treatment:** Mix 6 - 10 g of Trichoderma powder per Kg of seed before sowing.
2. **Nursery treatment:** Apply 10 - 25 g of Trichoderma powder per 100 m2 of nursery bed. Application of neem cake and FYM before treatment increases the efficacy.
3. **Cutting and seedling root dip:** Mix 10g of Trichoderma powder along with 100g of well rotten FYM per liter of water and dip the cuttings and seedlings for 10 minutes before planting.
4. **Soil treatment:** Apply 5 Kg of Trichoderma powder per hector after turning of sun hemp or dhainch into the soil for green manuring. Or Mix 1kg of Trichoderma formulation in 100 kg of farmyard manure and cover it for 7 days with polythene. Sprinkle the heap with water intermittently. Turn the mixture in every 3-4 days interval and then broadcast in the field.

5. **Plant Treatment:** Drench the soil near stem region with 10g Trichoderma powder mixed in a liter of water

Trichoderma formulations

Important commercial formulations are available in the name of Sanjibani, Guard, Niprot and Bioderma. These formulations contain 3x106 cfu per 1 g of carrier material. Talc is used as carrier for making powder formulation.

Uses

Used in Damping off caused by *Pythium* sp. *Phytophthora* sp., Root rot caused by *Pellicularis filamentosa*, Seedling blight caused by *Pythium*, Collar rot caused by *Pellicularia rolfsii*, Dry rot caused by *Macrophomina phaseoli*, Charcoal rot caused by *Macrophomina phaseoli*, Loose smut caused by Ustilago segetum, Karnal bunt diseases, Black scurf caused by *Rhizoctonia solani*, Foot rots of Pepper and betel vine and Capsule rot of several crops. Effective against silver leaf on plum, peach & nectarine, Dutch elm disease on elm's honey fungus (*Armillaria mellea*) on a range of tree species, Botrytis caused by *Botrytis cinerea*, Effective against rots on a wide range of crops, caused by *fusarium, Rhizoctonia*, and *pythium*, and sclerotium forming pathogens such as *Sclerotinia* & *Sclerotium*

Recommendation

Trichoderma is most useful for all types of Plants and Vegetables such as cauliflower, cotton, tobacco, soybean, sugarcane, sugarbeet, eggplant, red gram, Bengal gram, banana, tomato, chillies, potato, citrus, onion, groundnut, peas, sunflower, brinjal, coffee, tea, ginger, turmeric, pepper, betel vine, cardamom etc.

Precautions

- Don't use chemical fungicide after application of Trichoderma for 4-5 days.
- Don't use *Trichoderma* in dry soil. Moisture is an essential factor for its growth and survivability.
- Don't put the treated seeds in direct sun rays.
- Don't keep the treated FYM for longer duration.

Compatibility

Trichoderma is compatible with organic manure and biofertilizers like *Rhizobium, Azospirillum, Bacillus Subtilis* and *Phosphobacteria. Trichoderma* can be applied to seeds treated with metalaxyl or thiram but not mercurials. It can be mixed with chemical fungicides as tank mix.

(Ref:http://agropedia.iitk.ac.in/content/trichoderma-bio-control-agent-management-soil-born-diseases)

Growing disease/pest resistant crop varieties: There are good number of varieties and genotypes possessing tolerance and resistance against the biotic stresses. Growing a resistant crop variety is by far the safest check against the yield losses due to pests and diseases. Confining to only a single variety may create new races of pest and disease in time to come. Hence it is advisable to diversify the sources of pest / disease resistance or tolerance by growing more than one variety so as to prevent the break-down of resistance over a period of time.

Epilogue: Dryland agriculture being a fragile ecosystem, the incidence of biotic stresses though relatively of lesser magnitude, the yield losses inflicted are of immense proportion largely due to inherent limitations of the system. It is imperative that there is an urgent need to manage the biotic stresses in dryland ecosystem. The climate change adds a new dimension to the problems of dryland ecosystem including the intricacies of biotic stresses. Nevertheless, the management technologies to sustain reasonable levels of productivity in the face of biotic stresses should revolve around cultural and biocontrol options rather than depending totally upon chemical control for the reasons of higher cost involved and damage to the ecosystem. The choice of crop varieties resistant or tolerant to biotic stresses offers certainly sustainable mechanism in reducing the crop yield losses. Among the cultural methods, intercropping offers several advantages such as sustainable productivity levels apart from reducing the incidence of biotic stresses.

Molecular biological investigations are to be carried out at the genetic level to develop mechanisms in plants that prevent onslaught of biotic stresses. Unless responsive mechanisms are developed against biotic and abiotic stresses, crop plants will be continuously subjected to such stresses and ultimately will prove a great threat to world agriculture.

12

Biodiversity in Dryland Agriculture

Biodiversity encompasses the different kinds of life we find in one area—the variety of animals, plants, fungi, and even microorganisms like bacteria that make up our natural world. Each of these species and organisms work together in ecosystems, like an intricate web, to maintain balance and support life. Biodiversity refers to all species and living things on Earth or in a specific ecosystem.

Biodiversity brings together the different species and forms of life (animal, plant, entomological and other) and their variability, that is to say, their dynamics of evolution in their ecosystems. Traditionally, there are three levels of biodiversity: genetic diversity, species diversity, and ecosystem diversity. But what does each of these levels really mean?

Genetic Diversity

Genetic diversity is about the variability of genetic systems including the gene complexes in the existing vegetable kingdom. Between and within different species there are different genes and different expressions of genes. And this genetic variability contributes to the multiplicity of life forms, physical and biological characteristics and, depending on the interaction with the environment, phenotypes.

Species Diversity

We speak of species diversity to describe the diversity of living species. There are millions of living species on Earth. They're all different and divided into groups depending on their specificities (insects, animals, plants, fungi etc.). Species diversity is often separated into two categories:

Intraspecies Diversity

Intraspecific diversity refers to the genetic variation of individuals and populations of the same species. Humans having white or black skin, blond or brown hair, blue or green eyes is one of the many examples of intraspecific biodiversity.

Interspecies Diversity

On the contrary, interspecific biodiversity refers to the diversity of living species among themselves: by their number, their nature, and their relative importance. Let's say humans (or sapiens sapiens), a species with currently **7.7 billion** organisms, have a higher diversity than, for instance, the low number of **African elephants** that are currently facing extinction.

Ecosystem Biodiversity

Fig. 1: Representation of biodiversity of a typical Ecosystem
Source:https://www.cbd.int/article/biodiversityloss-climatechange. www.cbt.int 1920x1047

Ecosystem Biodiversity

Ecosystem biodiversity refers to the variety of ecosystems, by their nature and number, where living species interact with their environment and with each other. For example, on Earth, there are different ecosystems, each with their specificities like deserts, oceans, lakes, plains or forests. And even within these ecosystems, there are special details like cold or hot deserts, boreal or tropical forests, warm or cold water coastal regions. Each ecosystem has its own peculiarities, species, and ways of functioning.

At each level and between them, interaction is a primordial idea that emphasizes a constantly evolving reality. In fact, biodiversity isn't limited to the static inventory of living species at a given moment. The assessment of biodiversity and its interactions applies to all types of organisms (plants, animals, bacteria and others) and to the ecosystems of which they're part and in which they interact.

Biodiversity in danger: It was in the late 1980s that the concept of biodiversity became a major concern. This attention followed the increasingly obvious recognition of how human activities had been responsible for the degradation, fragmentation, and destruction of ecosystems and their biodiversity. Indeed, humans have been contributing to the disappearance of 60% of the world species since 1970. Not only because of our industrial, economic or urban activities but also because of how they all together contribute to climate change and its extreme events.

At the 1992 Earth Summit in Rio de Janeiro, an essential step towards the preservation of biodiversity has taken. Biodiversity was acknowledged as a crucial part of the sustainable development agenda. Therefore, in the Convention on Biological Diversity treaty, the signatory nations committed to protecting the diversity of life and restoring already degraded ecosystems.

Biodiversity loss: what are the consequences?

The gradual decline in biodiversity (we are even talking about the extinction of many species today) has many consequences for humankind. For example, the disappearance of certain pollinating insects such as butterflies or bees makes some agricultural productions more complicated. For instance, bees are very important pollinators of food crops and without them, the spread of biodiversity loses reach. The disappearance of some species can disrupt ecosystems and make them more fragile.

Biodiversity protection: how can we protect biodiversity?

https://youmatter.world/en/definition/definitions-biodiversity-what-is-it-definition-protection-loss-and-csr-commitments/#:~:text=Protecting%20biodiversity%20is,conna%C3%AEtre%20ce%20sujet

What is the kind of biodiversity on drylands used for agriculture?

Drylands, which encompass deserts, semi-deserts, grasslands and rangelands, occupy 41.3 per cent of the land surface on Earth, but are among the lesser researched ecologies with respect to agriculture and somewhat over-looked by decision- and policy-makers. Considering the fact that drylands are home to about 44% of area of all the world's cultivated systems and 50 per cent of the world's livestock and habitats for wildlife, it is imperative to give focussed attention on the role of agrobiodiversity in these regions to address the issues of food, nutrition and livelihood security of the nearly 2.1 billion people inhabiting these terrains, especially in the context of climate change threats.

Amongst the total 34 global hotspots, 9 are in the drylands and about 0.5 per cent of the plant species are endemic to the region. In terms of agriculture, plant species endemic to the drylands make up 30 per cent of the plants under cultivation today, including many ancestors and crop wild relatives (TAAS 2019).

Dryland biodiversity is of tremendous global importance, being central to the well-being and development of millions of people in developing countries. In June 2012, at the UN Conference on Sustainable Development (or "Rio+20"), global leaders from governments and civil society reaffirmed the intrinsic value of biological diversity and recognised the severity of global biodiversity loss and degradation of ecosystems. Although drylands were implicitly recognised, there continues to be inadequate attention to this major biome that covers such a vast part of our world's terrestrial surface. Yet, as this book conveys, conservation and sustainable management of drylands biodiversity offers a viable pathway to deliver international conservation and development targets. (Davies *et al*., 2012)

- Drylands are areas those which face great water scarcity. They cover over 40% of the earth's land surface, and are home to more than two billion people.
- They are highly adapted to climatic variability and water stress, but also extremely vulnerable to damaging human activities such as deforestation, overgrazing and unsustainable agricultural practices, which cause land degradation.
- Land degradation in drylands is known as desertification, and is the loss of the biological or economic productivity of land.
- Desertification reduces agricultural output, contributes to droughts and increases human vulnerability to climate change.
- The loss of biodiversity in drylands, including bacteria, fungi and insects living in the soil, is one of the major causes and outcomes of land degradation.
- Restoring rangelands and sustainable land management practices can preserve drylands biodiversity, restore ecosystem functions, and halt land degradation.
- Drylands (rangelands and agricultural lands) are highly resilient and dryland people have developed indigenous knowledge and know-how to manage these lands productively and sustainably. There is a need to integrated the indigenous know-how with modern techniques of

restoration of biodiversity. Otherwise, there is a hidden danger of losing the indigenous knowledge over time.

- Continuous mono-cropping on drylands of India is one of the important reasons for loss of biodiversity of dryland ecosystem

Drylands are places of water scarcity, where rainfall may be limited or may only be abundant for a short period. They experience high mean temperatures, leading to high rates of water loss to evaporation and transpiration. Drylands are also characterised by extremely high levels of climatic uncertainty, and many areas can experience varying amounts of annual precipitation for several years.

Drylands are found on all continents, and include grasslands, savannahs, shrub-lands and woodlands. They are most common in Africa and Asia – for example, in the Sahel region in Africa and almost all of the Middle East. Drylands cover over 40% of the earth's land surface, provide 44% of the world's cultivated systems and 50% of the world's livestock, and are home to more than two billion people.

Drylands are extremely vulnerable to climatic variations, and damaging human activities such as deforestation, overgrazing and unsustainable agricultural practices. The consequences of these include soil erosion, the loss of soil nutrients, changes to the amount of salt in the soil, and disruptions to the carbon, nitrogen and water cycles – collectively known as land degradation.

Loss of endemic biodiversity in the Thar Desert of Rajasthan in India

Farmers' reliance on use of biodiversity in ensuring stability of reasonable of farm production in desert ecosystem illustrates the basic relationship between genetic diversity in the farming systems and sustainable agriculture in such fragile systems. For example, human populations settled in arid zone make use of estimated 682 plant species belonging to 352 genera of 87 families of native flora for food, fibre, medicinal purposes and many other uses (Chopra and Rana, 1994). Due to its unique location at a biological crossroads of the Indian subcontinent, the Thar Desert harbours spectacular biodiversity. However, since the 1960s human population increase has led to steady expansion of cropping even on the most marginal lands. This has been facilitated by an exponential rise in irrigation from bore-holes and canals, especially from the Indira Gandhi Canal, during the last 30 years. Since the 1960s the area under irrigated crop fields has increased from 300,000 hectares to over 4 million hectares. Changing land use has significantly affected biodiversity, with desert-adapted species being replaced by species that demand more water. Increasing irrigation in the Thar Desert is leading to loss of numerous shrubs such as *Calligonum polygonoides, Haloxylon salicornicum* and *Dipterygium glaucum*. Reptiles that are adapted to survive in extremely dry environments are likely to be affected by the transformation of desert areas to irrigated crop fields. These include the lesser known lizards such as *Eumeces taeniolatus, Acanthodactylus* cantoris and *Cyrtodactylus kachhensis,* and the snakes *Leptotyphlops macrorhynchus* and *Lytorhyncus paradoxus*. At the same time, land-use changes create conditions that are favourable to a number of invasive species and other weeds that further undermine indigenous biodiversity. From a conservation perspective it is important to examine the presumed inevitability of agricultural expansion and the ways of mitigating the harmful impacts of land-use change. To some extent land-use changes can be moderated by greater recognition of the value of existing land-use practices. For example, studies in Kenya have shown that existing pastoral land use in the dry north of the country makes more efficient use of scarce water resources than irrigated crop farming16. Where land-use change is inevitable, different land-use options can be considered that impose less of an environmental cost, including more environmentally sensitive irrigation plans. Economic valuation of ecosystem services can be a strong tool for measuring such environmental costs in order to help planners avoid undesirable environmental outcomes. Five Village in the Thar Desert, India, are now military lands, or set aside for other purposes. Protected areas, as recognised by IUCN, give priority to nature conservation. Places where conservation outcomes are a beneficial secondary outcome of natural resource use, such as grazing or forest management, are therefore likely to be outside any official protected area system. In the Thar Desert conservation –cum- restoration activities include a) fencing of dune areas to protect against biotic interferences; b) creating micro-wind breaks in parallel stripe or checkerboard patterns by planting locally available brushwood and grass; c) afforestation on dune slopes by directly sowing grass seeds and transplanting seeds of indigenous and exotic species; d) planting grass slips, seeds of grasses or leguminous creepers on the leeward side of the micro-wind breaks; and e) continuous management of dunes till the input cost is recovered.

Leptadenia pyrotechnica (khimp), *Ziziphus nummularia* (pala), *Crotalaria burhia* (sinia) and *Panicum turgidum* (murath) were the species planted as brushwood to create micro-wind breaks. *Acacia tortilis*, *Prosopis* spp, *Acacia senegal*, *Parkinsonia articulata* and *Tamarix articulata* were planted as trees, and *Lasiurus sindicus* and *Cenchrus ciliaris* as grass. (Moharana, and Yadava, 2019)

Given the extent of such sustainable land management practices on drylands including arid deserts, it is likely that significant conservation outcomes are thus obtained. (Moharana, and Yadava, 2019)

(https://www.iucn.org/resources/issues-brief/drylands-and-land-degradation#:~:text=In%20 drylands%2C%20land%20degradation%20is,one%20hundred%20countries%20at%20risk.)

Land degradation leads to the reduction or loss of the biological or economic productivity and complexity of land. In dryland ares, land degradation is known as desertification. It is estimated that 25-35% of drylands are already degraded, with over 250 million people directly affected and about one billion people in over one hundred countries at risk.

Dryland Ecosystem

Drylands support an impressive array of biodiversity. This includes wild endemic species – such as the *Saiga* Antelope in the Asian steppe and American bison in the North American grasslands that do not occur anywhere else on earth – and cultivated plants and livestock varieties known as agrobiodiversity. Biodiversity in drylands also includes organisms which live in the soil, such as bacteria, fungi and insects – known as soil biodiversity – which are uniquely adapted to the conditions. Soil biodiversity comprises the largest variety of species on drylands – determining carbon, nitrogen and water cycles and thereby, the productivity and resilience of land.

Species and ecosystems in drylands are a result of distinctive evolutionary process, developing strategies to cope with environmental constraints such as water scarcity, extreme hot and cold temperatures, and unpredictable long drought periods with sporadic rainfall. In plants, these manifest into features such as short growth cycles, long roots, water storage in roots and trunks, and dormancy during dry seasons. Livestock species and breeds have adapted by optimizing the use of scarce vegetation and water, minimizing their water loss, being able to walk long distances over rough terrain, and other characteristics. Paradoxically, agricultural genetic resources, which are of fundamental importance for adaptation to climate change, also become a casualty under certain extreme edapho-climatic changes. Many dryland areas, especially mountain regions, which are the centres of origin and/ or diversity of domesticated plants and animals (including their wild relatives) are under threat. (TAAS 2019)

Domestication of plants and animals in these regions is the outcome of efforts of farmers and herders who bred and selected the innumerable varieties/breeds specifically adapted to these niche areas. These farmers and herders are the most extraordinary innovators and conservers of agrobiodiversity, as they managed to develop unique and highly technological agriculture and pastoral management systems – many of them still in use – adapted to very adverse and changing environments. (TAAS 2019).

Over millennia the availability of water, or rather the lack of it, has caused dryland organisms to adapt in many ways to survive. Some animals have the opportunity to move in response to water availability and long-range migrations are a common feature of drylands, but dryland fauna and flora display a tremendous variety of adaptations. Four broad categories of adaptation can be identified (SBSTTA, 1999)

Biological diversity of drylands, arid, semi-arid, savannah, grassland and Mediterranean ecosystems exhibit the following mechanisms of adaptation.

- Drought escapers: animals migrating in search of water or pasture, or insects 'escaping' into the egg or pupal stage until wet weather returns
- Drought evaders: plants like the salt bush with deep and efficient root systems or animals such as certain reptiles that avoid the heat by burying themselves underground
- Drought resistors: cacti that store the water in roots and trunks, or camels that minimise water loss
- Drought endurers: shrubs and trees that go dormant, or animals such as frogs that estivate during dry seasons

Historically, drylands have been the living basis for mankind. The first humans originated in the savannah grasslands of eastern and southern Africa. The origins of many of the earth's most important food crops are found in drylands. For example, maize, beans, tomato and potatoes originate from the drylands of Mexico, Peru, Bolivia and Chile. Millet and sorghum, come from the African drylands. The Mediterranean basin has given the world date palm and olive trees. And drylands continue to provide new food, as traditional food products are increasingly becoming commercialized globally in the age of health-consciousness (e.g. wild millet, wild rice, etc.).

Agricultural biodiversity is a vital subsect of biodiversity. It is the result of the careful selection and inventive development of farmers whose food and livelihood security depends on the sustained management of this biodiversity. Since the dawn of agriculture some 12,000 years ago, humans have selectively used and bred certain species of plants to provide food and other goods. These varieties of plants and breeds of animals, the species, and the agroecosystems that support them comprise agrobiodiversity (Mulvany, 1999).

A Team of Scientists of the ICAR-National Bureau of Plant Genetic Resource (NBPGR) New Delhi reported that the farmers of the north-eastern states of India carefully preserved and maintained the native land races of maize even after two decades of the release and spread of modern varieties and hybrids of maize (Chopra and Rana 1994). Tribal people and farmers engaged in crop production under stressed environments have known for a long time that their success and also their survival is linked to the genetic diversity in crop and forest resources available to them (Chopra and Rana 1994). This speaks volumes about the intent of farmers to conserve biodiversity. Biological diversity is obviously the key to sustainable agriculture in stressed environments like dryland agriculture, since it provides reasonable chances of success under threat of uncertainties and also promotes soil health status (Chopra and Rana, 1994).

The loss of biodiversity in drylands is one of the major causes and outcomes of land degradation.

Managing and Conserving Dryland Biodiversity

Although the conservation status of dryland biodiversity is not well monitored, many known drivers of biodiversity loss are present in the drylands. These drivers include rapid demographic shifts and urbanisation, agricultural expansion, land use change, weakening of governance arrangements and the introduction and spread of alien invasive species. Accelerating dryland development is anticipated to increase the rate of biodiversity loss. The combination of habitat loss and fragmentation will reduce the opportunities for dryland biodiversity to adapt and survive, with the additional impacts of climate change further exacerbating the problem. Despite approximately 9% of drylands receiving formal protection, the protected areas are not representative of all the dryland subtypes. For example, deserts are disproportionately represented whilst temperate grasslands have amongst the lowest level of protection of all biomes at 4-5%. To some extent this is because traditionally, areas with the lowest economic value were the ones designated as protected areas. Nevertheless, large areas of drylands are protected informally by the communities that inhabit the area, either consciously (for example as sacred sites) or as a by-product of sustainable management practices that evolved through generations (for example as seasonal grazing reserves). This indigenous protection is seldom recognised by government and is often undermined by government policies. Many traditional land management practices have proven to be more economically viable than more 'modern' alternatives, whilst simultaneously providing conservation benefits. The ecological rationale of these traditional strategies developed through a deep understanding by the indigenous communities of their surrounding natural environment ensures both economic and environmental sustainability. The drylands perhaps more than any other biome offer opportunities for achieving both conservation and development objectives simultaneously and in many cases have shown to do so for Biodiversity 2011-2020', which is a ten year framework for action by all countries and stakeholders to save biodiversity and enhance its benefits for people. The AICHI Biodiversity Targets are twenty actions adopted under 'Strategic Plan for Biodiversity 2011-2020', which is a ten-year framework for action by all countries and stakeholders to save biodiversity and enhance its benefits for people. The AICHI targets on protected areas could be more easily achieved, or even surpassed, in drylands by legitimising and supporting Indigenous and Community Conserved Areas, and traditional natural resource management strategies.

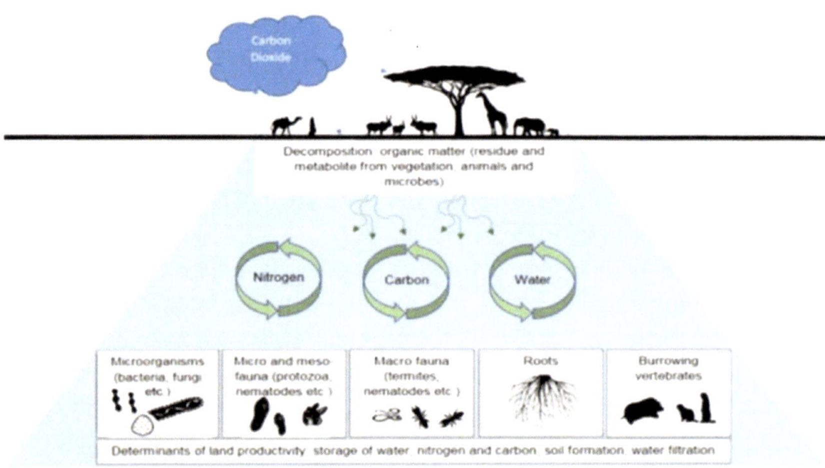

Fig. 2: Soil biodiversity and dryland ecosystem functions showing decomposition of organic matter food and water provision:

Low precipitation and prolonged dry seasons in drylands can lead to water scarcity, and limit agricultural productivity and output. Drylands biodiversity maintains soil fertility and moisture to ensure agricultural growth, and reduces the risk of drought and other environmental hazards. For example, vegetation is decomposed in the stomachs of large herbivores in the drylands, after which the dung is transformed into nutrients by bacteria in the soil, which are absorbed by plants. Bacteria and other microbes also break down plants and animals into decomposing residues – soil organic matter, which helps the soil easily absorb rainwater and retain moisture. Each gram of organic matter can increase soil moisture by 10-20 grams, and each millimetre of additional infiltration of water into the soil represents one million additional litres of water per square kilometre.

Poor crop and soil management, and habitat destruction undermine the ability of drylands biodiversity to perform nutrient recycling, and water storage and filtration services. On severely degraded land – devoid of biodiversity – as little as 5% of total rainfall may be used productively. An estimated 20 million hectares of fertile land is degraded every year, and in the next 25 years global food production could fall by up to 12% as a result of land degradation – threatening the food and water security of the rising human population.

Climate change mitigation and adaptation

The world's soils contain 1,500 billion tons of carbon in the form of organic matter – two to three times more carbon than is present in the atmosphere. The carbon stored in soil is released into the atmosphere when land is degraded, and about 60% of the earth's organic carbon has been lost through land degradation. This represents a significant contribution to man-made greenhouse gas emissions. Increasing the quantity of carbon contained in soil, for example through agriculture and pasture management practices which increase soil organic matter, can reduce the annual increase in carbon dioxide in the atmosphere. It is estimated that improved livestock rangeland management could potentially sequester a further 1,300-2,000 million metric tons of carbon dioxide by 2030.

Climate change will also impact drylands, with models predicting even more climatic variability and extreme temperatures. Biodiversity in drylands has adapted over millennia to the seasonality, scarcity and variability of rainfall, and can be useful in helping people adapt to climate change. For example, the unique species in drylands provide a genetic reservoir for new varieties of cultivated plants and livestock breeds, which are resilient to the climatic variations.

The drylands are different in a number of important ways from humid lands. However, development pathways for the drylands are often driven by a distorted idea of how drylands should or could exist, often modelled on more humid areas. Notions of "greening the desert" are developed on a misunderstanding of dryland ecology and have led to many harmful policies and investments. Furthermore, misrepresentation of drought and water scarcity in the drylands diverts attention away from sustainable and adaptive management, capable of being supported by limited resources, towards unsustainable practices that are ecologically harmful. Rather than adapting development strategies to fit the drylands, considerable effort is expended on trying to adapt drylands to fit development strategies (Davies,2012).

What can be done?

Conserving biodiversity in drylands, including soil biodiversity, ensures that vegetation for agriculture and livestock farming is maintained all year round, especially in between rainfall seasons. International Union for Conservation of Nature (IUCN) works with national governments, businesses and local communities to preserve and protect ecosystem functions in drylands by restoring rangelands for livestock and sustainable land management practices.

Rangelands restoration

Livestock farmers (pastoralists) depend on drylands resources such as grasslands and seasonal ponds to nourish their livestock. Sustainable pasture management through managed herd mobility can prevent degradation and sustain livelihoods. For example, the practice of Hima in Jordan takes into account the seasons and life cycle of grasses to prevent overgrazing by livestock herds, which also transport fertile seeds around the landscape. Governments can institute appropriate policies and grant rights to local communities to sustain these traditional practices. For example, grazing lands can be recognised as protected areas, to prevent their conversion to other land uses.

Sustainable land management practices

Biodiversity Conservation goals can be defined in many ways, including preventing species extinctions; sustaining the maximum diversity of species; maintaining intact and fully functional ecosystems; sustaining key ecosystem services (e.g. pollination services, water supplies, or carbon sequestration); and maintaining sustainable levels of harvestable or extracted resources (e.g. waterfowl, timber) (Stein and Shaw, 2013).

Strategies to resist the adverse impacts of climate change may work and be appropriate when *climate-forced impediments* are the modest or ecological resources are of very high value. Increasingly, though, conservationists will need to accept and manage rather than resist the change. Key to this shift will be finding ways to manage the change that would result in systems that retain functional processes (both ecological and evolutionary), and that sustain as much of the raw biological materials (i.e. genetic and species-level diversity) as possible. (Stein and Shaw, 2013)

Conservation of biodiversity and sustainable land management in the drylands is hampered by weak integration between sectors, which contribute to fragmentation of ecosystems, uncoordinated development of different resources and weak environmental-accountability. To protect ecosystems as the foundation of life and prosperity requires a concerted effort to manage them holistically. This can be achieved practically by mainstreaming environmental issues and a strong understanding of dryland ecology at all levels, from the farmer to the state and across borders, including in international treaties and through Intergovernmental Bodies. (Davies *et al.*, 2012)

Dryland ecosystems are highly vulnerable to climate change and land-use effects and therefore urgently require the implementation of standardised monitoring and assessment tools, which would allow timely prevention and adaptation actions at the various decision-making levels. However, changes in biodiversity and of species composition cannot be detected by remote sensing technology. It is therefore important to combine field analyses with data provided by remote sensing tools. The assessment of dryland biodiversity in different regions and under different land-use systems requires permanent terrestrial observation systems to be set up. These should operate on the basis of scientifically sound instruments and dedicated individuals, communities or other institutions that can ensure long-term regular monitoring (Davies, 2012)

Strategies are therefore needed to ensure more integrated and holistic planning of multiple resources, and to mainstream dryland biodiversity in rural development. This may be most feasible at the level of the land users, since they naturally operate in this way, and government institutions need to be tailored towards supporting and augmenting the skills of land managers. A priority is to greatly increase the use of tools that integrate the planning of land, water and biodiversity and to accelerate the adoption of effective participatory approaches by government services. Mainstreaming dryland biodiversity in key sector strategies such as agriculture, or in poverty reduction strategies, can contribute to conservation and more sustainable use of biodiversity (SCBD 2010). The ecosystem approach provides a framework for linking development and conservation and accommodating multiple agendas in a single landscape. It enables ecosystem services to be factored into planning in order to enhance the resilience and productivity of agro-ecosystems. National Biodiversity Strategy and Action Plans (NBSAPs) can play an important role in linking biodiversity and development priorities, including on dryland biodiversity issues (Bass *et al* 2010).

To reverse declines in dryland biodiversity and to ensure sustainable land management require greater attention to the condition of land and soils. A concept of land health should be factored into agricultural sector policies and investments to ensure sustainability. Incentives and policies or regulations are required to promote integrated farming systems in which biodiversity conservation is an explicit agricultural output, whether to produce marketable goods or to receive payments for ecosystem services. (Davies, 2012)

Sustainable land management practices often involve protecting biodiversity to boost soil organic matter and soil moisture. Traditional crop farming practices used by communities in drylands build up soil moisture and restore degraded land. For example, the zaï pits used by communities in the western Sahelian drylands (Burkina Faso, Niger and Mali) involve planting seeds in pits filled with organic manure to concentrate water and nutrients at the plant's base. Practices like agroforestry (planting trees together with agricultural crops) and low tillage agriculture (involving little or no ploughing of land) are based on traditional practices that have been revived and adapted to protect soil moisture and fertility of crop lands. Governments can encourage these traditional practices, and discourage less sustainable forms of land management such as prohibiting irrigation projects which intensely exploit water from small areas of land.

Crop diversification is an important aspect that adds to the biodiversity of dryland agriculture. According to Paroda (2022) for crop diversification, the future strategy must aim at: i) horizontal approach adopting crop intensification and crop substitution that are most suited to specific eco-regions, ii) vertical approach for enhancing productivity using genome editing and good agronomic practices aimed at judicious use and increased efficiency of costly inputs such as water, energy, fertilizers, pesticides, etc. iii) post-harvest processing, value addition, branding, packaging, etc. to enhance income, iv) water use efficiency through micro-irrigation especially in drylands, v) varietal diversification (introducing both high yielding varieties and hybrids for higher productivity), vi) incorporation of legumes, vii) adoption on large scale of integrated pest management, and viii) the risk management through inter-cropping, mixed cropping, a shift towards low volume high value crops and mixed farming.

In the paper entitled "the benefits and trade-offs of agricultural diversity for food security in low and middle income countries: A review of existing knowledge and evidence" published in the Journal of Global Food Security in June 2022, it was reported that the agricultural diversity consisting of diversity of crop types, domestic animals, fisheries and perennial trees had a positive effect on food security (Magzter 2022)

With global food prices surging due to Russia's invasion of Ukraine, South America has struggled. In a continent that relies heavily on wheat, corn and fertilizers, Brazil may have found the answer. According to The Economist, agricultural techniques of enhancement of biodiversity that combine crops and livestock with forestry practices can make a farm five times more productive than the current average across Brazilian agriculture. At present, however, such techniques are used on only about 5% of the country's farmland. (Sources:

Bloomberg, The Economist, The Wall Street Journal) News Item from Daily Dose Ozy_ May09, 2022

Endangered Plants in India

An endangered species is a native species that faces a significant risk of extinction in the near future throughout all or a significant portion of its range. Such species may be declining in number due to threats such as habitat destruction, climate change, or pressure from invasive species.

The term endangered species can be used either in a general or legal context. When used in a general sense, the term describes a species that faces a risk of extinction but does not necessarily indicate that the species is protected under any law.

When used in a legal context, the term refers specifically to a species that is listed on the US Endangered Species List and is defined legally as an animal or plant species in danger of extinction throughout all or a significant portion of its range.

The flora of India is one of the richest in the world due to the country's wide range of climate, topology, and environment. There are over 15,000 species of flowering plants in India which account for 6% of all plant species in the world. Many plant species are being destroyed, however, due to their prevalent removal.

International Union for Conservation of Nature (IUCN) comes up with a Red List from time to time. The list specifies the species of plants and animals that are threatened or are endangered in various parts of the world. The number of Indian plants and animals in the IUCN Red List has been steadily on the rise. In 2008, the Red List featured some 246 species of plants while in 2013 and 2014, this number went up to 325 and 332 respectively. In 2012, the Red List noted that some 62 species of plants in the country were critically Endangered, that is, in immediate danger of going extinct. Not only are the plants in our country endangered and threatened but a number of animals, insects and reptiles also face extinction. According to a World Bank report of endangered mammals, India is said to have the fourth largest number of threatened mammals' species in the world.

According to a report published in the scientific journal, 'Science,' between 22% and 47% of the world's plant species are endangered.

Roughly 1/4 of all plant species in the world are at risk of being endangered or going extinct. The combination of global warming and habitat destruction is the sole reason for the disappearance of many plants. Though there are thousands of interesting and unusual plants, here are some common plants which have become rare and endangered species in the past 30 years due to habitat destruction.

Plant	Also Known As	Region (Status)
Polygala irregularis	Milkwort	Gujarat (rare)
Lotus corniculatus	*Bird's foot*	*Gujarat (rare)*
Amentotaxus assamica	*Assam catkin yew*	*Arunachal Pradesh (threatened)*
Psilotum nudum	*Moa, skeleton, fork fern, and whisk fern*	*Karnataka (rare)*
Diospyros celibica	*Ebony tree*	*Karnataka (threatened)*
Actinodaphne lawsonii		*Kerala (threatened)*
Acacia planifrons	*Umbrella tree, kudai vel (Tamil)*	*Tamil Nadu (rare)*
Abutilon indicum	*Indian mallow, thuthi (Tamil) and athibalaa (Sanskrit)*	*Tamil Nadu (rare)*
Chlorophytum tuberosum	*Musli*	*Tamil Nadu*
Chlorophytum malabaricum	*Malabar lily*	*Tamil Nadu (threatened)*
Nymphaea tetragona		*Jammu (endangered), Kashmir (threatened)*
Belosynapsis vivipara	*Spider wort*	*Madhya Pradesh (rare and endangered)*
Colchicum luteum		*Himachal Pradesh (rare and threatened)*
Pterospermum reticulatum	*Malayuram, Malavuram*	*Kerala (rare), Tamil Nadu (threatened)*
Ceropegia odorata	*Jeemikanda (Gujarat)*	*Gujarat, Melghat Tiger, Rajasthan, and Salsette Island, (endangered)*

The above are some of the documented cases of endangered trees. There is every need to protect and enhance the population of the above trees to strengthen biodiversity.
(*Source*: (1) https://www.mapsofindia.com/my-india/india/endangered-plant-species-in-india and (2) https://owlcation.com/stem/Rare-and-Endangered-plants-of-India)

Fredrik Lähnn

Lotus corniculatus, one of the endangered plant species of Gujarat in India, belongs to the pea family. The plant bears pretty little yellow flowers that grow in a circle at the end of the stem. They are very bright and are easily spotted along the roadside. It is used in agriculture as a forage plant. It is also grown for pasture, hay and silage. The plants are perennial and herbaceous, similar to some clover. They are also called bird's foot, which refers to the appearance of the seed pods on their stalk.

Fig.3: *Allamanda cryptostegia* (Indian rubber wine) One of the critically endangered plant species of India

Strengthening Biodiversity of a farm

According to Nicole O' Malley the following five steps must be implemented to strengthen the Biodiversity of a farm (https://thatsfarming.com/beef/increase-biodiversity-on-land/) Increasing biodiversity of farm land is one simple measure that a farmer can take to contribute to the carbon-neutral pot.

- **Manage your hedgerows:** Each of the measures describe a way to increase biodiversity that also benefits soil fertility and grass output. The Rural Environmental Protection Scheme (REPS) began in 1994 and facilitated the establishment of around 10,000km of new hedgerows. If hedgerows are already on the farm, the farmer needs to promote the visits of the native birds by avoiding over-trimming.
- **Leave small, grassy areas unfertilised:** Similarly, leaving unfertilised, unsprayed grassy corners or zones encourage the competitive growth of nettles, thistles, and docks. These plants attract butterflies, bank voles, mice, shrews, linnet and meadow pipit. Plant these areas with a seed mix of grass, clover, herbs and wildflowers to make the most of awkward corners.
- **Planting catch crops:** Catch crops are sown between two main crops when land would otherwise lie idle, also known as fodder crops. They provide animal fodder, grow quickly, and fill the gap between autumn harvest time and spring. Kale, oilseed rape, vetches and stubble turnip are suitable crops that provide dry matter grazing, high in crude protein. Also, they prevent nitrogen leaching, reduce winter feed costs, and provide winter soil cover for microorganisms and small mammals.

- **Manage drainage effectively** Regularly waterlogged areas of a field and wet, compacted, poached areas deter organisms of all kinds, earthworms, beetles, birds, microbes. Effective drainage systems maintain habitats for a range of species so that they can carry out their natural function. In addition, this increases grass growth, improves soil aeration, reduces soil compaction and attracts organisms to the area.If soil air spaces are full of water due to a blockage, over time, this detracts organisms and soil fertility.
- **Leave a water buffer zone** When applying fertiliser, leave a 2m buffer zone if farm land is next to a water body. This prevents pollution and algal blooms, as unnatural levels of nutrients entering a water body upsets water quality. Leaving a buffer zone maintains the natural environment of earthworms, who contribute nutrient humus to soil. In addition, fish and insect species can suffer if their habitat receives abnormal levels of nutrients, depleting oxygen rapidly.

(*Source*: https://thatsfarming.com/beef/increase-biodiversity-on-land/)

13

Economy of A Small Farmer

About two-thirds of the developing world's three billion rural people live in about 475 million small farm households, working on land plots smaller than two hectares. Many are poor and food insecure and have limited access to markets and services. Their choices are constrained, but they cultivate their land and produce food for a substantial proportion of the world's population. Besides farming, they have multiple economic activities, often in the informal economy, to contribute towards their small incomes. These small farms depend predominantly on family labour. In China, nearly 98 percent of farmers cultivate farms smaller than 2 hectares – the country alone accounts for almost half the world's small farms. In India about 80 to 85 percent of farmers are small. In Ethiopia and Egypt, farms smaller than 2 hectares constitute nearly 90 percent of the total number of farms. In Mexico, 50 percent of the farmers are small; in Brazil smallholders make up for 20 percent of the total number of farmers (Rapsomanikis, 2019).

Indian agriculture too is beset with the above kind of structural limitations in terms of small sized holdings which culminated in the current size of the farm holding viz.,1.08 ha from 2.65 ha in 1960s (Sarah *et al.*, 2016). Small holdings face new challenges on integration of value chains, liberalization and globalization effects, market volatility and other risks and vulnerability, adaptation of climate change etc. (Thapa and Gaiha 2011).

Though there was increase in food-grain production in India from 257 million tons in 2012-2013 to 275 million tons during 2016-2017, the gain to small and marginal farmers is very low. The major problem in Indian Agriculture is the predominance of marginal and small farmers whose incomes are low, mainly due to the small sizes of their holdings. Though these category of farmers are fairly efficient in their agricultural skills, their income levels are very low due factors like disadvantages in marketing and high cost of credit (Subrahmanyam and Aparna,2019).

The marginal farmers derive 51% of their credit from private sources, while the middle level farmers get more than two-thirds of their credit from institutional source which has several advantages like low rate of interest and loan waiver facility (Subrahmanyam and Aparna, 2019).

Recent world-wide processes of farm change – commercialisation of increasing proportions of input and output, institutional developments such as super markets, privatization of key aspects of technical progress, and of output and process grades and standards – now indicate large farm focus (Lipton, 2006). Therefore, support is needed for small holdings in the context of these world-wide processes of farm change. There are also high returns from investments in agricultural R&D, rural roads and other infrastructure and knowledge generation. (Dev. 2014).

One of the paradoxes of the Indian economy is that the decline in the share of agricultural workers in total workers' has been slower than the decline in the share of agriculture in the GDP. For example, the share of agriculture and allied activities in the GDP declined from 57.7 per cent in 1950–1951 to 15.7 per cent in 2008 (Dev 2014). The share of agriculture in total workers, however, declined slowly from 75.9 per cent in 1961 to 56.4 per cent in 2004-05. Between 1961 and 2004-05, there was a decline of 34 percentage points in the share of agriculture in GDP while the decline in share of agriculture in employment was of only 19.5 percentage points. As a result, the gap between labour productivity in agriculture and non-agriculture increased rapidly (Dev, 2014)

There are an estimated 98 million small and marginal holdings out of around 120 million total land households in India. The share of marginal and small farmers accounted for around 81% of operational holdings in 2002-03 as compared to about 62% in 1960-61. Similarly, the area operated by small and marginal farmers has increased from about 19% to 44% during the same period. Recent data for 2005-06 shows that the share of small and marginal farmers in land holdings was 83% (Chand *et al*, 2011).

Consequently, the average size of holdings in India declined from 2.3 ha. in 1970-71 to 1.33 ha. in 2000-01. It may be noted that 63% of land holdings belong to marginal farmers with less than 1 ha. The average size of marginal holdings is only 0.24 at all India level (Dev 2014).

The vagaries of monsoon such as late onset, long dry spells, torrential rains, floods and droughts have rendered plans of many small farmers fail. Though farming has to contend with various kinds of aberrant weather throughout the world, the degree of severity is much higher in India, breaking the back of the economy of the small farmer. Small famers and agricultural labourers toil harder in India under the difficult environmental situations, despite lack of capital and any level of mechanisation. Labour force participation is lower among marginal land holders than in the cases of the higher categories.

Drawbacks regarding the economy of small farmers

The major throttle in the hands of the government is causing depressed prices of the agricultural commodities far below the boarder prices (NEWS-NCR 2021). Thus the farmer is denied the legitimate extra income to which he is entitled. Forced extraction of surplus from agriculture by taxation, confiscation, imposition of levies or arbitrarily kept low prices of agricultural products, can be the other measures taken by the government to transfer funds from the agricultural sector to the non-agricultural sectors. *(https://www.economicsdiscussion.net/economic-development/contribution-of-agriculture-to-economic-development/21407)*

Another anomaly is from the high protection accorded to industrial products by way of import tariffs and restrictions on the export of agricultural products. While this facility enabled the industries to enjoy high prices for their products, with higher rate of return on their investments, restrictions on the export of agricultural products and suppression of prices limit the returns on agricultural investments, caused flight of capital from agriculture to other sectors, impoverishing small farmers and starved agricultural sector of capital. The recent tightening of export restrictions has been limiting the supply of staple agricultural commodities available to global markets and pushing up prices (Cross, 2022). Small farmers bore the brunt of the above vicious factors much more, leading to their progressive impoverishment.

The top ten problems faced in Marketing Agricultural Goods are, low marketable surplus of Agricultural goods, producer not being permitted in the determination of the price, lack of storage, problems in transportation, long chain of middlemen, malpractices in the market, lack of Market Information, inelastic demand, lack of grading and bulky nature of produce. The marketable surplus of small farmer is too low due to the small size of holding, thereby causing their disability in meeting the financial needs of their families *(https://accountlearning.com/top-10-problems-faced-in-marketing-agricultural-goods/)*.

Literacy and mean years of education are lower for small holding farmers compared to medium and large farmers. For example, literacy among males and females for marginal farmers respectively were 62.5% and 31.2%, while the corresponding numbers for medium and large farmers were 72.9% and 39% (Dev.2004). This has implications in accessing the new farm production technological developments, inputs, farm business, government schemes and policy matters from time to time.

The NSS data also shows that across social groups, the indebtedness through formal sources is lower for STs as compared to others. Increasing globalization has added to the problems faced by the small holding agriculture. The policies of huge subsidies and protection policies by developed countries have negative effects on small holding farmers in developing countries (Dev 2014).

Added to their extremely low disposable incomes and poverty, there are the chronic problems of high indebtedness, recurrent crop failures, uncertain and erratic returns to his produce at the marketplace, supply of sub-standard and spurious inputs and the associated crop losses, breaking the back of small farmer. The more shocking is the incidence of suicides among farmers registered rapid rise in the recent past (average number of suicides/year over the last 3 years = 12000) (National Crime Records Bureau of India) (Ranga Rao, 2021).

Despite a host of national crop insurance schemes since 80s, risk protection through insurance remains a pipe dream even today; together they account for a negligible share of either total gross cropped area or number of farmers in the country (Raju and Chand, 2008, NABARD, 2016)

The basic problem with our agricultural strategy is aiming at achieving growth without improvement in income levels of small farmers who form the pivot of agricultural production. As such the small farmers are not able to get adequate income from crop production (Chandrasekhar and Mehrotra 2016).

The fact that the share of agriculture fell steeply, while the share of rural households depending on agriculture has not declined indicates that relative income levels of these households is falling continuously (Subrahmanyam and Aparna, 2019).

Hostile market forces

During the harvest season, crop prices drop drastically forcing the small farmers, who have no holding capacity, to succumb to distress sales, thereby losing heavily in the bargain (Ceballos *et al.*, 2021). This kind of lack of price support to the farmers discourages small farmers from producing more. Facing higher unit costs while offering their products to markets, is another impediment.

The Government of India announced the Minimum Support Price (MSP) to help farmers towards obtaining assured and better returns. During the decades of 1960s and 1970s market prices were higher than the announced MSP, which helped the farmers to a certain extent. Subsequently, due to globalization and invasion of the market by middle men, the prevailing market prices became much lower than MSP. While the Food Corporation of India procured rice and wheat at MSP, the coarse grains like sorghum (*jowar*), finger millet (*ragi*), Pearl millet (*bajra*) including pulses and oilseeds produced by small dryland farmers were left out of the governmental procurement, leaving them at the mercy of hostile middle men. The Farmers' Commission headed by Prof. M.S. Swaminthan recommended that farmer should be paid 50% over and above the cost of production in the market. However, the recommendations are not implemented. The traditional oilseed crops produced on drylands lost ground in the wake of liberal imports of vegetable oils from other countries. Despite the support provided by the way of MSP and some subsidies on fertilizers etc., the dryland farmer is virtually taxed due to the prevailing policies of free market and the tendencies of the market forces that kept the domestic prices of agricultural commodities far below their boarder prices through the flurry of export restrictions and taxes. (https://www.insightsonindia.com/2014/11/14/agriculture-issues-related-minimum-support-prices-msp-wto-subsidies/)

The Organization of Economic Cooperation and Development (OECD) estimated that Indian farmers are losing ₹ 345,845 crores per year by the way of depressed prices, while the government is supporting them to the extent of ₹181,576 crores through sector specific subsidies and MSP (OECD 2019).

There has been adverse impact of trade liberalization on the agricultural economy of the region's growing crops such as plantation, cotton and oil seeds in which foreign trade is important. With liberalization, India is not able to check import of a large number of commodities even at higher tariff (Dev 2014). This is true not only in the case of import from developed countries where agriculture is highly subsidized, but also in the case of products from developing countries. India is facing severe import competition in the case of items like palm oil from Malaysia and Indonesia, spices from Vietnam, China and Indonesia, tea from Sri Lanka and rice from Thailand and Vietnam (Planning Commission, 2007).

Intricacies in farm production

Lack of availability of recommended new productive technologies to small dryland farmers due to defective and inadequate machinery of agricultural extension network is forcing them to depend on local seed and fertilizer venders who exploit them often.

This is one of the main reasons for impoverishment of soils as a result of application of chemical fertilizers and plant protection chemicals as influenced by the venders, irrespective of the correct technical recommendations and decline of plant populations on the farms of small farmers. This unfortunate trend goes unnoticed and results in drastic decline in farm productivity and needless pushing-up of cost of production.

Water is the leading input in agriculture. Development of irrigation and water management are crucial for raising levels of living in rural areas. Agriculture has to compete for water with urbanization, drinking water and industrialization. As mentioned above, small holding agriculture depend more on ground water compared to large farmers who have more access to canal water. Ground water is depleting in many areas of India. Marginal and small farmers face more problems regarding water. (Dev 2014).

Climate change is a major challenge for agriculture, food security and rural livelihoods for millions of people including the poor in India. Adverse impact will be more on small holding farmers. Climate change is expected to have adverse impact on the living conditions of farmers, fisher-men and forest-dependent people who are already vulnerable and food insecure. Rural communities, particularly those living in already fragile environments, face an immediate and ever-growing risk of increased crop failure, loss of livestock, and reduced availability of marine, aquaculture and forest products. They would have adverse effects on food security and livelihoods of small farmers in particular (Cline, 2010; Climate Change, 2021)

Indebtedness of Dryland Farmers

According to Satyanarayana (2020), the debt of the farmers goes on increasing year after year and the situation leads the tragic phenomenon of rural indebtedness, which has turned chronic in nature. This depends on so many factors, which influence and reinforce others. Consequently, farmers more particularly, the dryland farmers remain in debt trap, due to frequent crop failures. The main reason behind all this tragic situation is chronic poverty of dryland farmers. Their income levels as well as their asset values are too low to sustain a decent life. While loans are taken, the incomes are not generated towards repayment. Indebtedness for the small and marginal farmers from formal institutional sources is much lower than large farmers and the reverse is true in the case of informal sources. This is because the small farmers have no access to institutional loans. The dependence on money lenders is the highest for sub-marginal and marginal farmers (Dev 2014), pushing them into the debt trap by local money lenders. Small and marginal farmers of Andhra Pradesh (India) have to depend on 73% to 83% of their loans on informal sources (Dev 2014)

The dryland farmers have no access to the loans offered by public sector banks, who flatly deny any loan facility to dryland farmers on the grounds of lack of credible means of repayment by them. Thus the farmer is forced to approach private money lenders who have their eye on farmers' landed property. That is how most of the dryland farmers are turning into agricultural labourers.

Indebtedness creates psychological problems like depression, loss of interest in agriculture and frustration resulting in low concentration on agriculture. Farmer's household becomes weak and economic transactions are always not beneficial to them. Rural indebtedness tells upon the overall personality of the borrowers. Its cancerous effects spread over the family members of the indebted household (Satyanarayana, 2020). This tragic development is the main reason for increasing number of farmers' suicides.

Farmer's Income

As per report of 70th round of NSSO survey 2002-03 to 2012-13, farmers in general seldom make on an average ₹6426 income per month only from farming, which is lower than legally laid out minimum wages for unskilled workers in agri-sector and not even 20-25% of either the country's average per capita income or what a lower strata Government employee earns today (Dalwai,2018 and NSSO,2014). What is worse still, a proportion of 61% segment, out of 90 million farm house-holds with one or lower than one hectare is reported to have a net negative monthly budget (income = ₹ 4718; consumption expenditure = ₹ 5701; net balance – ₹ 983) (RangaRao, 2021)

The average monthly income from different sources including those other than farming per agricultural household in July 2018 to June 2019 is reported to be only ₹10,218, where net receipt is obtained considering 'paid out expenses' approach. This amount of income further reduces to ₹8,337 when net receipt is obtained considering both the paid out and imputed expenses. This means that the per day income of farm households is only about ₹277, which is not much different from the minimum wage rate paid under the national employment guarantee scheme (Narayanamoorthy, 2021).

If that is the situation in the case of farmer growing irrigated crops etc., the average annual income per capita of a small dryland farmer in India as on 2014 was stated to be ₹11,000 to ₹16,000 as per the data available from the Ministry of Agriculture. According to a report entitled *Situation Assessment of Agricultural Households and Land and Livestock Holdings of Households in Rural India, 2019*, released by the Ministry of Statistics and Programme Implementation, most of small and marginal agricultural households now depend on wage work. (https://theprint.in/economy/indias-small-marginal-farmers-have-essentially-become-wage-labourers-data-shows/745390/)

There is increasing skepticism among farmers about the intentions of policy-makers, and realization among the farming community that they are being short-changed. Problems of inputs, water, extension services, agri-infrastructure, marketing and lack of capacity to face risks continue to characterize the farm sector (Chandrasekhar 2019).

Rays of Hope

There is a significant potential for small farmers to sequester soil carbon if appropriate policy reforms are implemented. The importance of collective action in climate change adaptation and mitigation is being recognized. Research and practice have shown that collective action by institutions are very important for technology transfer in agriculture and natural resource management among small holders and resource dependent communities. (Dev 2014)

Small and marginal farmers can get higher incomes with diversification of their farm production through the adoption of the technologies of the Integrated Farming System (RICAREA, 2017) adding high value crops, poultry, inland fisheries, diary, animal production and agro-forestry. But, there are risks in shifting to diversification as the current support systems are more for food grains. There is a need for extending financial support from financial systems like Banks for diversification to help the small holder farmers.

Public sector led improved technologies have been helping small farmers in India. The improved wheat varieties in Punjab and technology of single cross hybrid for maize have significantly contributed in enhancing yields in small farms.

Mysore' study (2021) on horticulture shows that one of the driving forces for horticulture development in India is due to 'easy-to-fit' technologies in the system. The study highlights a number of public sector driven improved technologies. These include: (a) productivity enhancing technologies; (b) input saving technologies; (c) nutrient balancing technologies; (d) value adding technologies. The small interventions reduced crop damage, increased production and raised income both in domestic and export markets. There are also technologies regarding efficient water management.

An initiative in Andhra Pradesh based on the Self Help Group (SHG) provides another example of an institutional platform for agriculture. The Community Managed Sustainable Agriculture (CMSA) programme was initiated by the Society for Elimination of Rural Poverty (SERP) in Andhra Pradesh in 2004. The mandate of the program is to eradicate poverty and to improve 18 family livelihoods of the poor (Joy Deshmukh-Ranadive, 2014).

In recent years, there has been some form of contract arrangements in several agricultural crops such as tomatoes, potatoes, chillies, gherkin, baby corn, rose, onions, cotton, wheat, basmati rice, groundnut, flowers, and medicinal plants. There is a silent revolution in institutions regarding non-cereal foods. New production –market linkages in the food supply chain are: spot or open market transactions, agricultural co-operatives and contract farming (Joshi and Gulati, 2003).

The Way Forward

The measures and strategies for enhancement of production, incomes and livelihoods of small farmers are outlined by Dev (2014) as follows.

The National Commission for Enterprises in the Unorganized Sector (NCEUS) has recommended a special programme for marginal and small farmers. The report of NCEUS analyses the status and constraints faced by marginal and small farmers and focuses on the need for a special programme which is aimed at capacity building of these farmers, both the farm and non-farm activities. As the marginal and small farmers suffer from market failures in agriculture in terms of credit, input supplies and marketing of output, access to new technologies etc., NCUS recommended the four measures.

These are

(a) Special programmes for marginal and small farmers;

(b) Emphasis on accelerated land and water management;

(c) Credit for marginal and small farmers and

(d) Farmers' debt relief commission.

The Commission strongly advocates that a strategy for marginal and small farmers must focus on group approaches in order to benefit from the economies of scale. A focused approach can be used to incentivize the formation of farmer's groups and apex organizations and government and other can facilitate in finding solutions to problems of irrigation, inputs, procurement, markets and risk. The Commission has considered four important models for group approach in the country.

These are: Co-operatives, Producer's Companies, Farmers' groups such as those in Andhra Pradesh and SEWA (Self Employed Women's Association) Farmers' model. Cooperatives and farmers' groups on the lines of Self Help Groups (SHGs) seem to hold greater promise for expansion. It may be noted that formation of marginal and small farmers' groups on the lines of SHGs has developed under agency structure such as *'Velugu'* or *Indira Kranti Pradham*

(IKP) or CMSA mentioned above in Andhra Pradesh, *'Kudumbashree'* in Kerala and SEWA in Gujarat. Such initiatives are being developed in Tamil Nadu, West Bengal, Orissa and Madhya Pradesh as well. As the Commission mentions that the '*main lesson of these experiences is the capacity building and group formation among the poor marginal and small farmers cannot be simply seen as an extension of routine departmental activity and as one of the many activities that a programme seek to promote*" (p.39). These groups under agency-approach can be promoted where farmers' cooperatives are not operating.

The elements of special programmes advocated by NCEUS are the following

- *Promotion of Marginal and Small Farmers' Groups:* In many states groups on the lines of self-help groups (SHGs) are few. Special efforts have to be made to facilitate formation of such groups. The special programme proposes setting up of Marginal and Small Farmers' 21 Development Society (MSFDS) for the promotion, capacity building and coordination of development of marginal and small famer's groups.
- *Enabling greater access to institutional credit*: Linking Marginal and small farmer's groups to banks is an essential step towards needed credit flow to these farmers.
- *Training and capacity building*: The special programme aims at motivating and enabling marginal and small farmers to acquire skills by establishing Community Resource Centres, by promoting marginal and small farmer activists at the village, cluster and block levels.
- *Support for strengthening and creation of non-farm activities*: This aims to bridge the farm activities and non-farm activities of small holding agriculture as income from small farming is hardly sufficient to meet the basic needs of the farm households.
- *Gender-focused activities:* It is known that the share of women is increasing in agriculture. This programme aims that the farmers' groups should have adequate representation of women farmers.
- *Planning for development of Marginal and Small Farmers:* The Marginal and Small Farmer's Development Society would develop a medium term development strategy for these farmers.
- *Encourage start-up ventures by young entrepreneurs to solve problems of dryland agriculture.* For example,

1. **Kamal Kisan** a Bengaluru-based social enterprise is helping small and marginal farmers reduce labour costs with its innovative agri-equipment.
2. **Ravgo**, a Punjab based start-up is an agri-equipment rental marketplace, which aims to bring access to modern technology for small farmers who cannot afford ownership of expensive machinery.
3. **FlyBird Farm Innovations**, a Bengaluru-based start-up uses sensors in the soil to detect moisture content and control moisture management in dryland agriculture. And
4. **Ninjacart** a Bengaluru-based start-up which uses technologically driven supply chain and price discovery platforms enabling retailers and merchants to source fruits and vegetables directly from farmers (Mashelkar, 2019)

According to the *Millennium Development Goals* (Kat Henrichs, 2013) the following additional steps are absolutely necessary to enhance the economy of dryland farmers

- *Protect and preserve the natural environment*: Without a healthy natural environment where native flora and fauna live productively, long-term sustainable agricultural practices will fail. Farms must be developed in conjunction and cooperation with local ecology, not at its expense.
- *Build and maintain soil productivity*: Healthy soil is the foundation of a healthy farm and leads to increased crop yields. Rebuilding soil after intensive cultivation is necessary to maintain soil productivity. Essential soil nutrients can be replenished through techniques such as fertilization, composting, inter-planting, and crop and crop rotation.
- *Supplement programs for new-borns and their mothers:* Even with an adequate food supply, pregnant and nursing mothers and their young children have unique nutritional needs. They need more protein, folate, calcium, and iron, as well as more calories.
- *End subsidies to wealthy farmers*: One Oxfam study showed that ending subsidies to wealthy cotton farmers would do more to help poor than the amount of aid they receive now. Farm subsidies drive down the prices of grown crops, making it impossible for small-scale farmers to compete.
- *Improve food security*: This means making sure that everyone in the community, including farmers, consistently has adequate calories and nutrition. Food security can be improved in many ways, including

building food storage facilities, providing access to fuel-efficient cook-stoves, and sourcing food locally, just to name a few.

- *Switching over to more remunerative tree culture on a less fertile part of the farm:* One important activity which doesn't find application with small farmers is the tree culture of more remunerative nature giving produce with much better storability to enhance the bargaining power of small farmers. Trees like *Pongmia pinnata (Karanj)* yielding pods and oil-rich kernels could be grown on less fertile patch of the farm. Considering the increasing demand for *Pongamia* oilseed bearing pods, with a better storability, the farmer will be at an advantageous position to earn better levels of income (Prasad, 2021A)

Along with the above strategies and schemes, every effort must be made to enhance the literacy of small farmers most of who are semi-literates and illiterates, despite their traditional skills in agriculture. In addition to the foregoing, the most important step to be taken to strengthen dryland agriculture lies in orienting the strategy including "Farmer First" approach (Paroda, 2022A). In other words, the economy of the dryland farmer needs to be augmented by ensuring timely supply of quality inputs, availability of farmers' loans at low interest rate and accessibility of farmers' to markets.

The small farmers should be stimulated to adopt diversification of farm production through farming system approach, instead of going with mono-cropping or simply confining to crop production alone. Many small farmers cannot afford to possess animals like cows, buffaloes, sheep and goat etc. Such of those farmers must be granted liberal loans at reasonable interest rates by the Public sector Banks.

References

Abbo, S., Redden, R.J. and Yadav, S.S. (2007) Utilization of wild relatives. In: Yadav, S.S., Redden, R.J., Chen, W. and Sharma, B. (eds.) Chickpea Breeding and Management

Abouziena, H.F., El-Saeid, H.M. and Amin, A.A.E. (2015) Water loss by weeds: a review, International Journal of Chem. Tech. Research CODEN (USA): IJCRGG ISSN: 0974-4290 Vol.7(01):323-336,2014-2015

Adam B.S and Richard W.K. (2013). "Smith A.B. and R. Katz, 2013: U.S. Billion-dollar weather and climate disasters: Data sources, trends, accuracy and biases" (PDF). Natural Hazards. 67 (2): 387–410. doi:10.1007/s11069-013-0566-5. S2CID 30742858. Retrieved 5 November 2017.

Afrin, S. , Latif, A. , Banu, N. , Kabir, M. , Haque, S. , Ahmed, M. , Tonu, N. and Ali, M. (2017) Intercropping Empower Reduces Insect Pests and Increases Biodiversity in Agro-Ecosystem. Agricultural Sciences, 8, 1120-1134. doi: 10.4236/as.2017.810082.

Agler M. T., Ruhe J., Kroll S., Morhenn C., Kim S.-T., Weigel D., et al. (2016). Microbial hub taxa link host and abiotic factors to plant microbiome variation. PLoS. Biol. 14:e1002352. 10.1371/journal.pbio.1002352

Agrawal, P.K. (2009). Global climate change and Indian agriculture; case studies from ICAR network project, Indian Council of Agricultural Research: 148.

Agriculture Profitable, July 26-27, 2013. Rajmata Vijayraje Scindia Krishi Vishwa Vidyalaya , Gwalior, pp 98-112.

Agropedia (2010) http://agropedia.iitk.ac.in/content/weed-control-strategies-sorghum

Ahloowalia, B. and Maluszynski, M. (2001). Induced mutations–A new paradigm in plant breeding. Euphytica, 118(2): 167–173.

Ahluwalia, V.K. and Malhotra, S. (2006) Environmental Science, Anne Books India, New Delhi

AICRIPDA-Arjia (2006) Annual Progress Report. Coordinated Research Project for Dryland Agriculture, ARJIA, Maharana Pratap University of Agriculture, Udaipur, Rajasthan.

AICRPDA (2014) Annual Progress Report of All India coordinated Research project for dryland agriculture

AICRPDA (2018) All India Coordinated Research Project for Dryland agriculture

AICRPDA,(2000 to 2006), Annual Progress Report of All India Coordinated Research Project for Dryland Agriculture Central Research Institute for Dry land Agriculture, Hyderabad, India,

AICRPDA, (2006), Annual Progress Report of All India coordinated Research project for dryland agriculture Central Research Institute for Dry land Agriculture, Hyderabad, India, 63 p

AICRPDA-TSP Report- 2012-15: All India Coordinated Research Project for Dryland Agriculture, ICAR- Central Research Institute for Dryland Agriculture, Hyderabad. p. 55.

AICRPDA.(1983). Improved agronomic practices for dry land crops in India, All India Coordinated Research Project for Dryland Agriculture, Central Research Institute for Dry land Agriculture, Hyderabad, India, 63 p

AICRPDA.(2000)All India Coordinated Research Project for Dryland Agriculture, Central Research Institute for Dry land Agriculture, Hyderabad, India, Central Research Institute for Dry land Agriculture, Hyderabad, India, 63 p

Ajit, Dhyani S.K., Handa A .K. , Sridhar, K. B., Jain, A. K., Uma, Sasindran, P., Kaza, M., Sah, R., Prasad, S. M. R. and Sriram K. (2014). Carbon sequestration assessment of block plantations at JSW Steels Limited. (In) Compendium of Abstracts, 3rd World Agroforestry Congress, organized by ICAR, WAC and ISAF at Delhi, 10-13 Feb 2014, pp 354–5.

Ajit,P., Chaturvedi, P.O., Singh, R.and Singh, U.P. (2009) Biomass production in multipurpose tree species in natural grasslands under semi arid conditions. Journal of Tropical forestry 25: 11-16.

Ajit DSK, Newaj R, Handa AK, Prasad R, Alam B, Rizvi RH, Gupta G, Pandey KK, Jain A, Uma (2013) Modeling analysis of potential carbon sequestration under existing agroforestry systems in three districts of Indo-Gangetic plains in India. Agrofor Syst 87(5):1129–1146

Alagirisamy, M. (2016) Groundnut, In: Surinder Kumar Gupta (Ed) Breeding Oilseed Crops for Sutainable Production: Opportuities and Constraints, Academic Press, Elsivier, Oxford Uk, Sa Diego, USA. Pp: 89-134

Alam, F.; Kim, T.Y.; Kim, S.; Alam, S.; Pramanik, P.; Kim, P.J.; Lee, Y.B.(2019) Effect of molybdenum on nodulation, plant yield and nitrogen uptake in hairy vetch (Vicia villosa Roth). J. Soil Sci. Plant Nutr. 61, 1–12.

Alamgir Hossain, Md and Sarder N. Uddin, S.N. (2011) Mechanisms of waterlogging tolerance in wheat: Morphological and metabolic adaptations under hypoxia or anoxia. Australian Journal of Crop Science, October 2011

Albrecht, A. and Kandji, S.T. (2003). Carbon sequestration in tropical agroforestry systems. Agriculture, Ecosystems and Environment 99: 15–27

Ali, M. and Gupta, S. 2012. Carrying capacity of Indian agriculture: pulse crops. Current Science 102(6): 874-881.

Ali, M. O., Alam, M. J., Alam, M. S., Islam, M. A. and Shahin-uz-zaman, M. (2007). Study on Mixed Cropping Mungbean with Sesame at Different Seeding Rates. Int. J. Sustain. Crop Prod. 2(5):74-77

Allison FE. 1973 Soil Organic Matter and its Role in Crop Production. New York: Elsevier;

Alok K. Sikka, Adlul Islam and Rao, K.V. (2017) Climate-Smart Land and Water Management for Sustainable Agriculture† Special Issue Paper, Irrigation and Drainage, https://doi.org/10.1002/ird.2162

Anita Kumawat, Devideen Yadav, Kala Samadharmam and Ittyamkandath Rashmi (2020) Soil and Water Conservation Measures for Agricultural Sustainability, FROM THE EDITED VOLUME Soil Moisture Importance. Edit. Ram Swaroop Meena and Rahul Datta

Annicchiarico, P. (2003). Breeding white clover for increased ability to compete with associated grasses. J. Agric. Sci. 140, 255–266. doi: 10.1017/S0021859603003198

Annicchiarico, P., and Filippi, L. (2007). A field pea ideotype for organic systems of northern Italy. J. Crop Improv. 20, 193–203. doi: 10.1300/J411v20n01_11

Annicchiarico, P., Collins, R. P., De Ron, A. M., Firmat, C., Litrico, I., and Hauggaard-Nielsen, H. (2019). Do we need specific breeding for legume-based mixtures? Adv. Agron. 157, 141–215. doi: 10.1016/bs.agron.2019.04.001

Annual Report (2006). Central Research Institute for Dryland Agriculture, Hyderabad

Anonymous (2007) Schemes, Programmes and Missions promoting agricultural development and farmer's welfare. MoA & FW. IPCC. 2007. Climate Change 2007: Synthesis Report.

Contribution of Working Group I, II and III to the Fourth Assessment Report of the Intergovernmental Panel on Climate Change. Fourth Assessment Report. Cambridge University Press, Cambridge, United Kingdom and New York, NY, USA

Anonymous (2012) Deforestation and Its Extreme Effect on Global Warming, Environment: Scientific American, November 13, 2012

Anonymous (2014) "Preparing for a poor monsoon". The Hindu. 26 April 2014. Retrieved 27 April 2014.

Anonymous (2014A) U.S. and African Leaders' Meet at a Leaders Summit for Food Security and Climate change at the National Academy of Sciences in Washington, D.C., on August 4, 2014.

Anonymous (2020) Weed Management, Knowledge Platform on Pulse Crops ICAR-Indian Institute of Pulses'Research, Kanpur (UP)

Anonymous (2021) Department of Primary Industries and Regional Development Govt. of Western Australia https://www.agric.wa.gov.au/soil-salinity/salinity-tolerance-plants-agriculture-and-revegetation-western-australia

Anonymous (2022) Mother Forest should prosper (Adavithalli, neevu challagundaala in Telugu language) Pages 9-11 Eenaadu, Sunday Weekly, June 05, 2022

APCAS. 2010. Asia and Pacific Commission on Agricultural Statistics, 23rd Session (APCAS10-28), 26-30 April 2010, Siem Reap, Cambodia.

Appiah, Bernard. (2012) Intercropping (Agroforestry) 'boosts maize yields by 50 per cent'. Flickr/M. DeFreese/CIMMYT. : http://www.scidev.net/en/sub-suharan-africa/news/intercropping-boosts-maize-yields-by-50-per-cent-.html)

Arachis pintoi . In: Kerridge, P.C. and Hardy,B. (eds) Biology and Agronomy of Forage Arachis CIAT, Cali, Columbia, pp.53–70

Arunahalam, A., Khan, M.L. and Arunachalam, K. (2002) Balancing traditional Jhum cultivation with modern agro-forestry in eastern Himalaya – A biodiversity hotspot. Curr. Sc.,83:117-118

Ashton, A.R. and Fischer, R.A. (1986) The effect of conventional cultivation, directdrilling, and crop residues on soil temperatures during the early growth of wheat at Murrumbaeman, New South Wales. Australian Journal of Soil Research.24:49-60.

Astaburuaga G., Ricardo (2004), "El agua en las zonas áridas de Chile", ARQ (Spanish), 57: 68–73

Atkinson, N. J.; Urwin, P. E. (2012). "The interaction of plant biotic and abiotic stresses: from genes to the field". Journal of Experimental Botany. 63 (10): 3523–3543. doi:10.1093/jxb/ers100. PMID 22467407. (22)

Awasthi, O. P. and Singh, I.S. (2010) Effect of ber and pomegranate plantation on soil nutrient status of typic torripsamments. Indian Journal of Horticulture 67(Special Issue): 138–142.

Ayyasamy, M. (2004). Tree borne oilseeds in agroforestry system. Strategies for Improvement and Utilization of Tree Borne Oilseeds. ICAR - Winter School (M. Paramathma, KT Parthiban, KS Neelakantan Eds). Forest College and Research Institute, Mettupalayam 641301. pp 110-117

Baishya, A., Kalita, M.C., Mazumdar, D.K., Hazarika, J.P. and Ahmed, S. (2007) Characterization of Farming Systems in Borpeta and Kamrup districts of Lower Brahmaputra Valley zone of Assam. J. Farming Systems Research and Development 13(2):168-175

Bajracharya, S.R., Mool, P.K. and Shrestha, B.R. (2007). Impact of Climate Change on Himalayan Glaciers and Glacial Lakes- Case Studies on GLOF and Associated Hazards in Nepal and Bhutan, International Centre for Integrated Mountain Development (ICIMD), Kathmandu.

Bal, S.K., Prasad, J.V.N.S and Singh, V.K. (2022). Heat wave 2022 Causes, impacts and way forward for Indian Agriculture. Technical Bulletin No. ICAR/CRIDA/ TB/01/2022, ICAR-Central Research Institute for Dryland Agriculture, Hyderabad,Telangana,India P.50

Bal, S.K., Sandeep, V.M., Vijaya Kumar, P., Subba Rao, A.V.M., Pramod, V.P., Srinivasa Rao, Ch., Singh, N.P., Manikandan, N., Bhaskar, S. (2022). Assessing impact of dry spells on the principal rainfed crops in major dryland regions of India. Agric. For. Meteorol. 313, 108768. https://doi.org/10.1016/j.agrformet.2021.108768.

Bal, S.K. and Minhas, P.S. (2017). Atmospheric Stressors: Challenges and Coping Strategies, In: P.S. Minhas et al.(eds) Abiotic Stress Management for Resilient Agriculture, Springers Nature Singapore Pte. Ltd., pp.9-50. DOI10.1007/978-981-10-5744-1

Balde, A.B.; Scope, L.E.; Affholder, F.; Corbeels, M.; DaSilva, F.A.M.; Xavier, J.H.V.; Wery, J.(2011) Agronomic performance of no-tillage relay intercropping with maize under smallholder conditions in Central Brazil. Field Crop Res. 124, 240–251

Balloli, SS Rajeshwar Rao,G Girija Veni, Visha Kumari ,V.and Osman,M (2018) Carbon Sequestration Potential in Agroforestry Systems. Agroforestry: Opportunities for Enhancing Resilience to Climate Change in Rainfed Areas.ICAR - Central Research Institute for Dryland Agriculture National Innovations in Climate Resilient Agriculture Hyderabad - 500 059, India

Banerjee, H., Dhara, P.K., Mazumdar, D. and Alipatra, A (2013) Development of Tree borne Oilseeds (TBOs) Plantation and Scope of Intercropping on degraded lands. Indian J. Agric. Res.., 47 (5): 453 – 456.

Banziger, M., Edmeades, G.O., Beck, D. and Bellon, M. (2000) Breeding for drought and nitrogen stress tolerance in maize. From theory to practice. CIMMYT, Mexico

Bapuji Rao, B., Santhibhushan Chowdary, P., Sandeep, V.M. Pramod, V.P and Rao, V.U.M (2015) Spatial analysis of the sensitivity of wheat yields to temperature in India. Agricultural and Forest Meteorology 200:192-202. 15 January 2015

Bapuji Rao B., Santhibhushan Chowdary, V.M. Sandeep, V.U.M. Rao and Venkateswarlu, B.(2014). Rising minimum temperature trends over India in recent decades: Implications for agricultural production. Global and planetary change, Vol. 117: 1-8.

Barik, K.C., Mohanty, D. Nath, S.K., Sahoo, S.K. aand Soren, L. (2010) Assessment of Integrated Faming System in Rainfed ecosystem in Deogarh district of Odisha. Poceedings of the Symposium on Resource Management Approaches towards Livelihood Security, Dec.2-4, 2010 Bangalore

Barillot, R., Escobar-Gutiérrez, A. J., Fournier, C., Huynh, P., and Combes, D. (2014). Assessing the effects of architectural variations on light partitioning within virtual wheat–pea mixtures. Ann. Bot. 114, 725–737. doi: 10.1093/aob/mcu099

Bass, S., Roe, D. and Smith, J. (2010). Look Both Ways: Mainstreaming Biodiversity and Poverty Reduction. Briefing Paper Series, IIED, London.

Batjes, N.H. (1996). Total carbon and nitrogen in the soils of the world. European Journal of Soil Science, 4

BBC (2004) "BBC NEWS - Asia-Pacific - Kazakh lake 'could dry up'". bbc.co.uk. 2004-01-15

Behera, U.K.,Yates, C.M., Kebreab, E. and France, J. (2008) Farming system methodology for efficient resource management at the farm level: An Indian Perspective. J. Agric. Sci. (Cantab)146:493-505.

Behera, U.K and Mahapatra, I.C. (1999) Income and employment of small and marginal farmers through integrated farming systems. Indian Journal of Agronomy 44(3):431-439

Bekavac, G., Stojakovic, M., Jockovic, D., Bocanski, J. and Purar, B. (1998) Path analysis of stay-green trait in maize. Cereal Research Communications 26: 161–167

Benjamin I.C., Justin S.M and Kevin J.A. (2018). "Climate Change and Drought: From Past to Future". Current Climate Change Reports. 4 (2): 164–179. doi.10.1007/s40641-018-0093-2. ISSN 2198-6061. S2CID 53624756.

Bertrand, A.R. (1966) Water conservation through improved practices. Pages 07-235 In: Plant Environment and efficient water use (Pierre, W.H., Kirkham, D. , Pesek, J. and Shaw, R.V. Eds.) Madison, Wisconsin, USA. American Society of Agronomy.

Bhadwal, S., Kelkar, U. and Bhandari P.M.(2007). "Impact on AgricultureHelp Reduce Vulnerability", in The Hindu Survey of the Environment, The Hindu, Special Issue,

Bhat, B.K., Dixit, S.K. and Darji, V.B. (2010) Monetary Evaluation of Sesame Based Intercropping Systems. Indian Journal of Agricultural Reseach,44(2)

Bhati, T.K. and Joshi, N.L. (2007). Farming systems for the sustainable agriculture in Indian Arid zone. In Dryland Ecosystem: Indian Perspective (Vittal KPR, Srivastava RL, Joshi NL, Kar A, Tewari VP and Kathju S Eds.), Central Arid Zone Research Institute and Arid Forest Research Institute, Jodhpur, pp 35-52

Bhatt, B.P. and Misra, L.K. (2003) Production potential and cost-benefit analysis of agri-horticultural and agro-forestry systems in Northeast India. J. Sustain. Agric.22:99-108

Bhatta, R., Saravanan, M., Baruah, L. and Prasad, C.S. (2015). Effects of graded levels of tannin-containing tropical tree leaves on in vitro rumen fermentation, total protozoa and methane production, Journal of Applied Microbiology, 118: 557- 564

Bhattacharya, R.C. and Singh, N.K. (2003) Transgenic for abiotic stress tolerance. Proceedings of the National Seminar in Oilseeds for attaining self-reliance in vegetable oils (Eds. Mangal Rai, Harvir Singh and D.M. Hegde) January 28-30. Indian Society of Oilseeds Research, Directorate of Oilseeds Research, Hyderabad, India. Pages 319-326.

Bhattacharyya R, Ghosh BN, Mishra PK, Mandal B, Rao CS, Sarkar D, et al. (2015) Soil degradation in India: Challenges and potential solutions. Sustainability. 7(4):3528-3570

Bheemaiah G., Subrahmanyam, M.V.R. and Ismail, S.(1992) Performance of arable crops with Acacia albida in drylands. Annals of Arid zone 31(4):303-304

Bhimaya, C.P, Kaul, R.N. and Gangul,i B.N. (1964). Studies on lopping intensities of Prosopis spicigera. Indian Forester. 90: 19-23.

Bidinger, F. R., Chandra, S., and Mahalakshmi, V. (2000). "Genetic improvement of tolerance to terminal drought stress in pearl millet (Pennisetum glaucum (L) R Br)," in Molecular Approaches for the Genetic Improvement of Cereals for Stable Production in Water-Limited Environments. A Strategic Planning Workshop held at CIMMYT, El Batan, Mexico, eds J.-M. Ribaut and D. Polland (Mexico: CIMMYT), 59–63.

Bieler, P., Fussel, L.K. and Bidinger, F.R. (1993) Grain growth of Pennisetum glaucum (L.) R. Br. under well-watered and drought-stressed conditions. Field Crops Research 31: 41–54

Birthal, P.S., Jha, A.K., Tiongco, M.M. and Narrod, C. (2008), "Improving Farm to Market Linkages through contract farming: A Case Study of Small Holder Dairying in India", IFPRI discussion paper 00814. Washington, D.C. B

Birthal, P.S. and Rao, P.P. (2002) Economic contributions of livestock sub-sector in India. IN: Technology Options for Sustainable Livestock Production in India. Proceedings of the ICAR-ICRISAT Collaborative Workshop on documentation, adoption and impact of Livestock Technologies in mixed Farming Systems in India (Eds. Birthal, P.S. and Rao, P.P) Jan.18-19, 2001 at ICRISAT India. pp12-19

Blanco H, Lal R. Principles of Soil Conservation and Management. Dordrecht: Springer; 2008. pp. 167-169

Blum, A., Mayer, J. and Gozlan, G. 1983.Associations between plant production and some physiological components of drought resistance in wheat. Plant Cell and Environment.6:219-225

Blum, A. and Sullivan, C.Y.1986. The comparative resistance of land races of sorghum and millet from dry and humid regions. Annals of Botany 57:835-846

Bohnert, H.J., Nelson, D.E., and Jensen, R.G. (1995) Adaptation to environmental stresses. Plant Cell, 7:1099

Bohra. H.C. and Ghosh. P.K. (1980). Nutritive value and digestibility of leaves of Khejri. In: Khejri (Prosopis cineraria) in the Indian desert-its role in agroforestry. (Eds. H.S. Mann and S.K. Saxena). Monograph No. 11, CAZRI, Jodhpur. pp 45-47

Bond, J.J., Army, T.J. and Lehman, O.R. (1964) Row spacing, plant population and moisture supply as factors in dryland grain sorghum production. Agronomy Journal 56: 3-6

Borland, A. M., Griffiths, H., Hartwell, J. and Smith, J. A. C. (2009) Exploiting the potential of plants with crassulacean acid metabolism for bioenergy production on marginal lands. J. Exp. Bot. 60, 2879–2896 (2009).

Borojevic, S. (1990). Principles and Methods of Plant Breeding. Elsevier, Amsterdam. ISBN 0-444-98832-7

Boyer J. S. (1982). Science 218: 443-448

Brahic, Catherine (2007) Sunshade› for global warming could cause drought 2 August 2007 New Scientist,

Branch, W. D. (2002). Variability among advanced gamma irradiation induced large-seeded mutant breeding lines in the 'Georgia Brown' peanut cultivar. Plant Breeding, 121: 275.

Brelivet, S. (2006). Wheat under the trees. South East NSW Private Forestry Newsletter, 4p.

Bresler E. (1987). Application of a conceptual model to irrigation water requirement and salt tolerance of crops. Soil Sci. Soc. Am. J. 51: 788-793.

Brim, C.A. 1966. A modified pedigree method of selection in soybean. Crop Sci.6:220

Brion, G. (2014) Controlling Pests with Plants: The power of intercropping, UVM Food Feed, Sustainable Food Systems &The University of Vermont, USA.

Brooker, R.W., Bennett, A. E., Cong, W.F., et al., (2015) Improving intercropping: a synthesis of research in agronomy, plant physiology and ecology. New Phytologist, 206(1):107-117. DOI: https://doi.org/10.1111/nph.13132., Zurich Open Repository and Archive University of Zurich Main Library Strickhofstrasse 39 CH-8057 Zurich www.zora.uzh.ch

Brovkin, V., Sitch, S., Von Bloh, W., Claussen,M., Bauer, E. and Cramer, W. (2004) Role of land cover changes for atmospheric CO2 increase and climate change during the last 150 years, Global Change Biology, 05 July 2004, https://doi.org/10.1111/j.1365-2486.2004.00812.x

Bruke, I., Pittman, D. and Tautges, N. (2014) Inter-row cultivation and intercropping for organic transition in dryland crop production systems Project ID: 120, Progress Reports 2012 & 2014, CSANR-Center for Sustaining Agriculture and Natural Resources, Washington State University, USA.

Buchanan F (2020) A Journey from Madras through the Countries of Mysore, Canara and Malabar, vol II. Blumar & Co, London

Bukhari YM (1998). Tree-Root Influence on Soil Physical Conditions, Seedling Establishment and Natural Thinning of Acacia seyal Var.Seyal On Clays of Central Sudan. Agroforestry Systems 42: 33–43,1998

Bunting, A.H. and Curtis, D.L. (1970). Local adaptation of sorghum varieties in northern Nigeria. Samaru Research Bulletin no. 106, Samaru, Zaria, Nigeria, Ahmadu Belo University, Nigeria.

Bunting, A.H. and Kassam, A.H. (1988) Principles of crop water use, dry matter production, and dry matter partitioning that governs choices of crops and systems. Pages: 43-61. In: In: Drought Research Priorities for the Dryland Tropics. (Eds: Bidinger, F.R. and Johansen, C.) ICRISAT (International Crops Research Institute for the Semi-Arid Tropics) Patancheru, India. 219pp.Millennium Development Goals

Bybee-Finley, K.A.; Ryan, M.R (2018). Advancing intercropping research and practices in industrialized agricultural landscapes. Agriculture 2018, 8: 80.

Caviedes (2001), p. 117 (Cross reference from Wikipedia 'Drought in India.)

Ceballos, F., Kannan, S. and Kramer, B. (2021) Crop prices, farm incomes, and food security during the COVID-19 pandemic in India: Phone-based producer survey evidence from Haryana State. Agric Econ. 2021 May; 52(3): 525–542. Published online 2021 May 25. doi: 10.1111/agec.12633 PMC8207062

CGIAR (2022) Grain legumes and dryland Cereals: Dryland Intercropping of Lentil and Quinoa is Sustainable and Profitable:

Ch. Srinivasarao , Deshpande, A.N. , Venkateswarlu,B., Rattan Lal, Anil Kumar Singh , Sumanta Kundu, Vittal K.P.R.,, Mishra, P.K., Prasad, J.V.N.S. , Mandal, U.K. and Sharma, K.L. (2012)Grain yield and carbon sequestration potential of post monsoon sorghum cultivation in Vertisols in the semi arid tropics of central India Geoderma 175-176 (2012) 90–97

Chand, R., Prasanna, P. A. L. and Singh, A. (2011). Farm size and productivity: Understanding the strengths of smallholders and improving their livelihoods, Economic and Political Weekly, 46(26 & 27).

Chandrasekhar, G. (2019) Farm crisis: Short-term palliatives are futile, BusinessLine. The Hindu. January 06, 2019

Chandrasekhar, S. and Mehrotra, N. (2016) Doubling Farmers' Income by 2022. What would It Take? Economic and Political Weekly 46 (26&27):5-11

Channelnewsasia.com (2010) is driest month for S'pore since records began in 1869". 3 March 2010. Archived from the original on 3 March 2010. Retrieved 5 November 2017

Chapter 5, Advances in Agroforestry, Springer India. Pages 117 to 154 DOI: 10.1007/978-81-322-1662-9_5,

Chaudary, M.R., Gajri, P.R., Prihar, S.S. and Khera, R. (1985) Effect of deep tillage on soil physical properties and maize yields on coarse textured soils. Soil and Tillage Research, 6: 31-44.

Chauhan, S.K., Sharma, S.C., Beri, V., Yadav, S. and Gupta, N. (2010). Yield and carbon sequestration potential of wheat (Triticum aestivum) – poplar (Populus deltoides) based agri-silvicultural system Indian. Journal of Agricultural Sciences. 80: 129- 35.

Chauhan, Sanjeev, Singh, Ateetpal. Sikka, S S., Tiwana, U., Sharma, Rajni and Saralch, Harmeet. (2013). Yield and quality assessment of annual and perennial fodder intercrops in Leucaena alley farming system. Range Management and Agroforestry. 35. 230-23

Chen, Q., Lu, D., Keller, M., dos-Santos, M.N., Bolfe, E.L., Feng, Y. and Wang, C. (2016). Modeling and Mapping Agroforestry aboveground biomass in the Brazilian Amazon using airborne lidar data. Remote Sensing 8(1): 21. doi:10.3390/rs8010021.

Choi, K.J., Chin, M.S., Park, K.Y., Lee, H.S., Seo, J.H. and Song, D.Y. (1995) Heterosis and heritability of stay-green characters. Maize Genetics Cooperative Newsletter 69: 122–123.

Chopra, V.L. and Rana, R.S. (1994) Genetic Diversity and Sustainability, In: Stressed Ecosystems and Sustainable Agriculture (Eds. Virmani, S.M., Katyal. J.., Eswaran, H. and Abrol, I.P.), Oxford &IBH Publishing Co. Pvt. Ltd. New Delhi, Bombay, Culcutta. Pages 173-179

Choudhary, V.K.; Choudhury, B.U. A staggered maize–legume intercrop arrangement influences yield, weed smothering and nutrient balance in the eastern himalayan region of india. Exp. Agric. 2016, 54, 181–200

Chowdhury, Q.M. and Monzur, K. 2016. "Impact Of Climate Change On Livestock In Bangladesh: A Review Of What We Know And What We Need To Know" (http://ajaset.e-palli.com/wpcontent/uploads/2013/12/impactof-climate-change-on-livestock-inbangladesh-a-review-of-what-weknow-and-what-we-need-toknow.pdf). American Journal of Agricultural Science Engineering and Technology. 3 (2): 18–25 – via e-palli.

Ciampitti I A, García F O, Picone L I and Rubio G. 2011. Soil carbon and phosphorus pools in field crop rotations in Pampean soils of Argentina. Soil Science Society of America Journal 75(2): 1–1

Clarke, L., Jiang, K., Akimoto, et al., (2014) Assessing Transformation Pathways. In: Edenhofer, O., Pichs-Madruga, R., Sokona, et al. (eds.) Climate Change 2014: Mitigation of Climate Change. Contribution of Working Group III to the Fifth Assessment Report of the Intergovernmental Panel on Climate Change. Cambridge, United Kingdom and New York, NY, USA: Cambridge University Press.

Climate Change (2021) The Physical Science Basis. Working Group I contribution to the WGI Sixth Assessment Report of the Intergovernmental Panel on Climate Change / Summary for Policymakers» (PDF)

Cline, William, R. (2010) Global Warming and Agriculture- Impact Estimates by Country. Center for Global Development. Peterson Institute for International Economics. Washington DC USA. Viva Books, New Delhi, Hyderabad, 186 pp.

Cole, T.G. and Ewel, J.J. (2006). Allometric equations for four valuable tropical tree species. For. Ecol. Manage. 229: 351–360

Cook, R.J. (1990) Diseases Caused by Root-Infecting Pathogens in Dryland Agriculture In: Dryland Agriculture, Strategies for Sustainability (Eds. Singh R.P., Par,J.F. and Stewart, B.A.) Advances in Soil Science,13. Springer-Verlag ISBN : 978-1-4613-8984-2

Cook, R.J. (2000) Advances in plant health management in twentieth century. Annual Review of Phytopathology, 38:95-116

Costanzo, A., and Bàrberi, P. (2014). Functional agrobiodiversity and agroecosystem services in sustainable wheat production. A review. Agron. Sustain. Dev. 34, 327–348. doi: 10.1007/s13593-013-0178-1

CRIDA (1996) Annual Report. Central Research Institute for Dryland Agriculture (CRIDA), Hyderabad, India

CRIDA (1997). Annual Progress Report of 1996-97, published by ICAR, Central Research Institute for Dryland Agriculture, Hyderabad, India.

CRIDA (1997) Central Research Institute for Dryland Agriculture,1997, CRIDA Perspective plan- Vision 2020, CRIDA,Hyderabad, India

CRIDA (2005) National Agricultural Technology Project-Rainfed Agro-ecosystem production system: Complete Report for the years 1994-2004. AgroecosystemDirectorate, CRIDA, Hyderabad, p 202

CRIDA (2007) Perspective Plan, Vision 2025. Central Research Institute for Dryland Agriculture, Hyderabad, Andhra Pradesh (Currently Telangana State)

CRIDA (2009-2010) Annual report. 2009-2010. NPCC Annual Report.

CRIDA (2010) Annual Reports cobined for the years from 1993 to 2020. Central Research Institute for Dryland Agriculture (CRIDA), Hyderabad,India

CRIDA (2011). Annual Report for the year 2010-2011, Central Research Institute for Dryland Agriculture (CRIDA) Hyderabad.162pp

CRIDA (2013). Annual Report, Central Research Institute for Dryland Agriculture (CRIDA) Hyderabad.162pp

CRIDA (2015) Eldoscope Electronic document. All India Co-ordinated Research Project for Dry land Agriculture (AICRPDA), Central Research institute for Dry land Agriculture (CRIDA), Hyderabad, India

CRIDA. (2002). Annual Report 2001-2002, Central Research Institute for Dryland Agriculture, Santoshnagar, Hyderabad. pp 111

Crispino Lobo, N., Chattopadhyay and K V Rao (2017) Integrated Agrometeorological Services. Making Smallholder Farming Climate-smart, Economic and Political Weekly, Vol. 52, Issue No. 1, 07Jan2017.

Cross, Olivia (2022) Risk of tighter agricultural export restrictions growing. Capital Economics, May 23, 2022 https://www.capitaleconomics.com/publications/commodities-overview

Culbertson, J.O. (1954). Breeding flax. Adv. Agron. 6: 174–178.

Curtis, D.L.1968. The relation between the date of heading of Nigerian sorghums and duration of the growing season. Journal of Applied Ecology, 5: 215-226

Cushman, J. C., Davis, S. C., Yang, X. & Borland, A. M (2015). Development and use of bioenergy feed-stocks for semi-arid and arid lands. J. Exp. Bot. 66: 4177–4193 (2015).

DAC. (2014). Agricultural Statistics at a Glance 2014. Department of Agriculture & Co-operation, Government of India, Ministry of Agriculture. Oxford University Press, New Delhi, India. 452 p.

Dalwai, Ashok. (2018). Report of the committee on doubling farmers' income, vol x 111 Structural reforms and governance framework "Strengthening the institutions, infrastructure and markets that govern agricultural growth". Department of Agriculture Cooperation and Farmers' Welfare, Ministry

Darthiya, M. and Subbalakshmi Lokanadhan (2020) Medicinal and Aromatic Plants in Drylands for Future Livelihood Security, International Journal of Agriculture Innovations and Research 9 (1) ISSN (Online) 2319-1473

Darwin, R.F. and Kennedy, D. (2000) Economic effects ofCO2 fertilization of crops. Transforming Yield into changes in supply. Environmental Modelling and Assessment 5, no.3: 157-168

Das, I.K. (2016). Diseases of sorghum. In: Dubey, H.C., Aggarwal, R., Patro, T.S.S.K., Sharma, P. (Eds.), Diseases of Field Crops and Their Management. Today & Tomorrow's Printers & Publishers, New Delhi, pp. 131e179.

Das, M., Nath, P.C., Reang, D., Nath, A.J. and Das, A.K. (2020) Tree Diversity and the Improved Estimate of Carbon Storage for Traditional Agroforestry Systems in North East India. Applied Ecology and Environmental Sciences, 8 (4) 154-159 DOI: 10.12691/aees-8-4-2

Dash, S., Chakravarty, A.K., Singh, A., Upadhyay, A., Singh, M. and Yousuf, S. 2016. Effect of heat stress on reproductive performances of dairy cattle and buffaloes: A review. Vet. World. 9(3): 235-244. FAO. 2012

Dass, A. and Sudhishir, S. (2010) Intercropping in finger-millet (Eleusine coracana) with pulses for enhanced productivity, resource conservation and soil fertility in uplands of southern Orissa. Indian J. Agron. 55:89–94.

Davies, F.S. 1986. Horticulture Review 8:120–129

Davies, J. et al. (2012). Conserving Dryland Biodiversity. iucn.org/drylands, IUCN's work on drylands

de la Paix Mupenzi, J., Li, L., Ge, J. et al. (2012)Water losses in arid and semi-arid zone: Evaporation, evapotranspiration and seepage. J. Mt. Sci. 9, 256–261 (2012). https://doi.org/10.1007/s11629-012-2186-z

Delauney, A.J. and Verma, D.P.S. (1993) Proline biosynthesis and osmoregulation in plants. Plant Journal, 4:215-223

Deng, W., Dong, X.F., Tong, J.M. and Zhang, Q. (2012). The probiotic Bacillus licheniformis ameliorates heat stress-induced impairment of egg production, gut morphology, and intestinal mucosal immunity in laying hens. Poultry Science. 91: 575–582

De Stefano A and Jacobson MG. (2017). Soil carbon sequestration in agroforestry systems: a meta-analysis. Agroforest Syst DOI 10.1007/s10457-017-0147-9

Dev, M. S. (2008) Inclusive growth in India: Agriculture, poverty and human development. Oxford University Press; 2008.

Dev, M. S. (2014) Small Farmers in India: Challenges and Opportunities. www.igidr.ac.in/pdf/publication/WO-P-2012 -2014.pdf .

Dhyani, S.K, Newaj, R, and Sharma AR. (2009). Agroforestry: its relation with agronomy, Challenges and opportunities. Indian Journal of Agronomy. 54: 249-266

Dhyani, S.K. (2012). Agroforestry interventions in India: Focus on environmental services and livelihood security. Indian J. Agroforestry. 13: 1-9.

Dhyani, S. K. (2014). National Agroforestry Policy 2014 and the need for area estimation under agroforestry. Current Science 107(1): 9–10

Dhyani,S.K., Asha-ram and Inder dev (2016) Potential of agroforestry systems in carbon sequestration in India Indian Journal of Agricultural Sciences 86 (9): 1103–12, September 2016/Review Articl

Dhyani, S.K., Vimala Devi, S. and Handa, A.K. (2015): A bulletin on Tree Borne Oilseeds for Oil and Biofuels. ICAR-CAFRI, Jhansi. India.

Dhyani, S K, Handa, A .K and Uma. 2013. Area under agroforestry in India: An Assessment for Present Status and Future Perspective. Indian Journal of Agroforestry 15(1): 1–11

Diaz, D., Hamilton, K., Johnson, E., (2011) State of the Forest Carbon Markets (2011): From Canopy to Currency. Ecosystem Marketplace Report, Forest Trends, Washington, DC, 2011.

Dillon, J.L. and J.B. Hardaker (1993). Farm Management Research for Small Farmer Development, FAO Farm Systems Management Series No. 6, Food and Agriculture Organization of the United Nations, Rome

Dimelu MU, Ogbonna SE, Enwelu IA. (2013) Soil conservation practices among arable crop farmers In Enugu–north agricultural zone, Nigeria: Implications for climate change. Journal of Agricultural Extension.;17(1):184-196

Dixon, R.K. (1995) Agroforestry systems: sources or sinks of greenhouse gases? Agrofor Syst 31:99–11

Dixon RK, Houghton RA, Solomon AM and Treler MC. (1994). Carbon pools and fluxes of global ecosystems, Science. 263: 185-190

Doraiswamy P.C. and Rsenberg, N.J. (1974) Reflectance induced modification in soybean canopy radiation. I. Preliminary tests with kaolinite reflectance. Agronomy Journal, 66 : 224-228

Dore, A.J., Mousavi-Baygi, M., Smith, R.I., Hall, J.,Fowler, D. Choularton, T.W. (2006). "A model of annual orographic precipitation and acid deposition and its application to Snowdonia".

Atmospheric Environment. 40 (18): 3316–3326. Bibcode:2006AtmEn..40.3316D. doi:10.1016/j.atmosenv.2006.01.043.

Drew, M.C. 1979. Plant responses to anaerobic conditions in soil and solution culture. Current Advances in Plant Science. 36: 1-14

Duncan, R.R., Blockholt, A. J. and Miller, F.R. 1981. Descriptive comparison of senescent and non-senescent sorghum genotypes. Agronomy Journal, 73: 849-853

Dwivedi, R.P., R.K. Tewari, K. Kareemulla, O.P. Chaturvedi and P. Rai. 2007. Agri horticultural system for household livelihood - A case study. Indian Res. J. Ext. Edu., 7 (1) : 22-26

Dwivedi, S.L., Nigam, S.N., Nageswara Rao, R.C. et al., (1996) Effect of drought on oil, fatty acids and protein content of groundnut (Arachis hypogaea, L.) seeds. Field Crops Res. 48: 125-133

Eck, H.V. and Unger, P.W. (1985) Soil profile modifications for increasing crop production. Advances in Soil Science 1: 65-100.

Ejeta, G. (2007) Breeding for Striga resistance in sorghum: exploitation of an intricate host–parasite biology Crop Sci., 47 (S3) (2007), pp. S-216-S-227, 10.2135/cropsci2007.04.0011IPBS

El-Swaify, S.A., Pathak, P., Rego, T.J. and Singh, S. (1985) Soil management for optimised productivity under rain-fed conditions in Semi-Arid Tropics. Advances in Soil Science 1: 1-64

Emerson, B.N. and Minor, H.C.1979. Response of soybean to high temperature during germination. Crop Science 19: 553-556

Emery, S.E., Anderson, P., Carlsson, G., Hanna Friberg, H., Larsson, M.C. Ann-Charlotte Wallenhammar, A.C. and Lundin,O. (2021) The Potential of Intercropping for Multifunctional Crop Protection in Oilseed Rape *(Brassica napus* L.), Frontiers in Agronomy, 3 .,2021 | https://doi.org/10.3389/fagro.2021.782686

Emmanouil N. Anagnostou (2004). "A convective/stratiform precipitation classification algorithm for volume scanning weather radar observations". Meteorological Applications. 11 (4): 291-300. Bibcode:2004MeApp..11..291A. doi:10.1017/S1350482704001409.

Envistats India (2018) (Website: http://mospi.nic.in/publication/envistats-india-2018)]

EOS (2020). Earth Observing System: Intercropping: Ergonomic and Efficient Farming, https://eos.com/

Ercisli,S. A. Esitken,A., Turkkal,C. and Orhan, E.(2005) The allelopathic effects of juglone and walnut leaf extracts on yield, growth, chemical and PNE compositions of strawberry cv. Fern PLANT SOIL ENVIRON., 51, 2005 (6): 283–287

Evans, L.T. (1983) Raising the yield potential: by selection or design. Pages 371-389 In Genetic Engineering of Plants (Eds: Kosuge, T., Meredith, C.P. and Hollaender, A.) Plenum Press, New York, USA.

Fand, B.B., Tonnang, H.E.Z., Bal, S.K., Dhawan, A.K. (2018). Shift in the Manifestations of Insect Pests Under Predicted Climatic Change Scenarios: Key Challenges and Adaptation Strategies. In: Bal, S., Mukherjee, J., Choudhury, B., Dhawan, A. (eds) Advances in Crop Environment Interaction. Springer, Singapore.. DOI: https://doi.org/10.1007/978-981-13-1861-0_15

Fang, Y. and Xiong, L. (2014) General mechanisms of drought response and their application in drought resistance improvement in plants, Cell Mol. Life Sci. 2015 72(4):673-89. doi: 10.1007/s00018-014-1767-0.

FAO (1995). https://www.fao.org/3/w7365e/w7365e05.htm

FAO (2007). http://www.fao.org/newsroom/en/news/2007/1000654/index.html

FAO (2015) Climate change and food security: risks and responses. Food and Agriculture Organization Of The United Nations. 2015. ISBN 978-92-5-108998-9 https://www.fao.org/3/i5188e/I5188E.pdf

FAO. (1984) Guidelines: Land Evaluation for Rainfed Agriculture. Rome: FAO Soils Bulletin; 19 p. 52

Faroda, A.S., Joshi, N.L., Singh, R. and Saxena, A. (2007). Resource management for sustainable crop production in arid zone – A review. Indian Journal of Agronomy 52 (3): 181–193

Fery, R.L.1980. Genetics of Vigna, Horticulture Reviews 2:311-394

Finch S and Collier RH (2000) Host-plant selection by insects - a theory based on 'appropriate/inappropriate landings' by pest insects of cruciferous plants. Entomol Exp Appl 96(2):91–102. https://doi.org/10.1046/j.1570-7458.2000.00684.x

Fisher, R.A. and Turner, N.C. (1978) Plant productivity in Arid and semi-arid zones. Annual Review of Plant Physiology. 29:277-317

Flor, H.H.1(955). Host-parasite interaction in flax rust—its genetics and other implications. Phytopathology 45: 680–685

Forest.org (2006) Kenya: Deforestation exacerbates droughts, floods, Source: Copyright 2006, IRIN Date: November 10, 2006 Original URL: Status ONLINE

Fraser, E.D.G., Termansen, M., Sun, N., Guan, D., Simelton, E., Dodds, P., Feng, K. and Yu, Y. 2008. Quantifying socioeconomic characteristics of drought sensitive regions: evidence from Chinese provincial agricultural data. Comptes Rendus Geosciences. 340:679-688. http://dx.doi.org/10.1016/j.crte.200 8.07.004

Fromme, D.A., Dotray, P.A., Grichar, W.J., Fernandez, C.J. (2012)"Weed Control and Grain Sorghum (Sorghum bicolor) Tolerance to Pyrasulfotole plus Bromoxynil", International Journal of Agronomy, vol. 2012, Article ID 951454, 10 pages, 2012. https://doi.org/10.1155/2012/951454

Front Plant Sci. (2018) Jul 2;9:905. doi: 10.3389/fpls.2018.00905. PMID: 30013587; PMCID: PMC6036327.

Fujita, K. and Ofosu-Budu, K.G. (1996) Significance of Intercropping in Cropping Systems. In: Ito, O., Johansen, C., Adu-Gyamfi, J.J., Katayama, K., Kumar, J.V.D.K. and Rego, T.J., Eds., Dynamics of Roots and Nitrogen in Cropping Systems of The Semi-Arid Tropics, Japan International Research Center for Agricultural Sciences, Ohwashi, Int. Agric. Series No. 3, 19-40.

Fukai, S.and Trenbath, B.R. (1993) Processes determining intercrop productivity and yields of component crops. Field Crops Res., 34, 247–271.

Fussell, L.K., Bidinger, F.R. and Bieler, P. (1991) Crop physiology and breeding for drought tolerance: research and development. Field Crops Research 27:183–199

Gabhane Vijay, Nagdeve Mahendra and Ganvir Mahipal (2013). Effect of long term integrated nutrient management on sustaining crop productivity and soil fertility under cotton and greengram intercropping in Vertisols under semi-arid agro-ecosystem of Maharashtra, India. Acta biologica indica 2013;2(1):284-291

Gahukar, R.T. (1989) Pest and disease incidence in pearl millet under different plant density and intercropping patterns. Agriculture, Ecosystems & Environment. Elsevier 26(1):69-74

Gajanan,G.N.., Hegde,B.R. Ganapathi, Panduranga, and Somasekhar,K.(1999) Organic manure for stabilizing productivity: Experience with dryland finger millet, All India coordinated Project for Dryland Agric.,Univ.of Agric,sci., Bangalore, India

Gangar, B. and Ravishankar, N. (2013b) Diversified cropping systems for food security Indian Farming 63 (9): 3-7

Gao, L., Huasen Xu, Huaxing Bi, Weimin Xi, Biao Bao, Xiaoyan Wang, Chao Bi, Yifang Chang (2013) Intercropping competition between apple trees and crops in agroforestry systems on the Loess Plateau of China. PLoS One 2013 Jul 25;8(7): e70739. doi: 10.1371/journal.pone.0070739. Print 2013

Garrity, D.P., Sullivan, C.Y. and Ross W.M. 1982. Alternative approaches to improving grain sorghum productivity under drought stress. Pages 339-356, In: Drought resistance in crops with emphasis on rice. International Rice Research Institute, Los Banos, Laguna, Philippines

Geraert, P.A., Padilha, J.C. and Guillaumin, S. 1996. Metabolic and endocrine changes induced by chronic heat exposure in broiler chickens: growth performance, body composition and energy retention. British Journal of Nutrition 75:195–204.

Getis, A., Getis, J. and Fellmann, J.D. (2000). Introduction to Geography, Seventh Edition. McGraw-Hill. p. 99. ISBN 978-0-697-38506-2.

Gill, H.A.K. and Robert, M. C.S. (2010) Effect of integrating soil solarisation and organic mulching on the soil surface insect community. Florida Entomologist. ;93(2):308-309

Gill, K.S. (1994) Efficient Plant Types. In Stressed Ecosystems and Sustainable Agriculture (Eds: Virmani, S.M., Katyal, J.C., Eswaran, H. and Abrol, I.P.) Oxford & IBH Publishing Co. Pvt. Ltd. New Delhi ISBN 81-204-0932-9. Pages: 307-332.

Gill, M.S. (2010) Farming System approach: Its principles and road towards sustainability. In: Extended summaries, 2010 XIX National Symposium on "Resource Management Approaches Towards Livelihood Security" Dec. 02-04, 2010Bangalore. Pp 155-159

Gill, M.S., Singh, J.P. and Gangwar, K.S. (2009). Integrated farming system and agriculture sustainability. Indian Journal of Agronomy 54(2): 128–139.

Giller and Wilson (1991) Giller, K.E. and Wilson, K.J. (1991) Nitrogen Fixation in Tropical Cropping Systems. CAB International, Wallingford, England, 167-237.

Gong X., Liu C., Li J., Luo Y., Yang Q., Zhang W., et al. (2019). Responses of rhizosphere soil properties, enzyme activities and microbial diversity to intercropping patterns on the Loess Plateau of China. Soil Tillage Res. 195:104355. 10.1016/j.still.2019.104355

Good, N.E. and Bell, D. H. (1980) Photosynthesis, plant productivity and crop yield. Pages 3-50 In: Section I. Interaction of Plants and Environment. In the Book: The Biology of Crop Productivity (Ed: Carlson, Peter S.) Academic Press, New York.471pp

Gopinath K.A., Dixit,S., Srinivasarao,Ch., Raju,M.K., Charry,G.R., M. Osman, M., Ramana, D.V., Nataraja K.C.,, Gayatri Devi, K., Venkatesh, G., Grover, M., M. Maheswari, M. and Venkateswarlu, B. (2012) Improving the existing rainfed farming systems of small and marginal farmers in Anantapur district, Andhra Pradesh.IndianJ. of Dryaland Agric Res and Dev.2012;27 (2): 43-47.

Gosal, S.S., Wani, S.H. and Kang, M.S (2009) Biotechnology and Drought Tolerance, Journal of Crop Improvement 23(1):19-54 DOI:10.1080/15427520802418251

Goulden, C.H. (1939). Problems in plant selection. Conference information. In: Proc. 7th Int. Genet. Congr. August 23–30, 1939, pp. 132–133.

Gowda, C.L.L. et al., (2009) Opportunities for improving crop water productivity through Genetic enhancement of Dryland Crops. In: Wani, S.P., Rockstrom, J. and Oweis, Th.(Eds) (2009) Rainfed Agriculture-Unlocking the potential. CAB International, ICRISAT, International Water Management Institute. © CAB International 2009. Cambridge, MA 02139 USA: Pages 133-163

Guertal, E.A. and Green, B.D. (2012) Evaluation of organic fertilizer sources for south-eastern (USA) turfgrass maintenance, Acta Agriculturae Scandinavica, Section B — Soil & Plant Science, 62:sup1, 130-138, DOI: 10.1080/09064710.2012.683201

Guhathakurta P. and Rajeevan M (2007) Trends in rainfall pattern over India. Int J Climatol; DOI: 10.1002/joc.1640 2007

Gulwa, U.; Mgujulwa, N.; Beyene, S.T (2017). Effect of Grass-legume Intercropping on Dry Matter Yield and Nutritive Value of Pastures in the Eastern Cape Province, South Africa. Univ. J. Agric. Res., 5: 355–362

Guo M, Zhang T, Li Z and Xu G. (2019) Investigation of runoff and sediment yields under different crop and tillage conditions by field artificial rainfall experiments. Water. ;11(5):1019

Gupta, B., Sarvade, S. and Mahmoud, A. (2015) Effects of selective tree species on phytosociology and production of under-storey vegetation in mid-Himalayan region of Himachal Pradesh. Range Management and Agroforestry. 2015;36(2):156-163

Gupta, J.P. (1997) Some alternative production systems and their management for sustainability. In: Gupta, J.P. and Sharma, B.M. (eds) Agroforestry for sustained produc- tivity in arid regions. Scientific publishers, Jodhpur, pp 31–39

Gupta, N., Kukal, S.S., Bawa, S.S, Dhaliwal and G.S. (2009) Soil organic carbon and aggregation under poplar based agroforestry system in relation to tree age and soil type. Agroforestry Systems 76: 27-35.

Gupta, S. K, Raina N.S and Sehgal S. (2007) Potential of silvi-pastoral systems in improving the forage production in the hills of Jammu and Kashmir. Journal of Research SKUAST-J. ;6(2):160-168

Gupta S.K. (1989). Choice of grasses and legume species for SAT area.--Subject Matter Workshop cum Seminar on Alternate Land use systems in Dryland Agriculture,12-24,Oct,CRIDA Hyderabad

Gurbachan Singh (2019) Recent Advances in Salinity Management in Agriculture—Indian Experience. 3rd World Irrigation Forum (WIF3) 1-7 September 2019, Bali, Indonesia ST-1.1 W.1.1.39 1

Gurung, T.R and Azad, A.K. (2013), Best practices and procedures of saline soil reclamation systems in SAARC countries

Haldar, S. (2014) Development of an organic integrated pest management (IPM) module against insect-pests of muskmelon in arid region of Rajasthan, India Journal of Experimental Biology and Agricultural Sciences, 2: 19-24

Hanson, A.D. and Nelsen, C.E. (1980) Chapter 3. Water: Adaptation of crops to drought-prone environment: Pages 77-152. In: Section I. Interaction of Plants and Environment. In the Book: The Biology of Crop Productivity (Ed: Carlson, Peter S.) Academic Press, New York.471pp

Hanumanthappa, M., Jayaprakash, R. and Nagaraj, R. (2018) Economic potentials of medicinal and aromatic plants in dryland and rainfed areas of India, Journal of Pharmacognosy and Phytochemistry 2018; SP3: 384-386

Hartman K., Heijden M., Roussely-Provent V., Walser J.-C. and Schlaeppi K. (2017). Deciphering composition and function of the root microbiome of a legume plant. Microbiome 5:2. 10.1186/s40168-016-0220-z

Hauggaard-Nielsen, H., and Jensen, E. S. (2001). Evaluating pea and barley cultivars for complementarity in intercropping at different levels of soil N availability. Field Crop Res. 72, 185–196. doi: 10.1016/S0378-4290(01)00176-9

Hay, R.K.M. (1995) Harvest index – a review of its use in plant breeding and crop physiology. Ann.Appl.Biol.126: 197-216. doi:10-1111/j.1744-7348. 1995.tb05015.x

Hayashi, H., Alia, Mustardy, L., Deshnium, P., Ida, M. and Murata, N. (1997) Transformation of Arabidopsis thaliana with codA gene for choline oxidase: accumulation of glycinebetaine and enhanced tolerance to salt and cold stress. Plant Journal,12: 132-142

Henry, A. (2013) Breeding Efficient Crops and Varieties for Dryland Conditions, Advances in Plant Physiology. Chapter 2. Pages 23-41 ICAR.Central Arid Zone Research Institute, Jodhpur-34203. file://C:/Users/Mvr%20Prasad/Downloads/123_Sample_Chapter%20(10).pdf

Henson, I.E., Mahalakshmi, V., Alagarswamy, G. and Bidinger, F.R. (1984) The effect of flowering on stomatal response to water stress in pearl millet (Pennisetum americanum (L.) Leeke). Journal of Experimental Botany 35: 219–226

Herridge, D.F., Marcellos, H.W., Felton, L., Turner, G.L. and Peoples,M.B.1(995) Chickpea increases soil-N fertility in cereal systems through nitrate sparing and N2 fixation Soil Biology and Biochemistry 27 (4–5), April–May 1995, Pages 545-551

Hiebsch, C.K.(1978) Interpretation of yields obtained in crop mixture. In Abstracts of American Society of Agronomy; Madison, Wisconsin, USA, 1978; p. 41.

Hirst, K. K (2016). "Plant Domestication – Table of Dates and Places". About.com.

Hirst, K.K. (2019) History of Animal and Plant Domestication: Mixed Cropping: The Co-Cultivation of Two or More Crops. Thought Co. https://www.thoughtco.com/mixed-cropping-history-171201

Hofman, V and Franzen, D. (1997). "Emergency Tillage to Control Wind Erosion". North Dakota State University Extension Service. Retrieved 2009-03-21

Holtum, J. A. M., Smith, J. A. C. and Neuhaus, H. E. (2005) Intracellular transport and pathways of carbon flow in plants with crassulacean acid metabolism. Funct. Plant Biol. 32: 429–449 (2005).

Hooda, N., Gera, M., Andrasko, K., Sathaye, J., Gupta, M.K, et al. (2007) Community and farm forestry climate mitigation projects: case studies from Uttaranchal, India. Mitigation and Adaptation Strategies for Global Change 12: 1099-1130.

Hooks CRR and Johnson MW (2003) Impact of agricultural diversification on the insect community of cruciferous crops. Crop Protect 22(2):223–238. https://doi.org/10.1016/s0261-2194(02)00172-2

Houze, Robert A., Jr. (1993). Cloud dynamics. San Diego: Academic Press. ISBN 9780080502106. OCLC 427392836.

http://gldc.cgiar.org/dryland-intercropping-of-lentil-and-quinoa-is-sustainable-and-profitable/#comment-957

https://horticulturedir.karnataka.gov.in/

https://www.researchgate.net/publication/342799551_Soil_and_Water_Conservation_Measures_for_Agricultural_Sustainability

https://www.slideserve.com/bien/self-help-groups-poverty-alleviation-and-empowerment-state-initiative-in-andhra-pradesh

Hubick, K.T., Farquhar, G.D. and Shorter, R., (1986). Correlation between water-use efficiency and carbon isotope discrimination in diverse peanut (Arachis) germplasm. Australian Journal of Plant Physiology 13, 803–816

Hull and Fred H. (1952). Recurrent selection and over dominance. In: John W. Gowen, (ed.), Heterosis. Hafner Publishing Co., N ew York, London. P.451-473

Hurd, (1974). Phenotype and drought tolerance in wheat. Agricultural Meteorology 14: 39-55.

IAEA Vienna (2003) Management of Crop Residues for Sustainable Crop Production © IAEA, 2003 Printed by the IAEA in Austria May 2003: 243pp

ICAR (2001) Indian council of Agricultural Research-Australian Center for International Agricultural Research.2001 Progress report of ICAR-ACIAR Project on tools and indicators for planning sustainable soil management on semi-arid farms and watersheds in India from Central Res. Inst. For Dryland Agric., Hyderabad

ICRISAT (2019) Annual Report. International Crops' Research Institute, Patancheru, India 2019. https://www.icrisat.org/2019/12/

Ilorkar, V.M., Suroshe, S.B., Jiotode, D.J. (2011) Agroforestry interventions across different agro-climatic zones in Maharashtra, India. Indian J Forestry 34(1):105–109

Immanuel, R., Ganapathy, M., Thiruppathi,M., Sudhagar Rao, G.B. and J. Nambi, J .(2019) Physiological responses of multipurpose tree seedlings to induced water stress January 2019, Plant Archives 19(Sup 1):444-447

Inamati, S.S., Patil, S.J. and Revathi, R. (2015) Performance of soybean intercropped under different seed sources of Pongamia pinnata agroforestry system. Indian Journal of Agroforestry, 17(2).

IPCC (1990) Inter Governmental Panel on Climate Change, 1990, UNFCCC-WMO- UNEP (UNO)

IPCC (1996) (Inter Governmental Panel on Climate Change, (1996). Climate Change, 1995: Impacts, Adaptations and Mitigation of Climate Change. Scientific-Technical Analyses. (Ed.)Robert T> Watson, Mrufu C.Zinyuwera and Richard H. Moss. IPCC Second Assessment Report: CLIMATE CHANGE 1995, Cambridge, UK. Cambridge University Press.

IPCC (2007) Intergovernmental Panel on Climate Change, 2007. Assessment report. T

Iqbal, N., Hussain, S., Ahmed, Z., Yang, F., Wang, X.; Liu, W.; Yong, T.; Du, J.; Shu, K.; Yang, W. (2018). Comparative analysis of maize–soybean strip intercropping systems: A review. Plant Prod. Sci., 22: 131–142.

IRRI (1975) Annual Report 1975. International Rice Research Institute, Las Banos Philippines.

Islam, A.K.M. and Nazrul. 2008. SIDR Cyclone in Bangladesh, 'End of the Course Project' Submitted to the National Institute of Disaster Management (NIDM), New Delhi, th for 4 Online Course on Comprehensive DM, online available at: http://siteresources. worldbank.org/ CMUDLP/Resources/AKM_Nazrul _Islam.pdf

Itnal.C.J.(1981).Water conservation measures for increased production in black soils of Bijapur, All India coordinated Project for Dryland Agriculture., Unv. of Agric, Sci.,Bangalore, India

Jabran K. (2019) Role of Mulching in Pest Management and Agricultural Sustainability. Cham: Springer;

James, B.K., Mishra, A., Mohanty, A., Rajeeb, K., Brahmanand, P.S., Nanda, P., Das, M. and Kannan, K, (2005) Management of excess rainwater in medium and low lands for sustainable productivities. Research. Bulletin No.24, ICAR-Water Technology Centre for Eastern Region

Jayanthi, C. (2017) Development of practices of successful crop based farming systems in Tamil Nadu. In: Integrated Farming Systems for Sustainable Agriculture and Enhancement of Rural Livelihoods (Eds. Muralidharan, K., Prasad, M.V.R., and Siddiq, E.A.) Proceedings of the Seminar. Retired ICAR Employees' Association (RICAREA). Hyderabad 500030 Pages144-157.

Jensen, E. S., Bedoussac, L., Carlsson, G., Journet, E. P., Justes, E., and Hauggaard-Nielsen, H. (2015). Enhanced yields in organic arable crop production by eco-functional intensification using intercropping. Sustain. Agr. Res. 4, 3–8. doi: 10.5539/sar.v4n3p42

Jeyabalan and Kuppuswamy (2001) The production of vermi-compost and its response in rice legume cropping system and soil fertility. European Journal of Agronomy, 15: 153-170. doi:10.1016/S1161-0301(00)00100-3

Jha, M.N., Gupta, M.K. and Raina, A.K. (2001) Carbon Sequestration: Forest soil and land use management. Annals of Forestry 9: 249-256

Jhariya, M.K., Surendra, S., Bargali, A. and Raj, S. (2015) Possibilities and perspectives of agroforestry in Chhattisgarh. In: Zlatic M, editor. Precious Soil Moisture Importance 22 Forests-Precious Earth. Croatia: Intech Open; 27-257. DOI: 10.5772/60841

Jones, J.W. and Zur, B. (1984). Simulation of positive adaptive mechanisms in crops subjected to water stress. Irrigation Science 5:251-264

Jordan, W.R. and Miller, F.R. (1980). Genetic variability in sorghum root systems: Implications for drought tolerance. Pages 383 – 399. In: Adaptation of plants to water and high temperature stress. (Eds.: Turner, N.C. and Kramer, P.J.) John Wiley, New York, USA.

Joseph, B., Rao, L.G.G., Sreemannarayana, B. (1999) Effect of pruning on yield of sunflower in Albizzia lebbeck based agri-silvi system. Indian Agrofor 1(2):129–133

Joseph S. D'Aleo; Pamela G. Grube (2002). The Oryx Resource Guide to El Niño and La Niña. Greenwood Publishing Group. pp. 48–49. ISBN 978-1573563789.

Joshi, P.K. and Tyagi, N.K. (2017). Assessment of Government policies and programmes on climate change- Adaptation, mitigation and resilience in South Asian Agriculture. Book Chapter: Agriculture under climate change threats, strategies and policies, section-VI: 431-434

Joshi P K and Gulati, A. (2003), 'From Plate to Plough: Agricultural Diversification in India', Paper presented at the Dragon and Elephant: A Comparative Study of Economic and Agricultural Reforms in China and India', New Delhi, India, March 25-26

Joy Deshmukh-Ranadive (2014) Self-Help Groups, Poverty Alleviation and Empowerment: State Initiative in Andhra Pradesh © 2022 Slide Serve | Powered By DigitalOfficePro

Kalaiselvi, K. and Arul Swaminathan, A. (2009). Alternate land use through cultivation of medicinal and aromatic plants - a review. Agricultural reviews; 30 (3): 176-183.

Kampen J. (1982). An approach to improved productivity on deep Vertisols. ICRISAT Information Bulletin II, Patancheru 502 324, Andhra Pradesh, India.

Kareem Ulla K, Shalander Kumar, Reddy K.S., Rama Rao C.A. and Venkateswarlu, B. (2010) impact of NREGS on Rural Livelihoods and Agricultural Capital Formation, Ind.J. Agri. Econ. 2010; 65 (3): S24-539.

Kariaga B.M., (2004). Intercropping maize with cowpeasand beans for soil and water management in westernKenya. 13th International soil conservation organization conference – Brisbanne

Kat Henrichs (2013) Millennium Development Goals. https://borgenproject.org/how-to-help-poor-farmers-and-their-communities-10-ways/

Katyal,J.C.(1997) Research and development in rainfed agriculture: Retrospective and perspective, plan 256-270. In Productivity of land and water. New Age Int. M. Virmani etal.,(ed) Stressed ecosystems and sustainable agriculture . Central Res. Inst. for Dryland Agric., Hyderabad. India

Katyal, J.C., Das, S.K., Korwar, G.R. .and Osman M. (1993). Technology for mitigating stresses: alternative land uses. (In Stressed Ecosystems and Sustainable Agriculture (eds. Virmani S.M.., KatyalJ.C.Eswaran H. and Abrol J.P.) Proceedings of Int. Symposium on Agroclimatology and Sustainable Agriculture in Stressed environments, Oxford and IBH291-305

Kaur, B., Gupta, S.R., Singh, G. (2002) Carbon storage and nitrogen cycling in silvipastoral system on sodic soil Northwestern India. Agroforestry Systems 54: 21-29.

Kaushik, N., Deswal, P., Tikko, A. and Kumari, S. (2016) Pongamia pinnata L. (Karanja) based agri-silviculture system under rainfed conditions of south-west Haryana Journal of Applied and Natural Science 8(1): 31-34.

Kaushik, N. ; Kumar, V. (2003)Range Management and Agroforestry 2003 Vol.24 No.1 pp.23-26 ref.7

Kavi-Kishore, P.B., Hong, Z., Miao, G.H., Hu, C.A.A. and Verma, D.P.S. (1995) Over expression of Δ synthetase increases proline production and confers osmotolerance in transgenic plants. Plant physiology 108: 1387-1394

Kenaschuk, E.O., (1975). In: Harpiak, J.T. (Ed.), Oilseed and Pulse Crops in Western Canada: A Symposium. Western Co-operative Fertilizers, Calgary, Alberta, pp. 203–221.

Kenya. Journal of Agricultural Science and Technology, 1, 372-384.

Keshwa, G.L. and Singh, M. (2004) Biomass production and soilfertility from Dichrostachys cinerea + Cenchrussilvipastoral system in arid and semi-arid regions. IndianJournal of Agronomy. 2004; 49(4): 293-29

Ketterings, Q.M., Coe. R., Vannoordwijk, M., Ambagau, K. and Palm, CA. (2001). Reducing uncertainty in the use of allometric biomass equations for predicting aboveground tree biomass in mixed secondary forests. Forest Ecology and Management. 146: 199–209.

Kharayat, B.S. and Singh,Y. (2013). Evaluation of inoculation techniques for screening sorghum genotypes against stalk rot caused by Erwinia chrysanthemi. Indian Phytopathology 2013.66 (4) : 400-402

Kiepe P.(1995) No runoff, no soil: Soil and water conservation in hedgerow barrier systems. Tropical Resource Management Papers No. 10; Wageningen; 1995. p. 158

Kinama, J.M., Ong, C.K., Stigter, C.J., & Ng, J.K., (2011). Hedgerow Intercropping Maize or Cowpea/Senna for Drymatter Production in Semi-Arid Eastern

Kinama, J.M., Stigter, C.J., Ong, C.K., & Ng, J.K., (2007). Arid Land Research and Management Contour Hedgerows and Grass Strips In: Erosion and Runoff Control on Sloping Land in Semi-Arid Kenya Contour Hedgerows and Grass Strips in Erosion and Runoff Control on Sloping Land in Semi-Arid Kenya 21, 1-19.

Knut V. (1993). Evaluation of components of the national biomass study, Uganda: Suggestions and recommendations for the biomass measurements, statistical procedures and strategies for Phase 111 of the project. Unpubl. Forest Department and Norwegian Forestry Society, Kampala.p

Kolarkar AS, Murthy KNK and Singh N. (1983). Khadin – a method of harvesting water for agriculture in the Thar Desert. Journal of Arid Environments 6 (1): 59-66.

Korwar, G.R. (1992) Alternate land use systems In: Dryland Agriculture in India: State of Art Research in India (Ed. L.L. Somani) Scientific Publishers, Jodhpur, India. pp 143-168.

Korwar, G.R. (2003). Agroforestry interventions for improvement of wastelands, A Compendium on wasteland development for rain-fed areas, CRIDA, Hyderabad. pp.157- 161

Korwar, G.R., Prasad, J.V.N.S., Rao. G.R., Venkatesh, G., Pratibha, G. and Venkateswarlu, B. (2014) Agroforestry Systems in India: Livelihood Security & Ecosystem Services,

Korwar, G.R. and Pratibha, G. (1999) Performance of short duration pulses with African winter-thorn (Faidherbia albida) in semiarid regions. Indian J Agric Sci 69(8):560–562

Korwar, G.R. and Radder, G.D. (1994) Influence of root pruning and cutting interval of Leucanea hedgegrows on performance of alley cropped rabi sorghum. Agrofor Syst. 25:95-109

Kumar, A., Natarajan, S., Biradar, N.B. and Trivedi, B.K. (2011). Evolution of sedentary pastoralism in south India: case study of the Kangayam grassland. Pastoralism: Research, Policy and Practice 1: 7

Kumar, A.K. (2010) Carbon sequestration: under-exploited environmental benefits of Tarai agroforestry. Indian Journal of Soil Conservation 38: 125-131.

Kumar, N., Nath, C.P., Hazra, K.K. and Sharma, A.R. (2016) Efficient weed management in pulses for higher productivity and profitability, Indian Journal of Agronomy 61(4th IAC Special Issue):S93-105 (2016)

Kumar, R. and Kumar, K.K. 2007. Managing physiological disorders in litchi. Indian Horticulture 52 (1): 22-24. Handbook. 2018.

Kumar, V.; Singh, R.P.; Kumar, S.; Shukla, U.N.; Kumar, K. (2018) Performance of pearlmillet + greengram intercropping as influenced by different planting techniques and integrated nitrogen management under rainfed condition. Int. J. Chem. Stud. , 6, 705–708.

Kumar D., Shivay Y. S., Dhar S., Kumar C. and Prasad R. (2013). Rhizospheric flora and the influence of agronomic practices on them: a review. Proc. Natl. Acad. Sci. India Sect. B Biol. Sci. 83 1–14. 10.1007/s40011-012-0059-4

Kumar KK; Rajagopalan B; Hoerling M; Bates G; Cane M (2006), Unraveling the Mystery of Indian Monsoon Failure During El Niño», Science, 314 (5796): 115–119, Bibcode:2006S ci 314..115K, doi:10.1126/science.1131152, PMID 16959975.

Kumar M. (2017). Carbon sequestration in a agroforestry system at Kurukshetra in Northern India. International Journal of Theoretical and Applied Sciences. 9(1): 43-46.

Kumar S, Kumar S and Choubey B K. (2009). Aonla based hortipastoral system for soil nutrient buildup and profitability. Annals of Arid Zone 48: 153–

Kumar Shalander, Venkateswarlu,B., Khem Chand, Murari Mohan Roy. (20/11/2013). Farm level rainwater harvesting for dryland agriculture in India: Performance assessment and institutional and policy needs. Harbin, China.

Kurukulasuriya, P. and Mendelsohn, R. (2006) A Ricardian Analysis of the Impact of Climate Change on African Cropland. CEEPA Discussion Paper No. 8, Centre for Environmental Economics.

Kushwah, S. K., Dotaniya, M. L., Upadhyay, A. K., Rajendiran, S., Coumar, M. V., Kundu, S. and Subba Rao A. 2014. Assessing carbon and nitrogen partition in kharif crops for their carbon sequestration potential. National Academy Science Letter37(3):213–7

Lai, R. (2004) Soil carbon sequestration in India. Climatic Change, 2004;65: 277-296.

Lal, M. (2001a) Climatic Change—Implications for India's Water Resources. Journal of Indian Water Resources Society, 21 : 101-119.

Lal, R. (2001b). Potential of desertification control to sequester carbon and mitigate the greenhouse effect. Climatic Change, 51: 35–72

Lal, R. (2009). Soil carbon sequestration for climate change mitigation and food security. (In) Souvenir, Platinum Jubilee Symposium on Soil Science in Meeting the Challenges to Food Security and Environmental Quality, Indian Society of Soil Science, New Delhi, pp 39–46.

Lal, R. (2010). Carbon sequestration potential of rainfed agriculture. Indian Journal of Dryland Agricultural Research and Development 25(1): 1–16.

Lal, R. (2016). Climate Change and Agriculture, Chapter 28. In Climate Change, 2nd ed.; Letcher, T.M., Ed.; Elsevier: Amsterdam, The Netherlands, 2016; pp. 465–489. ISBN 9780444635242.

Lal M. and Mishra S.K.(2015) Characterization of surface runoff, soil erosion, nutrient loss and their relationship for agricultural plots in India. Current World Environment: An International Research Journal of Environmental Sciences. 10(2):593-601

Lal R. (2009) Soil degradation as a reason for inadequate human nutrition. Food Security. 2: 1:45-57

Lal R. (2010) Managing soils and ecosystems for mitigating anthropogenic carbon emissions and advancing global food security. Bioscience. :59-82

Lal R. (2015) Restoring soil quality to mitigate soil degradation. Sustainability,7(5):5875-5895

Lal R. and Stewart BA.(1990) Soil Degradation. New York: Springer;

Landsberg, J.J. (1988) Interpretive Summary of Part I: Calculated Soil-Water Balances as Tools to Evaluate Crop Performance in Drought prone Regions. Pages 35-39, In: Drought Research Priorities for the Dryland Tropics. (Eds: Bidinger, F.R. and Johansen, C.) ICRISAT (International Crops Research Institute for the Semi-Arid Tropics) Patancheru, India. 219pp.

Land van Den Bergh, deWit, C.T. and van den Bergh, J.P. (1965) Competition between herbageplants. Netherlands Journal of Agricultural Science, 13, 212-221.

Laroche G, Domon G and Olivier A (2020) Exploring the social coherence of envisioned landscapes featuring agroforestry intercropping systems (AIS) using rural resident visual assessments and perceptions. Sust Sci. https://doi.org/10.1007/s11625-020-00837-3

Lawn, R.J. (1982) Response of four grain legumes to water stress in south-eastern Queensland. 1. Physiological response mechanisms. Australian Journal of Agricultural Research 33:481-496

Lawn, R.J. (1988) Breeding for Improved Plant Performance in Drought Prone Environments. Interpretive Summary of Part 4. Pages 213-219. In: Drought Research Priorities for the Dryland Tropics. (Eds: Bidinger, F.R. and Johansen, C.) ICRISAT (International Crops Research Institute for the Semi-Arid Tropics) Patancheru, India. 219pp.

Lee, J.K., Kwak, M.J., Jeong, S.G. and Woo, S.Y. (2022) Individual and Interactive Effects of Elevated Ozone and Temperature on Plant Responses. Horticulturae 2022, 8(3), 211; https://doi.org/10.3390/horticulturae8030211

Lencucha, R., Pal, N.E., Appau, A. et al. Government policy and agricultural production: a scoping review to inform research and policy on healthy agricultural commodities. Global Health 16, 11 (2020). https://doi.org/10.1186/s12992-020-0542-2

Lightfoot, C.W.F. and Tayler, R.S. (2008) Intercropping Sorghum with Cowpea in Dryland Farming Systems in Botswana. I. Field Experiments and Relative Advantages of Intercropping. Experimental Agriculture, 23 (4) : 425 – 434 DOI: https://doi.org/10.1017/S0014479700017385

Lin, H., Mertens, K., Kemps, B., Govaerts, T., De Ketelaere, B., De Baerdemaeker, J., Decuypere, E., and Buyse, J. (2004). New approach of testing the effect of heat stress on eggshell quality: Mechanical and material properties of eggshell and membrane. British Poultry Science 45: 476-482

Lindsey, R. (2017). Climate change: Atmospheric carbon dioxide. National Oceanic and Atmospheric Administration (NOAA). https://www.climate.gov/news-features/understanding-climate/climatechange-atmospheric-carbon-dioxide

Lipton, Michael (2006) Can Small Farmers Survive, Prosper, or be the Key Channel to Cut Mass Poverty? electronic Journal of Agricultural and Development Economics, 3(1): 58-85 2006 Agricultural and Development Economics Division (ESA) FAO available online at www.fao.org/es/esa/eJADE

Liu, J., Gao, G., Wang, S., Jiao. L., Wu. X. and Fu, B. (2018) The effects of vegetation on runoff and soil loss: Multidimensional structure analysis and scale 21 Soil and Water Conservation Measures for Agricultural Sustainability DOI: http://dx.doi.org/10.5772/intechopen.92895 characteristics. Journal of Geographical Sciences.28(1):59-78

Liu, Q. J., Zhang, H.Y.A. J. and Wu YZ. (2014) Soil erosion processes on row side-slopes within contour ridging systems. Catena.115:11-18

Long, S.P. et al (2006) Food for thought: lower-than-expected crop yield stimulation with rising CO2 concentrations. Science 2006 Jun 30;312(5782):1918-21. doi: 10.1126/science.1114722.

Lu, D. (2005). Aboveground biomass estimation using landsat TM data in the Brazilian Amazon. International Journal of Remote Sensing. 26: 2509-2525

Lu, D. (2006). The potential and challenge of remote sensing-based biomass estimation. International Journal of Remote Sensing. International Journal of Remote Sensing. 27(7): 1297–1328.

Lu. D., Chen, Q., Wang, G., Liu, L., Li, G. and Moran E. (2014). A survey of remote sensing-based aboveground biomass estimation methods in forest ecosystems. International Journal of Digital Earth, (December). 1-43. [Carbon Sequestration Potential 98 Agroforestry Opportunities for Enhancing Resilience to Climate Change in Rainfed Areas]

Ludlow, M.M. (1987) Contribution of osmatic adjustment to the maintenance of photosynthesis during water stress. Pages 161-168. In: Progress in Photosynthesis Research. Proceedings of the VII International Conference on Photosynthesis.Aug.16-23, 1986. Providence, Rhode Island, USA.

Ludlow, M.M., Fisher, M.J. and Wilson, J.R. (1985). Stomatal adjustments to water deficits in three tropical grasses and a tropical legume in the controlled conditions and in the field. Australian Journal of Plant Physiology 12: 131-149

Ludlow, M.M. and Muchow, R.C. (1988). Critical evaluation of possibilities for modifying crops for high production per unit of precipitation. In: Drought Research Priorities for the Dryland Tropics. Pages 179-211 (Eds.: Bidinger, F.R. and Johansen, C.) International Crops Research Institute for Semi-Arid Tropics, Patancheru 502324, India.

Maas E.V. (1986). Salt tolerance of plants. Applied Agricultural Research 1: 12-26.

Mader.P., Flieβbach, A. and Dubois, D. (2002) Soil fertility and biodiversity in organic farming. Science..296:1694-1697

Madhu, M., Hombegowda, H.C., Beer, K., Adhikary, P.P., Jakhar, P., Sahoo, D.C., Dash, C.J., Kumar, G. and Naik, G.B.(2019). Status of natural resources and resource conservation technologies in eastern region of India. In Resource Conservation in Eastern Region of India: Lead Papers of FFCSWR, 2019; Indian Association of Soil and Water Conservationists: Dehradun, India., pp. 60–82

Madsen, I and Pan. B. (2020) Water, Land, and Nutrient Use Efficiency for Intercropping Systems in the Dryland Pacific Northwest. Project ID: 120, Progress Reports 2019 2020, CSANR-Center for Sustaining Agriculture and Natural Resources, Washington State University, USA

Magzter (2022) The benefits and trade-offs of agricultural diversity for food security in low and middle income countries: A Review of existing knowledge and evidence. Journal of Global Food Security, June.2022, The Business Guardian, Aug,28, 2022

Mahalakshmi, V. and Bidinger, F.R. (1985) Water stress and time of floral initiation in pearl millet. Journal of Agricultural Science, Cambridge 105: 436–445

Maheswari, M., Sarkar, B., Vanaja, M., Srinivasa Rao, M., Prasad, JVNS., Prabhakar,M., Ravindra Chary,G., Venkateswarlu, B., Ray Choudhury, P., Yadava, D.K., Bhaskar, S. K and Alagusundaram (Eds.). (2019). Climate Resilient Crop Varieties for Sustainable Food Production under Aberrant Weather Conditions. ICAR-Central Research Institute for Dryland Agriculture, Hyderabad. P64.

Maheswari, M., Sarkar, B., Vanaja, M., Srinivasa Rao, M., Srinivasa Rao, Ch., Venkateswarlu, B., Sikka, A.K. (2015). Climate Resilient Crop Varieties for Sustainable Food Production under Aberrant Weather Conditions. CRIDA, Hyderabad. NICRA/2015/Bulletin No.4 P47.

Maikhuri, R.K., Semwa, R.L., Rao, K.S., Singh, K. and Saxena, K.G. (2000). Growth and ecological impacts of traditional agroforestry tree species in Central Himalya, India, Agroforestry Systems. 48: 257-272

Maitra, S.(2020) Intercropping of small millets for agricultural sustainability in drylands: A review. Crop Res. 55: 162–171

Maitra, S., Palai, J.N., Manasa, P. and Kumar, D.P. (2019) Potential of Intercropping in Sustaining Crop Productivity. International Journal of Agriculture, Environment and Biotechnology 12(1): 39-45, March 2019; DOI: 10.30954/0974-1712.03.2019.7

Maitra, S.; Hossain, A.; Brestic, M.; Skalicky, M.; Ondrisik, P., Gitari, H.; Brahmachari, K.; Shankar, T.; Bhadra, P.; Palai, J.B.; et al.(2021) Intercropping—A Low Input Agricultural Strategy for Food and Environmental Security. Agronomy 2021, 11:343.

Maitreya (2022) Environmental Protection and Conservation is the only recourse that can lead to Protection of Planet Earth (Article in Telugu Language, Paryavrana Parirakshna – Pudamiki Samrakshana) Eee Naadu (Telugu Daily) Hyderabad Edition No.49/96 of Nov[1]6, 2022 (www.eenadu.net)

Manasa, P., Maitra, S. and Reddy, M.D. (2018) Effect of summer maize-legume intercropping system on growth, productivity and competitive ability of crops. Int. J. Manag. Technol. Eng. 8: 2871–2875

Mandal, D.K., Rao, A.V.M.S., Singh, K. and Singh, S.P. 2002. Effects of macroclimatic factors on milk production in a Frieswal herd. Indian Journal of Dairy Science. 55(3):166–170.

Marcum, K.B. (1999) Salinity Tolerance Mechanisms of Grasses in the Subfamily Chloridoideae. Crop Science, 1999 https://doi.org/10.2135/cropsci1999.0011183X003900040034x

Markell,S. and Friskop, A. (2020) Diseases in Drought Years NDSU,NorthDakota State University, USDA, National Institute of Food&Agriculture https://www.ndsu.edu/agriculture/ag-hub/ag-topics/crop-production/diseasespests-and-weeds/plant-diseases/diseases-drought-years-07

Mashelkar, R. (2019) Leveraging Agritech Startups in Indian Agriculture Innovation Ecosystem. Dr. A.B. Joshi Memorial Lecture, 14th Agriculture Science Congress, 20 February, 2019, NASC Complex, New Delhi, India.

Mastouri, F., Bjorkman, T. and Harman, G. (2010) Seed Treatment with Trichoderma harzianum Alleviates Biotic, Abiotic, and Physiological Stresses in Germinating Seeds and Seedlings, Phytopathology,100 (11): 1213-21

Mathivanan, N., Srinivasan, K. and Chelliah, S.(2000) Biological control of soil borne diseases of cotton, eggplant, okra and sunflower by Trichoderma viridae. Z. Pfl.Krankh. Pfl. Schutz. 107:235-244

May, L.H. and Milthrope. F.L. (1962) Drought resistance of crop plants. Field Crop Abstracts 15: 171-179

Mayee, C.D. (2003) Management of diseases of oilseed crops. In: Mangala Rai, Harvir Singh and Hegde, D.M (Eds)Thematic Papers. National Seminar on Stress Management in Oilseeds for Attaining Self-Reliance in Vegetable Oils. ISOR, Directorate of Oilseeds Research, January 28-30.2003 Hyderabad. Pages 33-48.

Mc Cown, R.L., Jones, R.K. and Peake, D.C.I. (1980). Short term benefits of zero-tillage in tropical grain production. Page 220, In: Pathways to Productivity. Proceedings of the Australian Society of Agronomy: Meeting. April, 1980. Lawes, Queensland, Australia. Australian Society of Agronomy.

Mc Cown, R.L., Peake, D.C.I. and Jones, R.K.(1982). No tillage tropical legume by farming research in Northern Australia. Pages 239-247.In: Proceedings of the Monsanto Conservation Tillage Seminar, Nov.1981, Christ Church, New Zealand.

McCree. K.J. and Richardson, S.G. (1987). Stomatal closure vs. Osmatic adjustment: A comparison of stress responses. Crop Science 27 (3): 539-543

Meena R.S., Kumar, V., Yadav, G.S., and Mitran,T. (2018) Response and interaction of Bradyrhizobium japonicum and Arbuscular mycorrhizal fungi in the soybean rhizosphere: A review. Plant Growth Regulators. ;84:207-223

Mendelshon, R., Morrison. W., Schlesinger, M. E. and Andronova, N.G. (2000) Country specific market impacts of climate change. Climate Change 45: 553-69

Mendelsohn, R., Dinar, A. and Sanghi, A. (2001) The effect of Developmenton the climate sensitivity of Agriculture. Environment and Development Economics 6: 85-101

Mendelsohn, R. and Reinsborough, M. (2007) A Ricardian Analysis of US and Canadian Farmland. Yale University, New Haven, CT. Photocopy.

Mevada, K.D., Poonia, T.C., Piyush Saras, and Deshmukh, S.P. (2021) A Textbook of Rainfed Agriculture and Watershed Management Unknown Binding – 1 January 2021 Bhavya Books Edition : 1st 2021 Publishing Year : 2021 Total Pages : 325

Miles, L., Sonwa, D.J., et al.,(2015) UNEP Emissions Gap Report, Chapter 6: Mitigation potential from forest-related activities and incentives for enhanced ac on in developing countries, 2015.

Miller, D.E. and Aarstad, J.S. (1972) Effects of deep plowing on physical characteristics of Heze soil. Washington Agricultural Experiment Station Circular No. 556. Pullman, Washington, USA. Washington State University.

Miller, P.C. and Sunde, M.L. (1975). The effects of precise constant and cyclic environmentals on shell quality and other performance factors with Leghorn pullets. Poultry Science. 54: 36-46.

Mishra,A. and Mal, B.C.(2020) Soil And Water Conservation Engineering PDF Book In: Fundamental Of Soil Water Conservation & Engineering PDF. AGRIMOON.COM https://agrimoon.com/soil-and-water-conservation-engineering-pdf-book/

Mishra, A.K. (2002) Integration of livestock in land use diversification. In: Compendium of lectures. Summer School on Land-use Diversification in Rainfed Agro-ecosystem. April 15 to May 05, Central Research Institute for Dryland Agriculture (CRIDA)

Mishra, A. K.; Singh, V. P. (2011). "Drought modelling – A review". Journal of Hydrology. 403 (1–2): 157 175. Bibcode:2011JHyd..403..157M. doi:10.1016/j.jhydrol.2011.03.049.

Mishra, P.K. (2002). Indigenous technical knowledge on Soil and Water Conservation in semi-arid India (Eds: P.K. Mishra, G. Sastry, M. Osman, G.R. Maruthi Sankar and N. Babjee Rao). NATP, CRIDA, Hyderabd. 151p

Mittal SP and Singh P (1983) Studies on intercropping of field crops with fodder crops of subabul under rainfed conditions annual report. Central Soil & Water Conservation Research & Training Institute, India

Mittler R. (2006). Trends Plant Sci. 11: 15-19.

MOA (2021) Manual for Drought Management, Department of Agriculture Ministry of Agriculture & Farmers' Welfare, Government of India, New Delhi. https://agricoop.nic.in/sites/default/files/Updated%20Drought%20Manual.

Moharana, P.C. and Yadava, O.P (2019) What it takes to reclaim the Thar. Down to Earth December 04, 2019 | 10:19 AM

Molden, D. (ed.) (2007) Water for Food-Water for Life: a Comprehensive Assessment of Water Management in Agriculture Earth scan, London and International Water Management Institute (IWMI), Colombo, Sri-Lanka, pp.1–39

Montagnini, F. and Nair, P. K. R. (2004) Carbon sequestration: An underexploited environmental benefit of agroforestry systems; Agroforest Syst. 61: 281–295.

Morgan, I., McDonald, D.G., Wood, C.M., (2001). The cost of living for freshwater fish in a warmer, more polluted world. Global Change Biology. 7: 345–355.

Morgan, J.M., Hare, R.A. and Fletcher, R. J. (1986). Genetic variation in osmoregulation in bread and durum wheats and its relationship to grain yield in a range of field environments. Australian Journal of Agricultural Research, 37: 449-457

Morgan, J.M. and Codon, A.G. (1986). Water use, grain yield and osmoregulation in wheat. Australian Journal of Plant Physiology. 35: 299-319

Mucheru-Muna, M.; Pypers, P.; Mugendi, D.; Kungu, J.; Mugwe, J.; Merckx, R.; Vanlauwe, B. A. (2010). staggered maize legume intercrop arrangement robustly increases crop yields and economic returns in the highlands of Central Kenya. Field Crops Res. 115: 132–139

Muchow, R.C. and Sinclair, T.R. (1986) Water and Nitrogen limitations in soybean grain production, II Field and Model Analyses. Field Crop Research. 15: 143 -156.

Mukherjee, Sourav; Mishra, Ashok; Trenberth and Kevin E. (2018). "Climate Change and Drought: A Perspective on Drought Indices". Current Climate Change Reports. 4 (2): 145–163. doi:10.1007/s40641-018-0098-x. ISSN 2198-6061. S2CID 134811844.

Mukta, N., Sudhakara Babu, S.N., Nagaraj, G. and Ranganatha, A.R.G. (2000). Evaluation of oil yielding trees for their growth and utilization under agroforestry system. In: Extended Summaries Vol. 3. International Conference on Managing Natural Resources for Sustainable Agricultural Production in the 21st Century (Eds. JSP Yadav et al), Indian Society of Soil Science, New Delhi, India. pp 1010-1011

Mulvany, P. (1999) The year of farmer's rights: local to international governance of agricultural biodiversity. Paper presented at the National Workshop on Agricultural Biodiversity. Nairobi, Kenya.

Munns, R., James, R.A. and Läuchli, A. (2006) Approaches to increasing the salt tolerance of wheat and other cereals..Journal of Experimental Botany, 57(5): 1025–1043 March 2006, https://doi.org/10.1093/jxb/erj100

Murthy, I.K., Gupta, M., Tomar, S., Munsi, M. and Tiwari, R. (2013). Carbon sequestration potential of agroforestry systems in India. J Earth Sci. Climate Change. 4: 131. doi:10.4172/2157-7617.10001

Murthy, S.M. (1989) Biological suppression of pulse and oilseed pests. Proceedings of the Seminar-cum-seventh workshop on biological control of crop pests and weds. AICRP on BCCP&W, Biological Control Centre, Bangalore. Technical Document, No. 26:191pp

Musemwa, L., Muchenje, V., Mushunje, A. and Zhou, L. 2012. The Impact of Climate Change on Livestock Production amongst the Resource-Poor Farmers of Third World Countries: A Review. Asian Journal of Agriculture and Rural Development. 2(4): 621 – 631

Muthuraman Yuvaraj, Kasiviswanathan, Subash Chandra Bose, Prabakaran Elavarasi, and Eman Tawfik (2021) Soil Salinity and Its Management In: Recent Advances in Salinity Management in Agriculture. INDIAN EXPERIENCE Ed. Gurbachan Singh in 2019 Third World Irrigation Forum (WIF3) 1-7 September 2019, Bali, Indonesia ST-1.1 W.1.1.39 1

Mysore Study (2021) Opinion: Horticulture-Driving India's agricultural growth. https://government.economictimes.indiatimes.com/news/governance/opinion-horticulture-driving-indias-agricultural-growth/88256074

NABARD. (2016). Doubling farmers' income by 2022.Monograph of NABARD released on the eve of its 35th Foundation Day,12 July 2016

Nabel M, Schrey SD, Temperton VM, Harrison L and Jablonowski ND (2018) Legume Intercropping with the Bioenergy Crop Sida hermaphrodita on Marginal Soil.

Nadelhoffer, K.J. and Raich, J.W. (1992). Fine root production estimates and belowground carbon allocation in forest ecosystems. Ecology. 73: 1139–1147

Nagaraja, G.N., Rajeswari, K., Gowda, N. and Girish, M. (2004) Dryland based sustainable farming system models for eastern dry zone of Karnataka. Extension Education Unit. University of Agricultural Sciences (UAS), Bangalore.

Nagarajan, S., Bhavani, R.V. and Swaminathan, M.S. (2014) Operationalizing the concept of farming system for nutrition through promotion of nutrition-sensitive agriculture. Curr. Sci., 107(6).

Nageswara Rao, R.C., Talwar, H.S. and Wright, G.C. (2001) Rapid assessment of specific leaf area and leaf nitrogen in peanut (Arachis hypogaea L.).using chlorophyll meter. Journal of Agronomy and Crop Science, 186:175-182.

Nageswara Rao, R.C., Williams, J.H., Sivakumar, M.V.K., Wadia and K.D.R. (1988) Effect of water deficit at different growth phases of Peanut. II. Response to drought during pre-flowering phase. Agron.J. 80: 431-438

Nageswara Rao, R.C. and Wright, G.C. (1994) Stability of relationship between specific leaf area and carbon isotope discrimination across environments in peanut. Crop Science, 34:98-103.

Nair, P.K.R. (2012). Agroforest Syst. 86:243–253., DOI 10.1007/s10457-011-9434-z

Nair, P.K.R., Kumar, B.M. and Nair, V.D. (2009). Agroforestry as a strategy for carbon sequestration. J Plant Nutr Soil Sci. 172: 10–23

Nair, P.K.R., Nair, V.D., Kumar, B.M., Haile, S.G. (2009) Soil carbon sequestration in tropical Agroforestry systems: a feasibility appraisal. Environmental Science and Policy 12: 1099-1111

Nair, P.K.R., Nair, V.D., Kumar, B.M. and Showalter, J.M. (2010). Carbon sequestration in agroforestry systems. Adv Agron. 108: 237–307

Nair PKR, Nair VD, Kumar BM and Showalter JM. (2010) Carbon sequestration in agroforestry systems. Adv Agron. 108: 237–307

Nandini, D.U. and Somasundaram, E.(2020) Intercropping – A Substantial Component in Sustainable Organic Agriculture. Ind. J. Pure App. Biosci. (2020) 8(2), 133-143

Narain, P. (2008) Dryland management in arid ecosystem. Journal of the Indian Society of Soil Science 56: 337-347.

Narain, Sunita (2014) Breaking the impasse of 2013, Business Standard, January12, 2014

Narashima Reddy, K. (2006). Agriculture& Herbal Vision, Feb-March: 13-17.

Narayanamoorthy, A. (2021), Farm Income in India: Myths and Realities, Oxford University Press, New Delhi.

Narayanamoorthy, A. (2021) Why farm income in India is so low, Farm Trouble, Updated On: Oct 19, 2021 Business Line, The Hindu.

Natarajan, M. and R.W. Willey, 1985. Effects of rows arrangement on light interception and yield in Sorghum-pigeonpea intercropping. J. Agric. Sci., 104: 263-270.

Natarajan, M. and R.W. Willey,R.W. (1985). Effects of rows arrangement on light interception and yield in Sorghum-pigeonpea intercropping. J. Agric. Sci., 104: 263-270

NATP,(2004). Rainfed Agro -Ecosystem, Production Systems Research, Completion Report (199-2004) National Technology Project, Central Research Institute for Dryland Agriculture ,CRIDA, Hyderabad ,202 pp202

Nature Communications. 8: 1899 (2017). https://doi.org/10.1038/s41467-017-01491-7

Nazarov, P.A., Baleev,D.N., Ivanova,M.I., Sokolova, L.M. and Karakozova, M.V.(2020) Infectious Plant Diseases: Etiology, Current Status, Problems and Prospects in Plant Protection. Acta Naturae., 12(3): 46–59. doi: 10.32607/actanaturae.11026

Negi, M. S., Tandon, V. N. and Rawat H S. (1995). Biomass and nutrient distribution in young teak (Tectona grandis) plantation in Tarai Region of Uttar Pradesh. Indian Forester 121(6): 455– 63.

Nelson, S. C., and Robichaux, R. H. (2006). Identifying plant architectural traits associated with yield under intercropping: implications of genotype-cropping system interactions. Plant Breed. 116, 163–170. doi: 10.1111/j.1439-0523. 1997.tb02172.x

Neuendorf, K.E.K., Mehl, Jr. J.P. and Jackson, J.A. (2005). Glossary of Geology. Springer-Verlag, New York. p. 779. ISBN 978-3-540-27951-8

Newaj R and Dhyani SK. 2008. Agroforestry for carbon sequestration: Scope and present status. Indian Journal of Agroforestry. 10: 1-9.

NEWS-NCR (2021) Despite the rise in the international prices of agriculture BY Editorial News-Ncr, November 11, 2021.

NICRA (2014) Annual Report 2014-15NICRA-AICRPDA. All India Coordinated Research Project for Dryland Agriculture. ICAR-Central Research Institute for Dryland Agriculture, Indian Council of Agricultural Research, Hyderabad – 500 059, India. p. 336.

NICRA (2015-2019)Annual Reports, AICRPDA-NICRA, 2011 to 2015. Managing Weather aberrations through Real-time contingency planning. AICRPDA. CRIDA, Hyderabad.

Nigam, S.N., Nageswara Rao, R.C. and Wright, G.C. (2003) Breeding for increased water use efficiency in groundnut. Proceedings of the National Seminar in Oilseeds for attaining self-reliance in vegetable oils (Eds. Mangal Rai, Harvir Singh and D.M. Hegde) January 28-30. Indian Society of Oilseeds Research, Directorate of Oilseeds Research, Hyderabad, India. Pages 305-318

Nigude, V.K.,. Patil, S. A., Mohite, P.B. and Bagade, A.S. (2018) Seasonal Incidence of Tobacco Leaf Eating Caterpillar and Leaf Miner of Groundnut (Arachis hypogaea L.) Int.J.Curr. Microbiol.App.Sci (2018) 7(1): 562-565.

Nimblkar P.K., Chandrakant, A., Subhash, C. and Firoz, H. (2016) Review Article Multi-storeyed Cropping System in Horticulture-A Sustainable Land Use Approach .International Journal of Agriculture Sciences ISSN: 0975-3710&E-ISSN: 0975-9107, Volume 8, Issue 55, 2016, pp.-3016-3019.

NITI Ayog, (2021) Repot of the Committee Constituted for Formulation of Strategy for Flood Management Works in Entire Country and River Management Activities and works related to Border Areas (2001-2026). NITI Ayog, Government of India, New Delhi. 2021

NOAA (2002) National Oceanic and Atmospheric Administration (2002-05-16). "Warm Temperatures and Severe Drought Continued in April Throughout Parts of the United States; Global Temperature For April Second Warmest on Record"

NOAA (2007) National Oceanic and Atmospheric Administration. What is Drought? 2006. Retrieved 2007-04-10

NPCC (2007) Final report of Network Project on Climate Change (2007), Central Research Institute for Dryland Agriculture, ICAR, CRIDA, Hyderabad

NRAA. (2012) Prioritization of Rainfed Areas in India, Study Report 4, NRAA, New Delhi, India 100p

NRCAF. (2007). Perspective Plan Vision 2025 National Research Centre for Agroforestry, Jhansi, Uttar Pradesh.

NSSO. (2014). Key indicators of situation assessment survey in India, 70th round (2013), Ministry of statistics and uproar am implementation, National sample survey office, New Delhi.

NWDB (1987) Cost Models for Establishment of Silvi- pasture Forms in Different Regions. Monograph National Wasteland Development Board (NWDB), New Delhi

Nyarko, K.A., Seshu Reddy, K.V., Nyang'or, R.A. and Saxena, K.N. (1994) Reduction of insect pest attack on sorghum and cowpea by intercropping https://onlinelibrary.wiley.com/doi/abs/10.1111/j.1570-7458.1994.tb00745.x https://doi.org/10.1111/j.1570-7458.1994.tb00745.

Nyasani, J.O., Mayhofer, R., Subramanian, S. and Poehling, H.M. (2012) Effect of intercrops on thrips species composition and population abundance on French beans in Kenya. Entomologia Experimentalis et Applicata 142(3):236-246

Nyawade, S.O.; Gachene, C.K.; Karanja, N.N.; Gitari, H.I.; Schulte-Geldermann, E.; Parker, M.L. (2019) Controlling soil erosion in smallholder potato farming systems using legume intercrops. Geoderma Reg. 2019, 17: e00225.

Oad, F.C., Siddiqui, M.H. and Buriro, U.A. (2007). Growth and Yield Losses in Wheat Due to Different Weed Densities. Asian Journal of Plant Sciences, 6: 173-176. DOI: 10.3923/ajps.2007.173.176 URL: https://scialert.net/abstract/?doi=ajps.2007.173.176

OECD (2019) OECD Agricultural Policy Monitoring and Evaluation 2019 India's agro-food sector has made strong progress, but new policy approach needed to meet challenges. The Organization of Economic Cooperation and Development(OECD).

Okigbo, B.N. 1981. Alternatives to shifting cultivation. Ceres (FAO), 15 (6): 41-45. Rome.

Oladosu, Y., Rafii, M.Y. et al., (2016) Principle and application of plant mutagenesis in crop improvement: a review. Journal Biotechnology & Biotechnological Equipment 30 (1) : 1-16.

Oldeman, L.R.(1994) The global extent of land degradation. In: Greenland DJ, Szabolcs I, editors. Land Resilience and Sustainable Land Use. Wallingford UK: CAB International; 1994. pp. 99-118

Onwueme, I.C. and Adegoroye, S.A. (1975) Emergence of Seedlings from different depths following high temperature stress. Journal of Agricultural Science, Cambridge. 84: 525-528

Osman, M., Emmingham, W.H. and Sharrow, S.H. (1998). Growth and management. Annals of Forestry 9: 249-256.the International Technical Workshop on the Feasibility of Non-edible Oil Seed Crops for Biofuel Production May 25-27, 2007, Mae Fah Luang University, Chiang Rai, Thailand. pp 24-38.

Osman, M., Korwar, G.R. and Katyal, J.C. (1997). Composite Land Use Systems. In Resource Management in Rainfed Drylands, MYRADA, Bangalore, India and International Institute of Rural Reconstruction (IIRR), Silang, Cavite, Philippines. pp 195-198

Osman, M., Venkateswarlu, B. and Singh, R.P. (1989) Agroforestry systems for semi-arid tropics of India. – A Review. IN: Agroforestry Systems in India: Research and Development (Eds: Singh R.P., Ahlawat, I.P.S. and Gangasaran). Indian Society of Agronomy, New Delhi. India. pp 18-35

Ozede Igiehon, N. and Oluranti Babalola.O (2018) Rhizosphere Microbiome Modulators: Contributions of Nitrogen Fixing Bacteria towards Sustainable Agriculture, Int. J. Environ. Res. Public Health. 2018 Apr; 15(4): 574.

Pacey., A. and Cullis, A. (1986) Rainwater harvesting: the collection of rainfall and runoff in rural areas. Intermediate Technology Publications, London

Paepe, R., Fairbridge, R.W. and Jelgersma, S. (1990). Greenhouse Effect, Sea Level and Drought. Springer Science & Business Media. p. 22. ISBN 978-0792310174

Palaniappan SP and Sivaraman (976) cropping systems in the tropics Principles and Management,New Age International(P) publishers(formerly Wiley Eastern Limited)1-209

Pandey, D.N. (2007) Multi-functional Agro-forestry Systems in India. Curr. Sci.,92(4):455-463

Pandey D N. (2002) Carbon sequestration in agroforestry systems. Climate Policy 2: 367–77

Pang, H.C., Li, Y.Y., Yang, J.S. and Liang, Y.S. (2010). Effect of brackish water irrigation and straw mulching on soil salinity and crop yields under monsoonal climatic conditions. Agricultural ter Management.97:1971-1977

Pankaj, P.K., Ramana, D.B.V., Pourouchottamane, R. and Naskar, S. (2013) Livestock Management under Changing Climate Scenario in India, World Journal of Veterinary Science. 32 (39) DOI:10.12970/2310-0796.2013.01.01.5

Pareek, S.K., and Gupta R., (1993) Medicinal and aromatic improve profitability of cropping systems in India. In glimpses in plant research (J.W. Govil et al.,) pub Today and tomorrow printers and pub, New Delhi) Vol.11 pp 325-35.

Parmar, N., Singh, K.H., Sharma, D., Singh, L., Kumar, P., Nanjundan, J., Khan, Y.J., Chauhan, D.K. and Thakur, A. K. (2017) Genetic engineering strategies for biotic and abiotic stress tolerance and quality enhancement in horticultural crops: a comprehensive review, 3 Biotech. 2017 Aug; 7(4): 239. doi: 10.1007/s13205-017-0870-y

Paroda, Raj. (2022) Crop Diversification for Sustainable Agriculture. Publication of the Trust for Advancement of Agricultural Sciences (TAAS) New Delhi. 2022

Paroda, Raj. (2022A)Towards Secure and Sustainable Agriculture. StragyTrust for Advancement of Agricultural Sciences (TAAS) New Delhi. 2022 Strategy Paper.

Parry, M.A.; Reynolds, M.; Salvucci, M.E.; Raines, C.; Andralojc, P.J.; Zhu, X.G., Price, G.D.; Condon, A.G. and Furbank, R.T. (2011) Raising yield potential of wheat. II. Increasing photosynthetic capacity and efficiency. J. Exp. Bot. 2011, 62, 453–467

Passioura, J.B. (1976) Physiology of grain yield on wheat growing on stored water. Australian Journal of Plant Physiology, 3: 559-565

Passioura, J.B. (1977) Grain yield, harvest index and water use of wheat. Journal of Australian Institute of Agricultural Sciences. 43: 117-121.

Passioura, J.B. (1981) The interaction between the physiology and breeding of wheat. Pages 191-201. In: Wheat Science today and tomorrow. (Eds. Evans, L.T.E. and Peacock, W.J.) Cambridge University Press, Cambridge, UK.

Passioura, J.B. (1983) Roots and drought resistance. Agricultural Water management,7: 265-280

Pathak, P., Laryea, K.B. and Singh, S. (1989). A modified contour bunding system forAlfisols of the semi-arid tropics. Agricultural Water Management 16:187-199.

Pathak, P., Mishra, P.K., Rao, K.V., Suhas P Wani and Sudi, R. (2007)Best -bet options on soil and water conservation P75-94.Best-bet options for Integrated Watershed Management. Proceedings of the Comprehensive Assessment of watershed programs in India,25-27 july,2007 ICRISAT,Patancheru,India. (eds) Suhas P wani,B.Venkateswarlu.KL Sahrawat,KVRao and YS Ramakrihna.PP312

Pathak, P.S., Gupta, S.K. and Singh, P. (1996). IGFRI Approaches: Rehabilitation of Degraded Lands. IGFRI, Jhansi. pp 1-23

Pathak, P.S., Pateria, H.M. and Solanki, K.R. (2000) Agroforestry systems in India - A diagnosis and design approach. All India Coordinated Research Project on Agroforestry ICAR, National Research Centre fo r Agroforestry, Jhansi, India

Pathak, P.S. and Solanki, K.R. (2002) Agroforestry Technologies for Different Agro-climatic Regions of India. Indian Council of Agricultural Research (ICAR), p 41

Patidar M and Mathur B K. (2017). Enhancing forage production through a silvi-pastoral system in an arid environment. Agroforestry Systems 94: 713–27.

Patil, S.J., Mutanal, S.M., Patil, H.Y., Shahapurmath, G., and Maheswarappa, V. (2010). Performance of Sapota-Teak based agroforestry system in hill zone of Karnataka. Indian Journal of Agroforestry. 12(1): 27-34.

Patil EN, Bonde RD, Salunkhe CD, Borse RH (1999) Effect of fruit and fodder trees on rainfed arable crops.Indian J Dryland Agric Res Dev 14(2):81–83

Paudel, S., Sah,L., Devkota, M. and Podyl, V. (2020) Conservation Agriculture and Integrated Pest Management Practices Improve Yield and Income while Reducing Labor, Pests, Diseases and Chemical Pesticide Use in Smallholder Vegetable Farms in Nepal. Sustainability 12(16):6418 DOI:10.3390/su12166418

Paul Hawken, (2019) Project Drawdown A New Context for Global Warming. (https://www.theenergymix.com/2019/12/09/tree-intercropping-would-save-17-2-gigatons-of-carbon-by-2050-2/

Peacock, J.M. (1982). Response and tolerance od sorghum to temperature stress. Pages 143-159. In: Sorghum in Eighties. Proceedings of the International Symposium on Sorghum. Nov.2-7, 1981. ICRISAT Centre, Patancheru, A.P. India.

Pearce, R.P. (2002). Meteorology at the Millennium. Academic Press. p. 66. ISBN 978-0-12-548035-2. Retrieved 2009-01-02.

Pelzer E., Bazot M., Makowski D., Corre-Hellou G., Naudin C., Al Rifaï M., et al. (2012). Pea-wheat intercrops in low-input conditions combine high economic performances and low environmental impacts. Eur. J. Agron. 40 39–53. 10.1016/j.eja.2012.01.010

Pereira, A.M.F., Baccari Jr, F., Titto, E.A.L. and Almeida, J.A.A. 2008. Effect of thermal stress on physiological parameters, feed intake and plasma thyroid hormones concentration in Alentejana, Mertolenga, Frisian and Limousine cattle breeds. International Journal of Biochemistry. 52: 199-208

Perez, I. B.; Brown, P. J. (2014). "The role of ROS signaling in cross-tolerance: from model to crop". Front. Plant Sci. 5: 754. doi:10.3389/fpls.2014.00754. PMC 4274871. PMID 25566313. (20)

Peterson, G., Unger, P.W. and Payne, W.A. (2006). American Society of Agronomy, Crop Science Society of America. and Soil Science Society of America. Dryland agriculture. American Society of Agronomy: Crop Science Society of America: Soil Science Society of America, Madison, Wis.

Physiological Response of Multiple Tree Seedlings to Induced Water Stress. Plant Archives Vol. 19: 1, 2019 pp. 444-447 e-ISSN:2581-6063 (online), ISSN:0972-5210

Planning Commission (2007) Mid-term Report, 2007, Planning Commission, New-Delhi

Pranjit, P.K., Hazarika, M., Sarma, D., Saikia, P., Pankaj Kumar, N., Rajbongshi, R., Nikhita, K., Bhattacharjee, M. and Srinivasarao, Ch. (2015). Effect of foliar application of potassium on yield, drought tolerance and rain water use efficiency of toria under rainfed upland situation of Assam. Indian Journal of Dryland Agriculture Research and Development 30: 55-59

Prasad, J.V.N.S,, Korwar, G.R., Rao, K.V., Mandal, U.K., RamaRao,C.A., Rao, G.R., Venkateswarlu, B., Rao, S.N., Kulkarni, H.D., Rao, M.R. (2010a) Tree row spacing affected agronomic and economic performance of Eucalyptus -based agroforestry in Andhra Pradesh, Southern India Agrof. Syst. 78(3):253-267

Prasad, J.V.N.S., and Ramakrishna, B. (2018) Agroforestry as an adaptation and mitigation strategy for climate change ICAR-Sponsored Short Course on 'Assessment of vulnerability and adaptation to climate change in agriculture', November 28 to December 07, 2018, ICAR-CRIDA, Hyderabad.

Prasad, J.V.N.S., Korwar G.R., Rao, K.V., Srinivas, K., RamaRao, C.A.,, Srinivasa Rao, Ch., Venkateswarlu, B., Rao, S.N. and Kulkarni, H.D. (2010b) Effect of modification of tree density and geometry on intercrop yields andeconomic returns in Leucaena based agroforestrysystems for wood production in Andhra Pradesh, Southern India. Exp. Agric. 46(2):155-172

Prasad, J. V. N. S., Srinivas, K., Srinivasa Rao, Ch., Ramesh, C., Venkatravamma, K. and Venkateswarlu, B. (2012). Biomass productivity and carbon stocks of farm forestry and agroforestry systems of leucaena and eucalyptus in Andhra Pradesh, India. Current Science 103(5): 536–40

Prasad, JVNS 2011 Alternate land use systems for better soil and water resource utilization CRIDA, Hyderabad

Prasad, M.V.R. (1972). Use of induced mutants for grain characters in durum wheat breeding. Current Science, 41 (15):570-571.

Prasad, M.V.R. (1973) Some Considerations on Plant-types for Dryland Agriculture, Annals of Arid Zone, 12 (3&4) :124-134 September/December, 1973.

Prasad, M.V.R. (1985). Aerial podding in groundnut (Arachis hypogaea, L.). Indian J. Genetics. 45(1):89-91.

Prasad, M.V.R. (2018) Performance of castor and pigeon pea grown as intercrops in an elite clonal plantation of Pongamia on dryland near Hyderabad. (Unpublished).

Prasad, M.V.R. (2021) Oilseed Crops, section: Linseed. New India Publishing Agency, New Delhi.

Prasad, M.V.R. (2021A) Pongamia for Bioenergy and Better Environment, New India Publishing Agency, New Delhi.pp70

Prasad, M.V.R., Mamede, F.B.F. and De Silva, F.P. (1985). Mutational improvement of peanut (Arachis hypogaea, L.). Pesq. Agro. Brasileira 20 (4):439-445.

Prasad, M.V.R., Swarnalatha Kaul and Jain H.K. (1984) Induced Mutations of Peanut (Arachis hypogaea L.) for canopy and pod characters. Indian J. Genet and Plant Breeding. 44: 25-34

Prasad, M.V.R. and Reddy, B.N. (1992). Advances in oilseeds production under dryland situations. Chapter in the book "Sustainable Development of Dryland Agriculture in India", (R.P. Singh. Ed.) Scientific Publishers, Jodhpur, India. Pp 318-333

Prasad, M.V.R. and Sudhakar Babu, S.N. (1997). Technology for Dryland agriculture. Monograph in Agriculture and Environment Series, ICAR, New Delhi, India. 39pp

Prasad, Y.G., Venkateswarlu, B., Ravindra Chary, G., Srinivasarao, Ch., Rao, K.V., Ramana, D.B.V., Rao, V.U.M., Subba Reddy, G. and Singh, A.K. (2012). Contingency Crop Planning for 100 Districts in Peninsular India. Central Research Institute for Dryland Agriculture, Hyderabad 500 059, India. p. 302

Pratap Toppo and Abhishek R 2018 Role of agroforestry in climate change mitigation Journal of Pharmacognosy and Phytochemistry 2018; 7(2): 241-243

Pratibha G., and G.R. Korwar (2002). Crop diversification through medicinal, aromatic and dye yielding plants for sustainability in semi-arid regions. Proceedings of first national meet on medicinal & aromatic Plants. eds Mathur A.K., S. Dwivedi, DD Patra, GD

Prats, S.A., Wagenbrenner, J.W., Martins, M.A., Malvar, M.C. and Keizer, J.J.(2016) Hydrologic implications of post-fire mulching across different spatial scales. Land Degradation and Development. 27(5):1440-1452

Puangbut, D., Jogloy, S., Vorassot, N. et al., (2009) Variability in response of Peanut (Arachis hypogaea L.) genotypes under early season drought. Asian J. Plant Sciences, 8: 254-264.

Palacio- PabloGracia, María RemediosAlarcón, José LuisTenorio and Sara SánchezMoreno (2019) Ecological intensification of agriculture in drylands. Journal of Arid Environments 167: 101-105

Puri S, Singh S and Bhushan B (1994) Evaluation of fuel wood quality of indigenous and exotic tree species of India's semiarid region. Agrofor Syst 26:123–130

Rabindra, R. J. and Balasubrahmanian, M. (1980) Observations on an induced epizootic of nuclear polyhydrosis virus of Amsecta albistriga Wlk. Current Science, 49:279

Radhamani, S.(2001) Sustainable Integrated Farming System for dryland vertisol areas of Western Zone of Tamil Nadu. Ph.D. Thesis, TNAU,

Rai, M.P., Pathak, P.S., Deb Roy, R. (1983) Biomass production of *Sesbania grandiflora* in different habitats. Int. J Ecol. Environ Sci. 9:21-27

Rai, P., Ajit, Chaturvedi P.O., Singh, R. and Singh, U.P. (2009). Biomass production in multipurpose tree species in natural grasslands under semi-arid conditions. Journal of Tropical Forestry. 25: 11-16.

Rai, P., Deb Roy, R., Verma, N.C. and Rao, G.R. (1994) Performance of small ruminants on silvipasture and natural grassland under grazing condition. In: SinghP, Pathak PS, Roy MM (eds) Agroforestry Systems for Degraded Lands. Oxford and IBH publishing Co.Pvt. Ltd., New Delhi, pp 773-77

Raj, N. 2009. Air pollution-A threat in vegetable production. In: Sulladmath, U. V. and Swamy, K.R.M. International Conference on Horticulture (ICH-2009) Horticulture for Livelihood Society and Economic Growth. 158-159

Rajeshwar Rao G, Prabhakar M, Venkatesh G, Srinivas I and Sammi Reddy K (Eds.). (2018). Agroforestry Opportunities for Enhancing Resilience to Climate Change in Rainfed Areas, ICAR - Central Research Institute for Dryland Agriculture, Hyderabad, India. p. 224

Raj Kumar Gupta, Vijay Kumar, K.R., Sharma, Tejbir Singh Buttar, Gobinder Singh and Gowhar Mir. (2017). Carbon Sequestration Potential through Agroforestry: A Review. Int.J.Curr. Microbiol. App.Sci. 6(8): 211-220. Doi https://doi.org/10.20546/ijcmas.2017.608.029

Raju, N. J.; Gossel, W.; Ramanathan, A.; Sudhakar, M., eds. (2015). Management of Water, Energy and Bio-resources in the Era of Climate Change: Emerging Issues and Challenges. doi:10.1007/978-3-319-05969-3. ISBN 978-3319059686. S2CID 132165592. (22)

Raju, S.S and Chand, Ramesh (2008). A study on the performance of NAIS and suggestions to make it effective. Agricultural economics research review 21(1),11—19.

Raj Vir Singh. (2000). Watershed planning and management. Bikaner, India: Yash Publishing House. 470 pp.

Ram, N. and Gill, A.S. (1995) Effect of Leucaena and intercrop on anola (Emblica officinalis) under rainfed conditions. Van Anusnadhan 12(ii):51-55

Rama Rao C.A., Raju BMK, Subba Rao, AVM, Rao, K.V., Rao VUM, Kausalya Ramchandran, Venkateswarlu B. and Sikka, A.K. (2013). Atlas on vulnerability of Indian agriculture to climate change. Central Research Institute for dryland agriculture, Hyderabad: 116.

Ramasamy, C.S., Natarajan, C., Jayanthi and Suresh Kumar.D. (2007). Intensive integrated farming system to boost income of farmers. In: Proceedings of 32nd IAUA vice chancellors' annual convention on Diversification in Indian Agriculture, Birsa Agricultural University, December 20 - 21. pp. 28-4

Ram M. (2020) Effect of Intercropping on Productivity and Profitability of Sesame under Dryland Arid Conditions. Curr Agri Res 2020; 8(2). doi : http://dx.doi.org/10.12944/CARJ.8.2.1

Ram Mohan Rao, M.S., Chittaranjan, S., Selvarajan, S. and Krishnamurthy, K. (1981) Proceedings of the panel discussion on soil and water conservation in red and black soils, 20 March 1981, UAS, Bangalore, Karnataka: Central and Soil and Water Conservation Research and Training Institute, Research Center, Bellary, Karnataka and University of Agricultural Sciences, Bangalore, India. 127 pp

Ramrao, W.Y., Tiwari, S.P., Sarawat, S.S., Pathak, R. and Gupta, R. (2008) Integration of crop-livestock-poultry-duck farming system by large farmers in plain tribal areas of Chhattisgarh in Central India. Livestock Research for Rural Development 20. Art#42

Ramrao, W.Y., Tiwari, S.P. and Singh ,P. (2006)Crop-livestock integrated farming system for marginal farmers in rainfed regions of Chhattisgarh in Central India. Livestock Research for Rural Development. 18 Art.#102.

Rana S.S. and Rana, M.C. (2011) Cropping System. Department of Agronomy, College of Agriculture,CSK Himachal Pradesh Krishi Vishvavidyalaya, Palampur, 80 pages https://www.researchgate.net/publication/309211205_Cropping_System

Ranga Rao, V. (2021). Translating PM's Elusive Goal "Doubling Farmers' Incomes" into Ground Reality: Scenario, Challenges, Opportunities and the Way Forward In: RICAREA 2021. Environment and Agriculture.Souvenir,7th M S Swaminathan Award.2018-2019. Retired ICAR Employees' Association (RICAREA) Hyderabad. Pages 25-31

Ranga Rao G.V.., Desai S., Rupela O.P., Krishnappa K. and Wani S.P. (2009) Integrated Pest management options for better crop Production. IN: "Best-bet options aimed at integrated watershed management". Proceedings of comprehensive assessment of watershed programme in India,25-27 july 2007,I CRISAT.Patancheru,502324.Andhra Pradesh

Rao, A. V. R. K., Wani, S. P., Singh, K. K., Ahmed, M. I., Srinivas, K., Bairagi, S. D. and Ramadevi (2013) Increased arid and semi-arid areas in India with associated shifts during 1971-2004. Journal of Agrometeorology, 15 (1). pp. 11-18.

Rao, G.R., Prasad, Y.G., Prabhakar, M., Rao, J.V., Korwar, G.R. and Ramakrishna, Y.S. (2006). Agrotechniques for Biodiesel Plantations in Rainfed Areas. Technical Bulletin 2/2006. Central Research Institute for Dryland Agriculture, Hyderabad. p.16.

Rao, G.R., Sarkar, B., Reddy, P.S., Raju, B.M.K and Sharat, P.K. (2015) Identification of Pongamia (Pongamia pinnata L.) Based Suitable Intercropping Practices for Sustainable Production of Biofuel and Nutritional Security in Semi-Arid India. Indian Journal of Dryland Agricultural Research and Development. 30 (2).

Rao, J.V. (1999) Agri-horticultural systems in rainfed Alfisol watershed. Annual Report of CRIDA, Hyderabad,1998-99.

Rao, J.V. and Osman, M. (1994) Studies on silvi-pastoral systems in non-arable dry lands, In: Singh P, Pathak PS, Roy MM (eds) Agroforestry Systems for Degraded Lands. Oxford & IBH, New Delhi, pp 755-760

Rao, J. V. and Sujata, S. (2003). Alternate land use systems for marginal soils. A Compendium on Waste land development for rain-fed areas, CRIDA, Hyderabad. pp.188-194.

Rao, S.C. and Ryan, J. (2004). Challenges and strategies for dryland agriculture. In: The Symposium on "Challenges and strategies for dryland agriculture". Crop Science Society of America, Madison, Wis.USA

Rao K.V., Venkateswarlu,B., Vital,K.P.R., and Sharma, B.R. (2001) Water harvesting potential assessment in rainfed regions of India. In proc .National workshop on rain water harvesting and reuse through farm ponds: Experiences, Issues and strategies (Ed KV Rao et.al) CRIDA, Hyderabad, pp 67-74.

Rao MR, Ong CK, Pathak P, Sharma MM (1991) Productivity of annual cropping and agroforestry systems on a shallow Alfisol in Semi-arid India. Agroforest Syst 15:51–6

Raper, C.D. and Barber, S.A. (1970). Rooting systems of soybeans. 1. Differences in root morphology among varieties. Agronomy Journal, 62: 581-584.

Rapsomanikis, G. (2019) The economic lives of smallholder farmers, Food and Agriculture Organization of the United Nations (FAO) http://www.fao.org/3/a-i5251e.pdf

Ravindra Chary, G., Gopinath, K.A., Bhaskar, S., Prabhakar, M., Chaudhari, S.K. and Narsimlu, B. (Eds.). 2020. Resilient Crops and Cropping Systems to Cope with Weather Aberrations in Rainfed Agriculture. All India Coordinated Research Project for Dryland Agriculture (AICRPDA), ICAR-Central Research Institute for Dryland Agriculture, Hyderabad-500 059. 56 p.

Ravindra Chary, G., Venkateswarlu, B., Sharma, S.K., Mishra, J.S., Rana, D.S. and Ganesh Kute. (2012). Agronomic research in dryland farming in India: An Overview. Indian Journal of Agronomy 57 (3rd IAC special issue): 157-167.

Ravindra CharyG., Gopinath, K.A., Narsimlu, B. and Anantha Rao, D. (2017) Crops and cropping systems strategies in rainfed agriculture, In: K. Ravi Shankar, K. Nagasree, G. Nirmala, Jagriti Rohit, R. Nagarjuna Kumar, P.K. Pankaj and K. Sammi Reddy (Eds) 2017. Compendium of lectures on "Efficient Watershed Management in Rainfed Agriculture" ICAR- Central Research Institute for Dryland Agriculture, Hyderabad – 500 059. PP-91

Ravindranath, N.H. (2007). "Forests in India-Take Action Now", in The Hindu Survey of the Environment, The Hindu, Special Issue, New Delhi

Ravindra nathan, N.H., Chanakya, H.N. and Kerr, J.M. (1997). An ecosystem approach for analysis of biomass of energy, pp.561-587. In: Ker JM, Marothia DK, Singh K, Ramaswamy C and Bentley WR. (eds.) Natural Resource Economics Theory and Application in India. Oxford and IBH, New Delhi. p 156-163.

Ravindranath NH, Sukumar R (1998) Climate change and tropical forests in India. Clim Change 39:563–581

Reddy, G.S. (2009) Integrated farming system models for sustaining rural livelihoods in rainfed areas. In: Compendium of Lectures. Winter School on "Emerging concepts of soil and water management in drylands". February 10 to March 02, 2009. ICAR-Central Research Institute for Dryland Agriculture (CRIDA) Hyderabad, pp229-236.

Reddy, K.S. (2018) Farm pond Technology for Enhancing Resilience to Climate Change / Climate Vulnerability ICAR-Sponsored Short Course on 'Assessment of vulnerability and adaptation to climate change in agriculture', November 28 to December 07, 2018, ICAR-CRIDA, Hyderabadp.244-254

Reddy, M.D. (2005) Predominant farming systems and alternatives in Andhra Pradesh. Proceedings of the Seminar on Alternate Farming Systems. (Eds. Singh, A.K., Gangwar, B. and Sharma, A.K) Project Directorate of Cropping Systems Research, Modipuram, 16-18 September, 2004. Pp 217-227

Reddy, M.S. and Willey, R.W. (1981). Growth and resource use studies in an intercrop of pearl millet/groundnut. Field Crop Res., 4: 13-24.

Reddy, N.N. (2004). Develop agri-horticulture and agro forestry systems in Kharif sorghum area decreasing regions for overall sustainability of the production system-NATP RNPS

Reddy, N.N., Reddy, M.J.C., Reddy, M.V., Reddy, Y.V.R., Sastry, G. and Singh, H.P. (2002). Role of horticultural crops in watershed development programme under Semi – arid Sub tropical dryland conditions of Western India. Paper published in the 12th ISCO Conference held at Beijing, China from May 26- 31, 2002

Reddy, N.N. and Singh, H.P. (2002) Agri- horticultural and Horti pastoral Systems for Alternate Land Uses in Drylands of Indian Sub-Continent,459-462

Reddy,N.V. and Hampaiah, R.(1982) Alternate Land use in Dryland Agriculture in" A decade of dryland Agricultural Research in India,1971-80:AICRPDA,Hyderabad.

Reddy, Y.V.R., Sudha, M. (1988) Economics of different land use systems in dryland farming. Annual Report, CRIDA, Hyderabad. pp 62–63

Reddy S. (2011) Alternate land-based farming system in rainfed agriculture. In: Compendium of winter school organized by CRIDA, Hyderabad

Reinsborough, M.J. (2003) A Ricardian Model of climate change in Canada. Canadian Journal of Economics, 36 (1) February 2003: 21-40

Rejeb, I. B.; Pastor, V.; Mauch-Mani, B. (2014). "Plant Responses to Simultaneous Biotic and Abiotic Stress: Molecular Mechanisms". Plants. 3 (4): 458–475. doi:10.3390/plants3040458. PMC 4844285. PMID 27135514. (19)

Rhodes, D. and Hanson, A.D. (1993) Quaternary ammonium and tertiary sulfonium compounds in higher plants. Annual Review of Plant Physiology &Plant Molecular Biology.44: 357-384

RICAREA (2017) Integrated Farming Systems for Sustainable Agriculture and Enhancement of Rural Livelihoods. Proceedings of the RICAREA Seminar by Murlidharan, K., Prasad, M.V.R. and Siddiq, E.A.(Editors) Publisher: Retired ICAR Employees' Association(RICAREA) Hyderabad. India.280pp; ISBN978-[8]1-933823-0-1

Richards, R.A. (1982). Breeding and selecting for drought resistance in wheat. Pages 303-316. In: Drought resistance in crops with emphasis on rice. International Rice Research Institute, Las Banos, Laguna, Philippines.

Richthofen, F.V. (1882) "On the mode of origin of the loess". Geological Magazine (Decade II). 9 (7): 293–305. Bibcode:1882GeoM....9..293R. doi:10.1017/S001675680017164X.

Ricinus communis L.) totwo enhanced CO2 levels. Plant, Soil and Environment, 54 : 38–46

Rizvi RH, Dhyani SK, Yadav RS and Ramesh Singh. (2011). Biomass production and carbon stock of poplar agroforestry systems in Yamunanagar and Saharanpur districts of North western India. Current Science, 100 (5): 736-742

Rocheleau, D., Weber, F. and Field-Juma, A. (1988) Agroforestry in Dryland Africa. ICRAF, Kenya, p 18

Rogelj, J., Den Elzen, M., Höhne, N., et al.(2016). Paris Agreement climate proposals need a boost to keep warming well below 2°C, Nature, 534. 631-639.2016

Root, R. (1973). Organization of a plant-arthropod association in simple and diverse habitats. The fauna of collards (Brassica oleracea). Eco\. Monogr. 43: 95-124

Rosenow, D.T., Quisenberry, J.E., Wendt, C.W. and Clark, L.E. (1983). Drought tolerant sorghum and cotton germplasm. Pages 207-222. In: Plant production and management under drought conditions. (Ed. Stone, J.F. and Wills, W.O.) Amsterdam, Netherlands. Elsevier.

Roshetko, J.M., Lasco, R.D. and Angeles, M.S.D. (2007) Small holder agroforestry systems for carbon storage. Mitigation and Adaptation Strategies for Global Change 12: 219-242

Rotem, J.and Palti,J. (1969) Irrigation and Plant Diseases. Annual Review of Phytopathology. 7:267-288 https://doi.org/10.1146/annurev.py.07.090169.001411

Royal Society (2010) The Royal Society London: The Royal Society Science Policy Centre's Document Published in 2010,

Rudraradhya, M. (2002) A model of farming systems approach for sustainable agriculture. LEISA India (September) 13-14

Rukmani, R. and Manjula, M. (2009) Designing rural technology delivery systems for mitigating agricultural distress. A study of Wardha district. M.S.Swaminathan Reseaarch Foundation, Chennai and the Office of the Principal Scientific Adviser to the Government of India, New Delhi –2009

Russel, W.A. (1984) Agronomic performance of maize cultivars representing different years of maize breeding. Maydica 29: 375-390

Russel, W.A. (1995) Evaluation for plant, ear and grain traits of maize cultivars representing seven years of breeding. Maydica.30:85-96

Rockstrom, J., Louise Karlberg, Suhas P. Wani., Jennie Barron., Nuhu Hatibu., Theib Oweis., Bruggeman, A., Farahani, J. and Qiang, Z. (2010) Managing water in rain-fed agriculture—The need for a paradigm shift, Agricultural Water Management, 97 (4), April 2010, Pages 543-550

Saddington and David. (2016). "Small Islands, Big Impact: Marshall Islands Set Bold Carbon Targets." The Huffington Post. The Huffington Post.com, n.d. Web. 28 Jan.2016

Saharan, N., Korwar, G.R., Das, S.K., Osman, M. and Singh, R.P.. (1989). Silvipastoral system in marginal Alfisols for sustainable agriculture. Indian J. Dryland Agric. Res. Dev. 4: 41-47

Saharan N, Korwar GR, Das SK, Osman M and Singh RP. 1989. Silvipastoral system in marginal Alfisols for sustainable agriculture. Indian J. Dryland Agric. Res. Dev. 4: 41-47

Sahoo, D.C., Madhu, M., Adhikary, P.P., Dash, C. J., Sahu, S. S. and Devi, S. (2017) Adoption behaviour of different soil and water conservation measures among tribal farmers of Gajapati, Odisha. Indian Journal of Soil Conservation. 45(1):112-116

Salvagiotti, F., Cassman, K.G., Specht, J.E., Walters, D.T., Weiss, A. and Dobermann, A. (2008). Nitrogen uptake, fixation and response to fertilizer N in soybeans: A review. Field Crops Research, 108(1), pp. 1-13.

Samra, J.S. (2004) Resource management options in multi-enterprise agriculture (Eds: Acharya C.L., Gupta, R.K., Rao, D.L.N. and Subba Rao, A.) Proceedings of the Sixth Agricultural Science Congress Bhopal. Feb.13-15, 2003.NAAS, New Delhi.

Samra JS and Charan SS. (2000). Silvipasture systems for soil, water and nutrient conservation on degraded lands of Shivalik foot hills (subtropical northern India). Indian Journal of Soil Conservation, 28(1): 35-42

Samra JS and Sharma UC. (2002) Soil erosionand conservation. In: Sekhon GS,Chhonkar PK, Das DK, Goswami NN,Narayanasamy G, Poonia SR, Rattan RK,Sehgal JK, (editors). Fundamental of SoilScience. New Delhi: Indian Society SoilScience; 2002. pp. 159-170 (14) (PDF) Soil and Water Conservation Measures for Agricultural Sustainability

Sanchez-Bayo, F. and Goka, K. (2016) Impacts of Pesticides on Honey Bees. In: Emerson Dechechi Chambo, 2016. (Ed).Book on Beekeeping and Bee Conservation- Advances in Research Open Access Peer-Reviewed Chapter. https://www.intechopen.com/chapters/50073

Sanon, M.; Hoogenboom, G.; Traoré, S.B.; Sarr, B.; Garcia, A.; Garcia, Y.; Somé, L.; Roncoli, C. (2014) Photoperiod sensitivity of local millet and sorghum varieties in west africa. NJAS-Wagening. J. Life Sci. 2014, 68: 29–39

Sanon, M.; Hoogenboom, G.; Traoré, S.B.; Sarr, B.; Garcia, A.; Garcia, Y.; Somé, L.; Roncoli, C. (2014) Photoperiod sensitivity of local millet and sorghum varieties in west africa. NJAS-Wagening. J. Life Sci. 2014, 68: 29–39

Santamaria, J.M. (1986). Study of traits for drought resistance in Sorghum bicolor (L). Moench, with emphasis on osmotic adjustment. M.Sc. Thesis, University of Queensland, Brisbane, Queensland, Australia.

Santra,P., Mertia, R.S., Kumawat, R.N., Sinha, N. K. and Mahla, H. R. (2013). Loss of soil carbon and nitrogen through wind erosion in the Indian Thar Desert. Journal of Agricultural Physics.,13(1):13

Sarah, K., Lowder, J.S. and Terry, R. (2016) The number, size and distribution of farms.- Small holders farms and family farms worldwide. World Development, Elsevier Science Direct.87: 16-29

Sarma, J.S. (2004) Resource management options in multi enterprise agriculture. In: Multi-enterprise systems for viable agriculture (Eds: Acharya, C.L., Gupta, R.K., Rao, D.L.N. and Subba Rao, A.) Proceedings of the 6th Agricultural Science Congress, Bhopal, Feb.13-15.2003. NAAS, New Delhi.pp.19-39

Sarvade, S., Upadhyay, V.B., Kumar, M. and Khan, M.I. (2019). Soil and water conservation techniques for sustainable agriculture. In: Sustainable Agriculture, Forest and Environmental Management. Singapore: Springer; pp. 133-188

Sathaye, J. (2007). Policy Responses Adaptation and Mitigation", in The Hindu Survey of the Environment, The Hindu, Special Issue, New Delhi

Sathaye, J. and Ravindranath (1998) Climate change mitigation in the energy and forestry sectors of developing countries. Ann. Rev. Energy. Env. 23: 387-437

Satyanarayana, B.S., Hart, A., Milne, W.I. and Robertson, J. (1997) Field Emission from Tetrahedral Amorphous Carbon. Applied Physics Letters, 71, 1430-1435. http://dx.doi.org/10.1063/1.119915

Satyanarayana, D. (2020) A Study on Rural Indededness among Indian Farmers International Journal of Humanities and Social Sciences (IJHSS)ISSN(P): 2319–393X; ISSN(E): 2319–3948, 9(1), Dec–Jan 2020; 139–146© IASET

Saxena, K.B.; Chauhan, Y.S.; Kumar, C.V.S.; Hingane, A.J.; Kumar, R.V.; Saxena, R.K.; Rao, G.V.R. (2018) Developing improved varieties of pigeonpea. In Achieving Sustainable Cultivation of Grain Legumes Volume 2, Improving Cultivation of Particular Grain Legumes; Burleigh Dodds Science Publishing: Cambridge, UK, 2018; ISBN 9781786761408

SBSTTA, (1999). Biological diversity of drylands, arid, semi-arid, savannah, grassland and Mediterranean ecosystems. Draft recommendations to COP5. Montreal, Canada. 16.

SCBD (2010) (Secretariat of the Convention on Biological Diversity), 2010. Pastoralism, Nature Conservation and Development: A Good Practice Guide. Montreal, 40 + iii pages.

Schoeneberger, Michele; Domke, Grant. (2017). Chapter 3: Greenhouse gas mitigation and accounting. In: Schoeneberger, Michele M.; Bentrup, Gary; Patel-Weynand, Toral,(eds). 2017

Seetharam, A. and Virupakshappa, K. (1989) Package of practices for maximising production of sunflower. Directorate of Oilseeds Research, Hyderabad pp.45

Semere, T., and Froud-Williams, R. J. (2001). The effect of pea cultivar and water stress on root and shoot competition between vegetative plants of maize and pea. J. Appl. Ecol. 38, 137–145. doi: 10.1046/j.1365-2664.2001. 00570.x

Series (BDS) Publications are downloadable at: www.worldbank.org.bd/bds

Shankar, M.A., Aanda, N., Mahendra, R., Dhanapal, G.N., Umesh, M.R., Krishnamurthy, D. and Doreswamy, C. (2008). Strengthening of Farming System Enterprises to Enhance the Farm Productivity and Sustainability in Rainfed Condition, AICRIP for Dryland Agriculture, UAS, Bangalore.

Shankar, M.A., Dhanapal, G.N. and Umesh, M.R. (2007) Farming Systems in Rainfed Areas of Karnataka. MANAGE Ext. Res. Rev.8(2):1-19

Shao HB, Jiang SY, Li FM, Chu LY, Zhao CX. Shao MA, Zhao XN and Li F. (2007) Some advances in plant stress physiology and their implications in the systems biology era. Colloids and surfaces B. Bio-interfaces. 2007; 54 :33-36.

Sharda, V.N., Dogra, P. and Prakash, C. (2010) Assessment of production losses due to water erosion in rain fed areas of India. Journal of Soil and Water Conservation.,65(2):79-91

Sharda V.N. and Venkateswarlu,(2009) Proceeding of the Comprehensive Assessment of Crop Diversification and Alternate Land Use Systems in programs in India held during july 25-27,2007 ICRISAT ,Hyderabad ,published in Best -bet options for integrated watershed management(eds.,Suhas P Wani,B.Venkateswarlu,K.L.Saharwat,KV Rao and Ys Ramakrishna)PP111

Sharma, K.K. and Lavanya, M. (2002) Recent developments in transgenics for abiotic stress in legumes of semi-arid tropics. JIRCAS Working Report., 23:61-73

Sharma. P., Meena, R.S., Kumar, S., Gurjar, D.S., Yadav, G.S. and Kumar, S. (2019). Growth, yield and quality of cluster bean (Cyamopsis tetragonoloba) as influenced by integrated nutrient management under alley cropping system. Indian Journal of Agricultural Sciences. ;89(11):1876-1880

Sharmah, D., Debnath, B., Kandpal, B.K. and Das, D. (2019) Profitability of Integrated Farming System model for small and medium farmers of South Tripura, Tripura. Journal of Applied and Natural Science 11(3):587-589

Sharma R, Sharma E., and Purohit AN (1995) Dry matter production and nutrientcycling in agroforestry systems of mandarin grown in association with Albiziaanf mixed tree species. Agroforestry systems 29: 165-17

Sharma S K, Choudhury A, Sarkar P, Biswas S and Singh A et al. 2011 Greenhouse gas inventory estimates for India; Curr.

Shinde, R., Sarkar, P.K., Thombare,N. and Naik, S.K.(22019) Soil conservation: Today's need for sustainable development. Agriculture & Food: e-Newsletter. ;1(5):175-183

Shinde, S, Taneja, V.K. and Singh, A. 1990. Association of climatic variables and production and reproduction traits in crossbreds. Indian Journal of Animal Sciences. 60(1): 81–85

Shinozaki, K. and Kazuko Yamaguchi-Shinozaki (2007) Gene networks involved in drought stress response and tolerance. J Exp Bot 2007;58(2):221-7.

Shinozaki K. and Yamaguchi-Shinozaki K. (2007). J. Exp. Bot. 58:

Sinclair, T.R. (1988) Interpretive Summary od Part 2: Selecting Crops and Cropping Systems for Water limited Environments. In: Bidinger, F.R. and Johansen, C.(Eds.) (1988) Drought Research Priorities for the Dryland Tropics., International Crops Resear[ch] Institute for the Semi-Arid Tropics, Patancheru-502324 India. Pages 87-94.

Sinclair, T.R., Tanner, C.B. and Bennett, J.M. (1984) Water use efficiency in crop production. Bioscience 34: 36-40.

Sinclair, T.R. and Ludlow, M.M. (1986) Influence of soil-water supplyon the plant water balance of tropical grain legumes. Australian Journal of Plant Physiology, 13:329-341.

Singh, A. and Verma, S.K. (2018) Management of ravines through anicuts and afforestation2018. In: Ravine Lands: Greening for Livelihood and Environmental Security. Singapore: Springer; pp. 477-504

Singh, A. K., Kumar AK, Katiyar, V.S., Singh, K.D. and Singh, U.S.(1997) Soil and water conservation measures in semi-arid regions of South-Eastern Rajasthan. Indian Journal of Soil Conservation. ;25(3):186-189

Singh, G., Singh, H., Dagar, J.C., Singh, N.T. and Sharma, V.P. (1997) Evaluation of Agriculture, forestry and agroforestry practices in moderately alkali soil in north-western India. Agrofor.Syst.37: 279-295

Singh, H.P., Sharma, K.D., and Subba Reddy, G. (2002) Rainfed agriculture: Challenges of the 21st century-National Perspective.p.3-4. In Proc. Of the 89th Indian Sci. Congr- Part IV. Lucknow, India.

Singh, H.P., Sharma, K.L., Venkateswarlu, B., and Neelaveni, K.1(999). Fertilizer use in rainfed areas. Problems and Potentials. Fert. News 44(1):27-38.

Singh, H.P., Sharma, K.P., Reddy, G.S. and Sharma, K.L. (2004). Dryland Agriculture in India. p. 67-92. In: Challenges and strategies. Crop Science Society of [S.I.]

Singh, H.P., Venkateswarlu, B., Vittal, K.P.R.and Ramachandran (2000) Management of rainfed agro-eco system p.669-774. In Natural resource management for agricultural production in India. Indian Soc.of Soil.Sci., New Delhi

Singh, J.P. and Ravishankar, N. (2017). Integrated farming systems for sustainable agricultural growth: Strategy and Experience from Research. In: Integrated Farming Systems for Sustainable Agriculture and Enhancement of Rural Livelihoods (Eds. Muralidharan, K., Prasad, M.V.R., and Siddiq, E.A.) Proceedings of the Seminar. Retired ICAR Employees' Association (RICAREA). Hyderabad 500030. Pages 18-31

Singh, R.P. (1986) Farm ponds. Project bulletin No.6, Central Research Institute for Dryland Agriculture. Hyderabad, India

Singh, R.P. (1995) Problems and prospects of dryland agriculture in India; In: Sustainable development of dryland agriculture in India (Ed. Singh, R.P.) Scientific Publishers, Jodhpur, India. Pages 13 to 23

Singh, R.P., Reddy, Y.V.R. (1986) Degradation and rehabilitation of drylands some technical and economic aspects. Paper presented in Regional Workshop for 9 countries of South and Southeast Asia on Economics of Dry and Degradation and Rehabilitation held from August 25–29, 1986 at New Delhi

Singh, R.P. and Khan. M.A. (1999) Rainwater management: water harvesting and its efficient utilization.p.301-313.In:H.P. Singh et al.,(eds.)Fifty years of dryland agricultural research in India. Central Research Institute for dryland Agriculture, Hyderabad

Singh, R.P. and Reddy, G.S. (1988) Identifying crops and cropping systems with greater production stability in water deficit conditions. In "Drought Research Priorities for Dryland Crops (Bidinger, F.R and Johansen, eds.) ICRISAT (International Crops Research Institute for the Semi-Arid Tropics, Patancheru, India, ICRISAT, 86

Singh, R.P. and Subba Reddy, G.(1988) Drought research priorities for dryland tropics (Bidinger,F.R.,and Jphansen,C., eds.), ICRISAT(International Crops Research Institute for semi-arid tropics),Patancheru,A.P. 502324,India.P.85

Singh, S.K., Meena, H.R., Kolekar, D.V. and Singh, Y.P. (2012). Climate change impacts on livestock and adaptation strategies to sustain livestock Production. Journal of Veterinary Advances. 2(7): 407- 412.

Singh. H.P. (2002). Farming systems and best practices for drought prone areas of India. In farming systems and best practices for drought prone area in Asia and the Pacific. Central Res. Inst. for Dryland Agric., India Hyderabad. India

Singh K, Chauhan HS, Rajput DK, Singh DV (1989) Report of a 60 month study on fitter production, changes in soil chemical properties and productivity under Poplar (P.deltoides) and Eucalyptus (E. hybrid) interplanted with aromatic grasses. Agroforestry Systems 9: 37-45.

Singh N.R., Jhariya, M.K. (2016) Agroforestry and agrihorticulture for higher income and resource conservation. In: Narain S, Rawat SK, editors. Innovative Technology for Sustainable Agriculture Development. New Delhi: Biotech Books; . pp. 125-145

Sivamani, E., Bahieldin, A., Wraith, J.M. et al., (2000) Improved biomass productivity and water use efficiency under water deficit conditions in transgenic wheat constitutively expressing the barley HVA1 gene. Plant Science, 155: 1-9

Slenning, B.D. 2010. Global climate change and implications for disease emergence. Veterinary Pathology. 47: 28-33.

Smith, M.R., Rao, I.M. and Merchant, A. (2018) Source-Sink relationship in crop plants and their influence on yield development and nutritional quality. Front. Plant Sc., 20. December 2018. https://doi.org/10.3389/fpls.2018.01889

Solanki KR, Ram-newaz AK (1999) Agroforestry, an alternate land use system for dryland agriculture. In: Singh HP, Ramakrishna YS, Sharma KL, Venkateswarlu B (Eds.) Fifty Years of Dryland Agricultural Research in India CRIDA. Hyderabad, pp 463–473

Solanki M. K , Wang F.-Y., Wang Z., Li C.-N., Lan T.J., Singh R. K., et al. (2019). Rhizospheric and endospheric diazotrophs mediated soil fertility intensification in sugarcane-legume intercropping systems. J. Soils Sediments 19 :1911–1927. 10.1007/s11368-018 2156-3

Squire, G.R., Ong, C.K. and Monteith, J.L. (1986) Crop growth in semi-arid environments. Page 219-231 In Proceedings of the International Pearl-millet Workshop, April 7-11, 1986, ICRISAT Centre, Patancheru, A.P. India.

Srinivasan, S. and Rao, D.V.S. (1987) New report of parasites of groundnut leaf webber Aproaerema modicella Deventer. Entomon, 12:117-119

Srinivasarao, Ch., Girija Veni, V., Prasad, J.V.N.S., Sharma, K.L., Chandrasekhar, Ch., Rohilla, P.P. and Singh, Y.V. (2017). Improving carbon balance with climate resilient management practices in tropical agro-ecosystems of Western India, Carbon Management. DOI: 10.1080/17583004.2017.1309202

Srinivasarao, Ch., Gopinath, K.A., Prasad, J.V.N.S., Prasanna, K. and Singh, A.K. (2016a). Climate resilient villages for sustainable food security in tropical India: Concept, process, technologies, institutions, and impacts. Advances in Agronomy. 140(3): 101-214

Srinivasarao, Ch., Rao, K.V., Gopinadh, KA., Prasad, Y.G. Arunachalam, A., Ramana, D.B.V., Chary, G.R., Gangaiah, B., Venkateswarlu, B. and Mohapatra, T.(2019). Agriculture Contingency Plans for Managing Weather Aberrations and Extreme Climatic Events: Development, Implementation and Impacts in India. Advances in Agronomy. https://doi.org/10.1016/bs.agron. 2019.08.002

Srinivasarao, Ch., Ravindra Chary, G., Mishra, P.K., Maruthi Sankar, G.R. Nagarjuna Kumar, R., Venkateswarlu, B. and Sikka, A.K. (2013). Real Time Contingency Planning: Initial Experiences form AICRPDA. AICRPDA, CRIDA, Hyderabad, 73 p

Srinivasarao,Ch., Subha Lakshmi, C. and Sumanta Kundu (2019) Critical Role of Nutrient Management in Climate Adaptive Dryland Farming Systems in India, Indian Journal of Fertilisers 15 (11) : 1220-1233, November 1220-1234

Srinivasarao, ch., Sudha Rani, Y., Girija Veni, V., Sharma, K.L., Maruthi Shankar, G.R., Jrasad,J.V.N.S., Prasad, Y.G. and Sahrawat, K.L. (2016). Assessing village-level carbon balance due to greenhouse gas mitigation interventions using EX-ACT model. Int. J. Environ. Sci. Technol, 13: 97-112.

Srinivasarao, Ch., Venkateswarlu, B., Dixit, S., Wani, S.P., Sahrawat, K.L., Kundu, S., Gayatri Devi, K., Rajesh, C. and Pardasaradhi, G. (2010). Productivity enhancement and improved livelihoods through participatory soil fertility management in tribal districts of Andhra Pradesh. Indian Journal of Dryland Agriculture Research Development 25(2), 23– 32.

Srinivasarao, Ch. and Manjunath, M. (2017). Potential of beneficial bacteria as eco-friendly options for chemical-free alternative agriculture. In Plant-Microbe Interactions in Agro-Ecological Perspectives (D. Singh, H. Singh and R. Prabha, Eds.), pp. 473-493. Springer, Singapore.

Srinivasa Rao, M., Manimanjari, D., Vanaja, M.,CA Rama Rao, C.A. Srinivas, Vum Rao,K. and Venkateswarlu, B. (2012) Impact of Elevated CO2 on Tobacco Caterpillar, Spodoptera litura on Peanut, Arachis hypogea, J Insect Sci. 2012; 12: 103.

Srinivas I., Sanjeeva Reddy B., Adake R.V., Mayande V.M., Thyagaraj C.R., Pratibha G., Rao K.V., Sammi Reddy K., and Srinivasa Rao Ch., (2014). Farm Mechanization in Rainfed Regions: Farm Implements Developed and Commercialized. Central Research Institute for Dryland Agriculture, Hyderabad, Telangana State. 31p.

Srinivas Rao Ch., Gopinath, K.A., Prasad yen and Alok Sikka (2016b). Climate Resilient Rainfed Agriculture: Status and Strategies2016, https://www.researchgate.net/publication/304131389

Srinivas RaoCh.,Gopinath K.A., prasad yen., and Alok Sikka (2016): Climate Resilient Rainfed Agriculture: Status and Strategies2016, https://www.researchgate.net/publication/304131389

Stahlman, P.W. and Wicks, G.A., (2000). Weeds and their Control in Grain Sorghum. In: [Smith, C.W., Frederiksen, R.A. (Eds.)], Sorghum Origin, History, Technology and Production. John Wiley and Sons, Inc, New York. 535-590p

Stahlman PW, Wicks GA. (2000) Weeds and their control in grain sorghum. in C. W. Smith, ed. Sorghum: Origin, History, Technology, and Production. New York: Wiley; c2000. p. 535-582. 1

Stein, B.A. and Shaw, M.R. (2013) Biodiversity conservation for a climate-altered future, In Book: Successful Adaptation: Linking Science and Practice in Managing Climate Change Impacts (pp.50 – 66) Editors: Susanne C. Moser, Maxwell T. Boykoff. Publisher: Routledge

Steiner, J.L. (1986) Dryland grain sorghum water use, light interception, and growth responses to planting geometry. Agronomy Journal 78: 720-726

Stern (2006). Review: The Economics of Climate Change, Cambridge. University Press, London IPCC (Intergovernmental Panel on Climatic Change) 2007. Climate Change: The Physical Science Basis. Extracts from the IV Assessment Report. Survey of the Environment 2007, The Hindu, 147-155.

Strange, N.R. and Scott, P.R. (2005) Plant Disease: A Threat to Global Food Security. Annual Review of Phytopathology 43:83-116 https://doi.org/ 10.1146/annurev.phyto.43.113004.133839

Subba Rao, I.V. 2002 Land use diversification in rainfed agriculture. CRIDA Foundation Day, Lecture. Central Res. Inst. For Dryland Agric., Hyderabad

Subba Rao A, and Saha R (2014) Agroforestry for soil quality maintenance, climate change mitigation and ecosystem services. Indian Farm 63(11):26–29

Subba Reddy, G., Gangadhar Rao,D., Venkareswarlu, S., and Maruthi V.(1996). Drought management options for rainfed castor grown in Alfisols.J. Oil seeds.Res.13 (2):200-207

Subba Reddy, G., Ramakrishna, Y.S., Ravindra Chary, G. and Maruthi Sankar, G.R. 2008. Crop and Contingency Planning for Rainfed Regions of India- a compendium, All India Coordinated Research Project for Dryland Agriculture, CRIDA, Hyderabad, India,174 p.

Subba Reddy,G and Maruthi,V.(2006) Efficient crops and cropping systems in rainfed agriculture. In Land use diversification for Sustainable Rainfed Agriculture. (Eds). Sharma K.D. and Soni B., PP132-148

Subba Reddy. G. (2002) Potentials of green manuring in rainfed conditions. In Potentials of green manure crops in rainfed agriculture. Univ. of Agric. Sci., Bangalore, India

Subrahmanyam, S. and Aparna, P. (2019) Recent Trends in Rural Sector and Impact on Small Farmer. In: (Eds) Purnachnadra Rao, K., Prasad, M.V.R., Muralidharudu, Y. and Damodaram, T. (2019) Sustainability of Small Farmers in Changing Agricultural Scenario. Seminar Proceedings, Retired ICAR Employees' Association, RICAREA. Hyderabad. Pages 54-61

Subramani, Mancombu (2020) Farmers' Protests: 3 Reasons why the prices of agricultural commodities have not spiked despite the stir. Money Control, December 10, 2020. https://www.moneycontrol.com/news/business/markets

Sudhakara Babu, S.N, Sujatha, M. and Rao, G.R. (2008). Non-edible oil-seed crops for biofuel production: Prospects and challenges In: J. Keith Syers, David Wood, and Pongmanee Thongbai (Eds) Proceedings of the International Technical Workshop on the Feasibility of Non-edible Oil Seed Crops for Biofuel

Sudhakara Babu, S.N., Mukta, M., Ranganatha, A.R.G. and Hegde, D.M. (2006.) Compatibility of annual crops for intercropping under tree borne oilseeds. Technical papers. Regional Workshop on Deserts and Desertification. AMR-AP Academy of Rural Development, Hyderabad. pp 126-128.

Sudhakar Reddy, L. (2019) Dryland Practices in AP. https://www.scribd.com/document/403045666/Dryland-practices-AP-pdf

Sullivan, C.Y. and Ross, W.M (1979). Selecting for heat and drought resistance in sorghum. Pages 263-281.In: Stress physiology in Crop plants. (Ed. Mussel, H. and Staples, R.C.) New York, USA. John Wiley.

Summerfield, R.J., Huxley, P.A. and Stele, W. (1974) Cowpea (Vigna unguiculata, L. Walp). Field Crop Abstracts, 27: 301-312

Sundarajan and Palaniappan (1979) Agricultural History Vol. 53, No. 1, Jan., 1979

Sundaramurthy, V.T., Jayaraj, S. and Swamiappan, M. (1976) Telenomous Nixon, a potential egg parasite of groundnut red hairy caterpillar. Amsecta albistriga Walker. Current Science,45:775

Suresh, G., Shankar, N. and Varaprasad, K.S. (2017) Successful oilseeds and pulses based Farming Systems: IIOR's Field Level Experiences In: Integrated Farming Systems for Sustainable Agriculture and Enhancement of Rural Livelihoods (Eds. Muralidharan, K., Prasad, M.V.R., and Siddiq, E.A.) Proceedings of the Seminar. Retired ICAR Employees' Association (RICAREA). Hyderabad 500030, pages 137-143.

Sushil Kumar (2006) Climate change and crop Breeding objectives in the twenty first century. Curr. Sci.2006; 90:1053-1054

Sutcliffe, J.F. (1968) Plants and water. Institute of Biology's Studies in Biology No.14.London, UK:Edward Arnold.

Swain, S. et al. (2017). "Application of SPI, EDI and PNPI using MSWEP precipitation data over Marathwada, India". IEEE International Geoscience and Remote Sensing Symposium (IGARSS). 2017: 5505–5507. doi:10.1109/IGARSS.2017.8128250. ISBN 978-1-5090-4951-6. S2CID 26920225.

Swami, S.L., Puri, S. and Singh, A.K. (2003). Growth, biomass, carbon storage and nutrient distribution in Gmelina arborea Roxb. Stands on red lateritic soils in central India. Bioresource Technology. 90: 100-120 Tumwebaze SB, Bevilacqua E, Briggs R and Volk T. 2013. Allometric biomass equations for tree species used in agroforestry systems in Uganda. Agroforest. Syst. 87: 781-795.

Swaminathan, M.S. (2012) Combating hunger, Science 338:1009

TAAS (2019) Dryland Agrobiodiversity for Adaptation to Climate Change, (Satellite Symposium: Jodhpur, 13 February, 2019). Accelerating Science-Led Growth in Agriculture: Two Decades of TAAS, Page 72, Trust for Advancement of Agricultural Sciences (TAAS) New Delhi.

Tanner, C.B. and Sinclair, T.R. (1983) Efficient water use in crop production: research of re-search? Pages1-27 In: Limitations to efficient water use in crop production. (Eds.: Tylor, H.M. Jordan, W.R. and Sinclair, T.R.) Madison, Wisconsin, USA. American Society of Agronomy, Crop Science Society of America and Soil Science Society of America.

Tanwar, S.P.S., Singh, A., Patidar, M., Mathur B.K. and Lal, K. (2016). Integrated farming system with alternate land uses for achieving economic resilience in arid zone farming. (In) Extended Summaries of Fourth International Agronomy Congress 22–26 November 2016, IARI, New Delhi, pp. 246–247.

TARP-IVLP (2003) Annual Report on Technology Assessment and Refinement through Institute-Village Linkage Program (TARP-IVLP) under Rainfed Agro-ecosystem. Central Research Institute for Dryland Agriculture, Hyderabad, 94pp.

TARP-IVLP (2005) Technology Assessment and Refinement of Nutritious Cereal based Production System through Institute-Village Linkage Program for Northern Karnataka, University of Agricultural Sciences (UAS) Dharwad, pp:36-39.

Termansen, M., Sun, N., Guan, D., Simelton, E., Dodds, P., Feng, K. and Yu, Y. (2008). Quantifying socioeconomic characteristics of drought sensitive regions: evidence from Chinese provincial agricultural data. Comptes Rendus Geosciences. 340:679-688. http://dx.doi.org/10.1016/j.crte.200 8.07.004

Tewari, J.C. (2007). Agroforestry systems in arid regions of India. In: Puri S and Panwar P (eds) Agroforestry – Systems and Practices. New India Publishing Agency, New Delhi: 175-189

Tewari, J.C., Bohra, M.D., Harsh, L.N. (1999) Structure and production function of traditional extensive agrofor- estry systems and scope of agroforestry in Thar desert. Indian J Agroforestry. 1(1):81–94

Tewari, J. C., Ram, M. and Roy, M. M.(2014), Livelihood improvements and climate change adaptations through agroforestry in hot arid environments. In Agroforestry Systems in India: Livelihood & Ecosystem Services (eds Dagar et al.), Advances in Agroforest., Springer, India, 2014; doi:10.1007/978-81-322-1662-9_6.

Tewari, J.C., Ram M, Roy, M.M. and Dagar, J.C. (2013). Livelihood improvements and climate change adaptation through agroforestry in hot arid environment. In: Dagar J, Singh A and Arunachalam A. (eds) Agroforestry Systems in India: Livelihood Security and Ecosystem Services. Advances in Agroforestry, Springer, New Delhi. 10: 155-183

Thapa, G. and R. Gaiha (2011), "Smallholder farming in Asia and the Pacific: Challenges and Opportunities", paper presented at the Conference on new directions for small holder agriculture, 24-25 January 2011, Rome, IFAD

Thapa, S. and Stewart, B.A. (2016) Dryland Farming: Concept, Origin and Brief History

Thornthwaite, C.W. (1948) An approach toward a rational classification of climate. Geographical Review, 38: 85-94

Thornthwaite, C.W. and Mather, J.R. (1955) The Water Balance. Laboratory of Climatology, Centerton. Arkansas USA.

Thyagraj.,. Vittal, K.P.R., Mayande, V.M.and Sharma, K.L. (1999) Tillage and soil management for higher productivity in drylands.p.329-344. In: Fifty years of dryland agricultural research in India. Central.Res.Inst.for Dryland Agric. Hyderabad, India

Thyamini (2010). Thiamine in plants: Aspects of its metabolism and functions. Phytochemistry,71, Issues 14–15, October 2010, Pages 1615-162

Tiwari, R., Murthy, I.K., Killi, J., Kandula, K., Bhat, P. R., et al. (2010) Land use dynamics in select village ecosystems of southern India: drivers and implications. Journal of Land Use Science 5.

Tollenaar, M. (1991) Physiological basis of grain improvement in maize hybrids in Ontario from 1959 to 1988. Crop Science 31:119-124

Tompe, A.A., HoleU.B.,,Kulkarni, S.R., Chaudhari, C.S.and Chavan, S.K. (2020) Studies on seasonal incidence of leaf eating caterpillar, Spodoptera litura (Fab.) infesting capsicum. Journal of Entomology and Zoology Studies 2020; 8(1): 761-7

Trenbath, B.R. (1976) Plant Interactions in Mixed Crop Communities, In: Multiple Cropping, Volume 27(Eds.R.I. Papendick,P.A. Sanchez,G.B. Triplett) Book Series: ASA Special Publications, https://doi.org/10.2134/asaspecpub27.c8

Trenbath, B.R. (1993) Intercropping systems for the management of pests and diseases. Field Crops Research 34 (3&4): 381-405

Treut, L.H., Somerviolle, R., Cubasch, U., Ding, Y., Mauritzen, C., Mokssit, A., Peterson, T. and Prather, N. (2007) Historical over view and Climate Change. In: Climate Change 2007, The Physical Science Basis and Contribution of Working Group –I to the Fourth (IV) Assessment Report.

Trexler, M. C. (1993). Mitigating global warming through forestry: A Partial literature review report to GTZ for Enquette Commission, German.

Tuhan, N. C., Pawar, A.D. and Singh, B. (1987) Preliminary studies on Trichogramma brasiliensis Ashmead against the red hairy caterpillar , Amsecta moore Butler. Journal of Biological Control,1: 72-73

Tumwebaze, S.B., Bevilacqua, E., Briggs, R. and Volk, T. (2013). Allometric biomass equations for tree species used in agroforestry systems in Uganda. Agroforest. Syst. 87: 781-795

Turner, N.C., and Nicolas, M.E. (1987). Drought resistance of wheat for light textured soils in a Mediterranean climate. Pages 203-216. In: Drought tolerance in winter cereals. (Eds. Srivastava, J.P. et al.), New York, USA, John Wiley.

Turner, N.C. and Ward, P.R. (2002). The role of agroforestry and perennial pasture in mitigating water logging and secondary salinity: summary, Agricultural Water Management, 53: 271-275.

Udawatta, R.P., Krstansky, J.J., Henderson, G.S., Garrett, H.E. (2002) Agroforestry practices, runoff, and nutrient loss: A paired watershed comparison. Journal of Environmental Quality. 31:1214-1225

UK Met Office (2011) Warming: A Guide to Climate Change, Exerter, UK Met Office, Hadley Centre, 2011.

Umezawa et al. (2006). Curr. Op. Biotech.17: 113-122.

Undie, U.L., Uwah, D.F. and Attoe, E.E.(2012) Effect of intercropping and crop arrangement on yield and productivity of late season maize/soybean mixtures in the humid environment of south southern Nigeria. J. Agric. Res., 4, 37

UNEP (2017) The Emissions Gap Report, United Nations Environment Programme, Nairobi. UNEP. 2017.

UNFCCC (2015) 'The Paris Agreement', The Conference of the Parties. 2015, https://unfccc.int/resource/docs/2015/cop21/ eng/l09r01.pdf [Accessed 15 October 2017].

Unger, P.W. (1979) Effects of deep tillage and profile modification on soil properties, root growth, and crop yields in USA and Canada. Geoderma, 22: 275-295.

Unger, P.W., Jones, O.R. and Steiner, J.L. (1988) Principles of crop and soil management procedures for maximizing production per unit of rainfall. In "Drought Research Priorities for Dryland Crops (Bidinger, F.R and Johansen, eds.) ICRISAT (International Crops Research Institute for the Semi-Arid Tropics, Patancheru, India, ICRISAT, Pages 97-112

Unger, P.W., Steiner, J.L. and Jones, O.R. (1986) Response of conservation tillage sorghum to growing season precipitation. Soil and Tillage Research.7:291-300

Unger, P.W. and Parker, J.J. (1976). Evaporation reduction from soil with wheat , sorghum ad cotton residues. Soil Science Society of America Journal 40: 938-942.

UNISDR (2018) United Nations Office for Disaster Risk Reduction Annual Report, 2018. © 2019 United Nations 9-11 Rue de Varembé, 1202 Geneva, Switzerland, Tel: +41 22 917 89 08 unisdr-annual-report-2018-e-version.pdf

Upadhyay, R.C., Sirohi, S., Ashutosh, Singh, S.V., A. Kumar and Gupta, S.K. (2009). Impact of climate change on milk production of dairy animals in India. In: Global climate change and Indian agriculture case studies from the ICAR network project (Ed. PK Aggrawal). ICAR Pub.: 104-106.

USDA (2016) U. Sagricult URE And for Estry Greenhouse Gas Inventory 1990–2013 Technical Bulletin Number 1943. September 2016. Office of the Chief Economist | Climate Change Program Office. USDA, USA.

USDS (2005) U S Department of State: Nigerian Scholar Links Drought, Climate Change to Conflict in Africa - US Department of State". state.gov. Archived from the original on 28 October 2005.

U S G S (2004). United States Geological Survey 2004 «Dunes – Getting Started». Archived from the original on 2012-06-22. Retrieved 2009-03-21.

Uthappa, A.R., Chavan, S.B., Dhyani, S.K., Handa, A.K. and Newaj, R. (2015). Trees for soil health and sustainable agriculture. Indian Farming. 65: 2-5..,

Vanaja M., Jyothi M., Ratnakumar P., Raghuram Reddy P., JyothiLakshmi N., Yadav S.K., Maheshwari M. and Venkateshwarlu B. (2008) Growth and yield response on castor bean Plant Soil Environ., 54, 2008 (1): 38–46

Vanaja M., Maheswari M., Ratnakumar P. and Ramakrishna Y.S.(2006): Monitoring and controlling CO2 concentrations in open top chambers for better understanding of plants response to elevated CO2levels. Indian Journal of Radio and Space-Physics, 35: 193–197

van Den Bergh (1965) Wit, C.T. de; Bergh, J.P. van den 1965Competition between herbage plants The Journal of Agricultural Science 13 (1965). - ISSN 1916-9752 - p. 212 – 221

Van den Bossche, P. and Coetzer, J. A. W. (2008). Climate change and animal health in Africa Review Science Technology Off. int. Epiz. 27 (2): 551-562.

van der Maesen, L.J.G. (1986) Cajanus DC and Atylosia W&A (Leguminosae). Agricultural University Wageningen Papers 85–4 (1985) Agricultural University, Wageningen, the Netherlands.

Vanwalleghem, T. (2016). Soil erosion and conservation. International Encyclopedia of Geography: People, the Earth, Environment and Technology: People, the Earth, Environment and Technology.,12:1-10

Varshney, R.K. et al., (2021) Breeding custom-designed crops for improved drought adaptation, Review. Advanced Genetics, Wiley online Library, https://doi.org/10.1002/ggn2.202100017

Vasal, S.K., Dhillon, B.S. and Pandey, S. (2010) Recurrent Selection Methods Based on Evaluation-Cum-Recombination Block, Plant Breeding Reviews (June 2010) 139-163; SN 97804706 50073; DO 10. 1002/9780470650073.ch5

Vayugrid (2014) Progress Report 2014. Board of Directors, Vayugrid Market Place Services Pvt. Ltd. Bangalore, India. 21pp.

Velayutham, M. (1999). Crop and land use planning in dryland agriculture. pp. 73-80. (In) Fifty years of dryland agricultural research in India (Singh, H.P., Ramakrishna, Y.S., Sharma, K.L. and Venkateswarlu, B., Eds.), Central Research Institute for Dryland Agriculture, Hyderabad, India.

Venkareswarlu, B., Osman M., Padmanabhan M.V.., Karemulla K., Korwar G.R. and Rao K.V.(2013) Field Mannual on Watershed Management, Central Research Institute for Dryland Agriculture,Santoshnagar,Hyderabad161-176

Venkateswaralu, J. (2010) Rainfed agriculture in India: Research and development scenario. Indian Council of Agricultural Research, New Delhi

Venkateswarlu, B. (2017). Climate smart agriculture: Are we poised to outsmart climate change impacts? Current Science, Vol. 112 (5): 891-892.

Venkateswarlu, B. (2019) Small-holder Agriculture and climate change: Vulnerability and Adaptation. In: (Eds) Purnachnadra Rao, K., Prasad, M.V.R., Muralidharudu, Y. and Damodaram, T. (2019) Sustainability of Small Farmers in Changing Agricultural Scenario. Seminar Proceedings, Retired ICAR Employees' Association, RICAREA. Hyderabad. Pages 28-33

Venkateswarlu, B., Mishra, P.K., Ravindra Chary, G., Maruthi Sankar, G.R. and Subba Reddy, G. 2009. Rainfed farming – A compendium of improved technologies. All India Coordinated Research Project for Dryland Agriculture, CRIDA, Hyderabad. 132 p

Venkateswarlu, B., Singh, A.K., Prasad, Y.G., Ravindra Chary, G., Srinivasa Rao, Ch., Rao, K.V., Ramana, D.B.V. and Rao, V.U.M. (2011). District level contingency plans for weather aberration in India. Central Research Institute for Dryland Agriculture, Natural Resource Management Division. ICAR, Hyderabad, pp.136.

Venkateswarlu, B. and Gopinath K.A. (2017) Sustainable farming systems for dryland agricultural ecosystem with special reference to small farmers. In: Integrated Farming Systems for Sustainable Agriculture and Enhancement of Rural Livelihoods In: (Eds). Muralidharan, K., Prasad, M.V.R., and Siddiq, E.A. (2017)Integrated Farming Systems for Sustainble Agriculture and Enhancement of Rural Livelihoods. Proceedings of the Seminar. Retired ICAR Employees' Association (RICAREA). Hyderabad 500030. Pages. 89-100.

Venkateswarlu, B. and Shanker, A.K. (2009). Climate change and agriculture: Adaptation and mitigation strategies. Indian Journal of Agronomy. 54 (2): 226-230.

Vernon, D.M., Tarczynski, M.C., Jansen, R.G. and Bohnert, H.J. (1993) Cyclitol production in transgenic tobacco, Plant Journal, 4: 199-205

Vijayan, Roshni (2016).Dryland agriculture in India – problems and solutions. Asian J. Environ. Sci., 11(2): 171-177, DOI: 10.15740/HAS/AJES/11.2/171-177.

Vijendra K. Boken; Arthur P. Cracknell; Ronald L. Heathcote (2005). Monitoring and Predicting Agricultural Drought : A Global Study: A Global Study. Oxford University Press. p. 349. ISBN 978-0198036784

Vinod Gupta , Pradeep Kumar Rai and Risam, K.S. (2012) Integrated Crop-Livestock Farming Systems: A Strategy for Resource Conservation and Environmental Sustainability, Indian Research Journal of Extension Education, Special Issue (Volume II), 201 Indian Research Journal of Extension Education, Special Issue (Volume II), 2012,p49-54

Virmani,S.M., Pathak, P.and Singh.R (1991). Soil related constraints in dryland crop production in vertisols,Alfisols and Entisols of India,p.80-95.In Soil related constraints in crop production. Indian. Soc. of Soil Sci., New Delhi. India.

Vision 2020(1997): Perspective Plan. Central Research Institute for Dryland Agriculture (CRIDA). Hyderabad. Indian Council of Agricultural Research (ICAR), pp.80

Vision 2050 (2015) Central Research Institute for Dryland Agriculture (CRIDA). Hyderabad. Indian Council of Agricultural Research (ICAR), pp.36.

Vittal.K.P.R., Vijayalakshmi, K. and Rao.,U.M.B. (1983).Effect of deep tillage on dryland crop production in red soils of India. Soil Tillage Res.3: 377-384

Vivekanandan, E., Ratheesan, K., Manjusha, U., Remya, R. and Ambrose, T.V. (2009). Temporal changes in the climatic and oceanographic variables off Kerala. In: Vivekanandan, E. et al. (eds.), Marine Ecosystems Challenges and Opportunities. Book of Abstracts, Marine Biological Association of India, Cochin, 260-261.

Vohra, C.P. (2005) GLACIERS OF INDIA—A Brief Overview of Glaciers in the Indian Himalaya in the 1970s and at the End of the 20th Century https://pubs.usgs.gov/pp/p1386f/pdf/F5_India.pdf

Von Cossel, M., Wagner, M, Lask, J., Magenau, E., Bauerle, A., Von Cossel, V.; Warrach-Sagi, K. and Winkler, B. (2019). Prospects of Bioenergy Cropping Systems for A More Social-Ecologically Sound Bio-economy. Agronomy 9, 605.

Wagg C., Schlaeppi K., Banerjee S., Kuramae E. E., van der Heijden M. G. A. (2019). Fungal-bacterial diversity and microbiome complexity predict ecosystem functioning. Nat. Commun. 10:4841. 10.1038/s41467-019-12798-

Wang, B. (2006). The Asian Monsoon. Springer Science & Business Media. p. 206. ISBN 978-3540406105

Wang, Y., Fan, J., Cao, L. and Liang, Y.(2016) Infiltration and runoff generation under various cropping patterns in the red soil region of China. Land Degradation and Development., 27:83-91

Wang-Qianfeng, et al. (2015) The alleviating trend of drought in the Huang-Huai-Hai Plain of China based on the daily SPEI. International Journal of Climatology.2015. . doi:10.1002/joc.4244

Wang et al. (2005). Plant J 43: 413-424.

Wani, S.P., Rockstrom. J., Sahrawat, KL., (2011) Integrated Watershed Management in Rainfed Agriculture, CRC Press, NewYork, pp. 1 -447

Wani, S. P., Rockstrom. J. and Oweis, T. (2009) Preface, Rainfed Agriculture:

Wani, S.P. and Kumar, M.S. (2002). On-farm generation of N-rich organic material. In: A Training Manual on Integrated Management of Watersheds. SP Wani, PPathak and TJ Rego, ICRISAT, (eds.) Patancheru, Andhra Pradesh, India. 30 pp

Wani, S. P. and Sreedevi, T K and Marimuthu, S and Rao, A V R K and Vineela, C (2009) Harnessing the Potential of Jatropha and Pongamia Plantations for Improving Livelihoods and Rehabilitating Degraded Lands. In: 6th International Biofuels Conference, 4-5 March 2009, New Delhi, India.

Wani S.P., Venkateswarlu, B., Sahrawat, K.L., Rao, K.V. and Ramakrishna, Y.S.(eds.). (2009). Best-bet Options for Integrated Watershed Management. Proceedings of the Comprehensive Assessment of Watershed Programs in India, 25-27 July 2007,ICRISAT, Patancheru 502 324, Andhra Pradesh, India. Patancheru 502 324, AndhraPradesh, India: International Crops Research Institute for the Semi-Arid Tropics. 312 pp.ISBN 978-92-9066-526-7: Order code: CPE 167

Wasti, S., Sah, N. and Mishra, B. (2020)Impact of Heat Stress on Poultry Health and Performances, and Potential Mitigation Strategies. Animals 2020, 10(8), 1266; https://doi.org/10.3390/ani10081266

Wasti,S., Sah, N. and Mishra, B. (2020) Impact of Heat Stress on Poultry Health and Performances, and Potential Mitigation Strategies Animals 2020, 10(8), 1266; https://doi.org/10.3390/ani10081266

Weih, M., Adam,E., Vico, G. and Rubiales,D.(2002) Application of Crop Growth Models to assist breeding for Intercropping: Opportunities and Challenges. Front. Plant Sci., 04 February 2022 | https://doi.org/10.3389/fpls.2022.720486

Weißhuhn, P., Moritz Reckling, M., Stachow, U. and Wiggering, H. 2017 Supporting Agricultural Ecosystem Services through the Integration of Perennial Polycultures into Crop Rotations. Sustainability 2: 9

West, J.W. 2003. Effect of heat stress on production in dairy cattle. Journal of Dairy Science. 86: 2131- 2144. World Bank. 2008. World Bank's Approach to Climate Change in South Asia: An Overview, Bank Information Center, online available at: www.bicusa.org

Wiese, A.F. (1983) Weed Control. Pages 463-488 In: Dryland Agriculture. (Dregne, H.E. and Wills, W.O. Eds.) Agronomy Monograph No.23, Madison, Wisconsin, USA, American Society of Agronomy, Crop science Society of America and Soil Science Society of America.

Wiggs and Giles F.S. (2011). "Geomorphological hazards in drylands". In Thomas, David S.G. (ed.). Arid Zone Geomorphology: Process, Form and Change in Drylands. John Wiley & Sons. p. 588. ISBN 978-0-470-71076-0

Willey, R.W. (1979) Intercropping, Its Importance and Research Needs. Part 1. Competition and Yield Advantages. Agronomy and Research Approaches. Field Crop Abstract, 32, 1-10

Willey, R.W., Natarajan, N., Reddy, M.S., Rao, M.R., Nambiar. P.T.C., Kannayyan, J. and Bhatnagar, V.S. (2014) Intercropping studies with annual crops, International Crops Research Institute for the Semi-Arid Tropics (ICRISAT) CP0140 Patancheru P.O. 502 324, India.

Willey, R.W. and Rao, M.R. (1980) A Competitive Ratio for Quantifying Competition between Intercrops. Experimental Agriculture, 16, 117-125. http://dx.doi.org/10.1017/S0014479700010802

Wilson, G.L., Raju, P.S. and Peacock, J.M. (1982). Effect of soil temperature on seedling emergence of sorghum. Indian Journal of Agricultural Science, 52: 848-851.

Winkel, T. and Do, F. (1992) Caracteres morphologiques et physiologiques de resistance du mil (Pennisetum glaucum (L.) R. Br. a la secheresse. L' Agronomie Tropicale 46: 339–350

Wischmeier, W.H. and Smith, D.D.(1978) Predicting rainfall erosion losses: a guide to conservation planning. In: USDA, Agriculture Handbook. Washington, DC: U.S. Government Printing Office; pp. 537

World Bank (2011) The Cost of Adapting to Extreme Weather Events in a Changing Climate.

Wortmann, C. S., Mamo, M., Mburu, C., Letayo, E., Abebe, G., Kayuki, K. C., et al. (2009). Atlas of Sorghum (Sorghum bicolor (L.) Moench): Production in Eastern and Southern Africa. University of Nebraska-Lincoln, USA: INTSORMIL CRSP.

Wright, G.C., Nageswara Rao, R.C. and Farquhar, G.D. (1994) Water use efficiency and carbon isotope discrimination in peanut under water deficit conditions. Crop Science,34:92-97

Wright, G.C., Smith, R.C.G. and Morgan, J.M. (1983). Differences between two grain sorghum genotypes in adaptation to drought stress. 3. Physiological responses. Australian Journal of Agricultural Research.34:637-651

Wright, G.C. and Nageswara Rao, R.C. (Eds.) (1984) Selection for water use efficiency in grain legumes. Report of workshop held at ICRISAT Centre, Andhra Pradesh, India May 05-07, 1993.ACIAR Technical Reports No.27. 70pp

Wright, J.C. and Smith, R.C.G. (1983) Differences between two grain sorghum genotypes in adaptation to drought stress. II Root water uptake and Water Use. Australian Journal of Agricultural Research.34: 627-636.

Wright A L, and Hons F M. 2005. Tillage impacts on soil aggregation and carbon and nitrogen sequestration under wheat cropping sequences. Soil and Tillage Research 84: 67–75 www.OECD.org

Xiaohan Yang, Rongbin Hu et al (2017) The Kalanchoë genome provides insights into convergent evolution and building blocks of crassulacean acid metabolism.

Xu, D., Duan, X.,Wang, B., Hong, B. et al (1996) Expression of late embryogenesis abundant protein gene, HVA1 from Barley confers tolerance to water deficit and salt stress in transgenic rice. Plant Physiology, 110:249-257

Xue, Y.; Xia, H.; Christie, P.; Zhang, Z.; Li, L. and Tang, C. (2016) Crop acquisition of phosphorus, iron and zinc from soil in cereal/legume intercropping systems: A critical review. Ann. Bot. Lond. 2016, 117: 363–377

Yadav, R.S., Bidinger, F.R., Hash, C.T., Yadav, Y.P., Yadav, O.P., Bhatnagar, S.K. and Howarth, C.J. (2003) Mapping and characterization of QTLE interactions for traits determining grain and stover yield in pearl millet. Theoretical and Applied Genetics 106: 512–520

Yadav, R.S., Hash, C.T., Bidinger, F.R., Cavan, G.P. and Howarth, C.J. (2002) Quantitative trait loci associated with traits determining grain and stover yield in pearl millet under terminal drought stress conditions. Theoretical and Applied Genetics 104: 67–83.

Yadav, R. S., Yadav, B.L. and Chhipa, B.R. (2008) Litter dynamics and soil properties under different tree species in a semi-arid region of Rajasthan, India, Agroforestry Systems, 73(1):1-12

Yadav, S. and Rathee (2020) M Sucking Pests of Crops, In the book: Sucking Pests of Crops (pp.193-238) (Ed. Onkar) Publisher: Springer

Yadav R S, Yadav B L, Chhipa B R, Dhyani S K and Ram M. (2011). Soil biological properties under different tree based traditional agroforestry systems in a semi-arid region of Rajasthan, India. Agroforestry Systems 81: 195–202.

YuHong, NicoHeerink, ShuqinJin, PaulBerentsen, LizhenZhang and Wopkevan der Werf (2017) Intercropping and agroforestry in China – Current state and trends Agriculture, Ecosystems & Environment 244: 52-61

Zhang, Z., Xutian Chai, Akash Tariq, Fanjiang Zeng, Xiangyi Li a[n]d Corina Graciano (2021) I[n]tercropping Sys[t]ems Modify Desert Plant-Associated Microbial Communities and Weaken Host Effects in a Hyper-Arid Desert, Front Microbiol. 12: 754453. doi: 10.3389/fmicb.2021.754453

Zhang J. Z. et al. (2004). Plant Physiol. 135: 615-621

Zheng, D., Rademacher, J., Chen, J., Crow, T., Bresee M, Le Moine J and Ryu S. (2004). Estimating aboveground biomass using Landsat 7 ETM+ data across a managed landscape in northern Wisconsin, USA. Remote Sensing of Environment. 93: 402–411.

Zomer, R., et al.,(2016) Global Tree Cover and Biomass Carbon on Agricultural Land: The contribution of agroforestry to global and national carbon budgets. Nature Scientific Reports, 2016., 6.

Živanov D., Savić A., Katanski S., Karagić Đ., Milošević B., Milić D., Đorđević V., Vujić S., Krstić Đ. and Ćupina B.(2018). Intercropping of field pea with annual legumes for increasing grain yield production. Zemdirbyste-Agriculture, 105 (3): 235–242 DOI 10.13080/z-a.2018.105.030